THE DIALECTICS OF ECOLOGY

The Dialectics
of Ecology

SOCIALISM AND NATURE

John Bellamy Foster

MONTHLY REVIEW PRESS
New York

Library of Congress Cataloging-in-Publication Data
available from the publisher

ISBN: 978-1-68590-046-5 paper
ISBN: 978-1-68590-047-2 cloth

Typeset in Minion Pro

MONTHLY REVIEW PRESS, NEW YORK
monthlyreview.org

5 4 3 2 1

Contents

To Carrie Ann Naumoff

Preface

THE DIALECTICS OF ECOLOGY represents an attempt to transcend in both theory and practice what Richard Levins and Richard Lewontin in *The Dialectical Biologist* called "the *alienated* world," characteristic of modern thought, where mind is seen as alienated from body, society from nature, part from whole, cause from effect, subject from object, and past from present. The alienated world, in this sense, reflects the real material alienation of humanity associated with capital's all-encompassing expropriation of labor and the earth. The dire consequences of this now extend to the entire planet as a place of human habitation.[1]

More specifically, the present work represents an attempt to carry forward a project conceived almost a quarter-century ago in *Marx's Ecology*, in which I sought to retrieve Karl Marx's materialist ecological critique, including his now famous theory of metabolic rift.[2] Implicit in that book but never fully explored was the wider question of the *dialectics of nature*, which has long divided the Marxist tradition, with official Marxism reducing this core element of classical historical materialism to mere dogma, while the Western Marxist philosophical tradition simply rejected it altogether.

A way of transcending the theoretical chasm represented by these two one-sided perspectives was tentatively presented in the epilogue to *Marx's Ecology*, where I argued that numerous radical thinkers within the sciences and cultural theory had sought to investigate the relations between nature and society (and the relations of nature independent of society), inspired especially by Engels's later work, but also by Marx's own dialectical naturalism.[3] In the process, socialist theorists, primarily in the natural sciences, I suggested, had developed a critical-ecological analysis directly confronting the alienated world of capitalism.[4] Nevertheless, it took two decades to complete the research necessary to carry this argument forward, which finally materialized in 2020 in the form of *The Return of Nature*.[5]

With the publication of *The Return of Nature*, it finally became possible to reflect in a more systematic way on the implications of the dialectics of nature for the development of socialist ecology and its relevance for our time. For the classical historical materialist tradition, represented by the work of Marx and Engels, the human understanding of the world around us is a product of the *dialectics of nature and society*, that is, the intersections of these internally heterogenous, ever-changing realms. Society itself here is to be understood as an emergent form of nature with its own laws, but ultimately dependent on the "universal metabolism of nature" of which it remains a part. It is by carrying out labor and production, that is, through our "social metabolism" with nature, Marx argued, that we come to understand the "human essence of nature" and the "natural essence of man," and through which we are also able to infer relations and processes that extend beyond human experience.[6] It was this developed understanding uniting the political-economic and ecological critiques in *Capital* that was to be Marx's most enduring contribution, to be seen in tandem with Engels's *Dialectics of Nature*.

Still, it is important to understand that the ecological and social critique offered by Marx's metabolism argument, even when united with Engels's (and Marx's) dialectics of nature, does not provide direct practical solutions to the ecological contradictions of our time, 150 years or more later. However much one may think that "degrowth communism" is the solution to our present planetary predicament, there is no evidence that suggests that such an outlook was present in Marx's work itself, beyond a broad commitment to a process of sustainable human development controlled by the associated producers.[7] Rather, it is the *method* of materialist dialectics that is Marx and Engels's chief legacy to us today as we confront the twenty-first-century planetary emergency. It is this overarching dialectical critique applied to our own time that demands we take seriously the issue of planned degrowth, together with ecological civilization. The theoretical critique of our alienated world takes on practical significance as transformative praxis only by means of the concrete struggles carried out in relation to ever-changing historical conditions. The very fact that the future is not determined points to the necessity of the exercise of freedom.[8]

The Dialectics of Ecology is, therefore, aimed at the integration of the ecological and political-economic critiques of capitalism with the conditions of the global struggle conceived in the broadest terms. In the Anthropocene Epoch, the alienated world of capitalism has turned deadly on a planetary scale. Not only is capitalism crossing planetary boundaries, creating a habitability crisis for humanity, but the extreme logic of the accumulation system is rapidly accelerating the planetary emergency by offering the financialization of nature as its ultimate solution. Opposing this alienated-world logic are two emerging socialist projects full of contradictions and hope: the struggle to create an ecological civilization, principally emanating from China, and the strategy of planned degrowth in the rich economies, coupled with sustainable human development in the world as a whole. Ultimately, these approaches will need to converge if they are to be successful, generating a society of substantive equality and ecological sustainability on a planetary level. This, however, will depend on the emergence of a global environmental proletariat animated by a dialectical ecology and opposed to imperialism, extractivism, and class (and other) forms of exploitation, representing a new communal age of the earth.

I wrote *The Dialectics of Ecology* between March 2022 and September 2023. Each individual chapter, apart from the Introduction, which provides a comprehensive historical background, was published separately during that period of a year and a half, though revised for publication in this book, and thus each chapter can be approached as a "totality" in itself. However, each of the ten chapters was also the product of the development of an overall project on Marxian ecology. Hence, *The Dialectics of Ecology*, taken as a whole, is to be viewed as much more than the sum of its parts.

I could not have written this book without the help and encouragement of others, who have inspired me at every stage. John Mage has played a critical role in my thinking by recognizing those qualitative leaps in our understanding that have already been made, along with others that still need to be made. Brett Clark is my close collaborator and coauthor of chapter 9, on "Socialism and Ecological Survival." Fred Magdoff has advised me throughout and has frequently guided me in relation to ecological science. Joseph Fracchia and I have engaged in numerous conversations, in which he shared his deep insights into Marx's *corporeal* materialism, aiding me at times with wider issues with respect to Marxian theory and translation. Brian Napoletano and I have been in constant correspondence so that ideas constantly flow back and forth dialogically. Ian Angus has helped me face the Anthropocene, continually crossing the bridge between natural and social science. Helena Sheehan, whose work has inspired much of my own, has been steadfast in supporting my attempts to connect the dialectics of nature and society. Victor Wallis has provided useful suggestions in relation to nearly every chapter. Hannah Holleman has continually widened my vistas and worldview. Chris Shambaugh has provided much needed research support. Oscar A. Ralda has helped with fact-checking parts of the book and in providing helpful perspectives.

At *Monthly Review* and Monthly Review Press Martin Paddio, Michael D. Yates, Sarah Kramer, Jamil Jonna, Camila Valle, and Rebecca Manski, and the late John J. Simon have helped with this book in too many ways to enumerate.

Others I would like to thank who have significantly influenced my thinking and doing in this period are Intan Suwandi, Robert W. McChesney, Desmond Crooks, Richard York, Mauricio Betancourt, Michael Dreiling, Jason Hickel, Kenny Knowlton, Sue Dockstader, Alejandro Pedregal, Andy Ryan, Saul Foster, Carles Soriano, Pedro Urquijo, Linda Berentsen, Tobiah Moshier, Keri Bartow, Mira Castano, and Fray Castano.

The book is dedicated to Carrie Ann Naumoff, with whom I share an unbreakable dialectical relation.

— EUGENE, OREGON
OCTOBER 1, 2023

Introduction

> All nature is in a perpetual state of flux. . . .
> There is nothing clearly defined in nature. . . . Everything is
> bound up with everything else.
>   —DENIS DIDEROT

AS HARVARD ECOLOGIST and Marxian theorist Richard Levins observed, "perhaps the first investigation of a complex object as a system was the masterwork of Karl Marx, *Das Kapital*," which explored both the economic and ecological bases of capitalism as a social-metabolic system.[1] The premise of the *dialectics of ecology*, as it is addressed in this book, is that it is above all in classical historical materialism/dialectical naturalism that we find the method and analysis that allows us to connect "the history of labor and capitalism" to that of the "Earth and the planet," enabling us to investigate from a materialist standpoint the Anthropocene crisis of our times.[2] In Marx's words, humanity is both "a part of nature" and itself "a force of nature."[3] There was, in his conception, no rigid division between natural history and social history. Rather, "The history of nature and the history of men [humanity]" were seen as "dependent on each other as long as men exist."[4]

In this view, the relation of labor and capitalism to the earth's metabolism is at the center of the critique of the existing order. "Labour," Marx wrote, "is, first of all, a process between man and nature, a process by which man, through his own actions, mediates, regulates and controls the metabolism between himself and nature. He confronts the materials of nature as a force of nature."[5] However, with the advent of "capitalist production," a systematic disturbance and displacement occurs in "the metabolic interaction between man and the earth," creating a metabolic rift, or ecological crisis, severing essential natural relations and not only "robbing the worker but . . . robbing the soil."[6]

Today, this ecological rift in the metabolism of society and nature can be seen as having reached an Earth System level, creating what scientists have called an "anthropogenic rift" in the biogeochemical cycles of the entire planet, resulting in what Frederick Engels referred to metaphorically as the "revenge" of nature.[7] In the classical historical-materialist perspective, this contradiction can only be resolved by reconciling humanity and nature. Such a reconciliation requires overcoming not simply the alienation of nature but the self-alienation of humanity itself, manifested most fully in today's destructive, commodified society. What is necessary in such an analysis is recognition from the start of the "corporeal" nature of human existence itself, which is tied to production. Hence, if a "new universal history of the human" is necessary in our time, it is here, within the historical materialist tradition, that the necessary materialist, dialectical, and ecological method is to be found. For Marx, "Universally developed individuals, whose social relations, as their own communal relations, are hence also subordinated to their own communal control, are no product of nature, but of history."[8] However, human history is never detached from "the universal metabolism of nature," of which the social metabolism based in the labor and production process is an emergent part.[9]

In such a dialectical-ecological perspective, there are no fixed answers applicable to all of history, since everything around us in natural history and social history—constituting, as Marx said, the "two sides" of a single material reality—can be seen as in a state of constant flux.[10] Nevertheless, it will be argued here that the method of dialectical ecology, rooted in historical materialism and aimed at transcending the alienation of humanity and nature, provides a basis for uniting theory and practice in new, revolutionary ways. This constitutes the necessary dialectical negation or overcoming of the material conditions of our current alienated, divided, and dangerous world, itself the product of human historical development. Such a view assumes that there is a contingent, ever-changing historical process in which each new emergent reality bears within it an incompleteness and various contradictory relations, leading to further transformative developments. As Corrina Lotz indicates, dialectical negation properly embraces "absenting (Roy Bhaskar's term), removal, loss, conflict, interruption, leaps and breaks," often understood in terms of the general concept of *emergence*, or the qualitative shift to higher organizational levels, which, as Engels said, always carries within it the potential for *annihilation*.[11] The structure of history, including natural history, thus always contains within it crises and catastrophes, along with the possibility of something qualitatively new, drawn from a combination of residuals of the past (previously negated realities) interacting in contingent ways with the present as history and generating transformative change. History, whether natural or human history, is thus not linear, but rather manifests itself as a spiral form of development.

The notion of human historical development, a relatively recent conception that scarcely precedes the capitalist era, is a product of the changing relation of human beings to nature as a whole. As Marx recognized, Epicurus in Hellenistic antiquity saw the origins of natural philosophy or natural science as tied to an overriding sense of danger that the natural world represented in the daily lives of human beings.[12] In Epicurean philosophy, there was no rational answer to be found to this existential condition, other than reconciliation with the world through forms of contemplative self-consciousness and the development of a sense of oneness with nature, or *ataraxia*, by means of enlightenment/science.

The enormous historical development of the productive forces, separating antiquity from the modern world, and the emergence of modern science in this context was to alter fundamentally the relation between humanity and its natural environment. Bourgeois society, as a result of this "progress" and the scientific revolution of the seventeenth century, would revel in the "domination of nature" provided by Enlightenment science. The realm of natural necessity was seen in this conception as being forever pushed back and even transcended.[13] This, however, gave rise to the conceit, as Engels noted, of "human victories over nature" in the manner of "a conqueror over a foreign people," a view that, because of its lack of foresight and its narrow objectives, led to human-generated ecological catastrophes.[14]

As a result of the historical process, humanity finds itself once again confronted with an overarching sense of danger emanating from the forces of nature. Yet, behind this existential threat to humanity and life lies human labor, itself a *force of nature*, now generating planetary-level catastrophe. The alienation of nature under capitalism is such that money is fetishistically mistaken for existence, while private extraction and expropriation, the robbery of the earth, is confused with real wealth. In the historical-materialist view, the contradiction between humanity and the earth can be transcended before it proves fatal, but only if the two sides of human self-alienation—alienation from humanity and alienation from nature—are transcended through the "revolutionary reconstitution of society as a whole" and the creation of a world of substantive equality and ecological sustainability.[15]

The development of such an approach based on classical historical-materialist grounds cannot consist simply of a theoretical reconstruction of the analysis of Marx and Engels in this area, involving a synthesis of their contributions to an ecological-materialist dialectics. At best, the only thing such an approach can generate is a more critical method in analyzing the present, although it is the actual overcoming of the present as history that is the overriding concern. Above all, it is necessary to address the rapidly developing ecological crisis of the Anthropocene Epoch in human history, which marks the rise of anthropogenic, as opposed to non-anthropogenic, factors as the main driving force of Earth System change. Here we must confront the current financialization of nature, the new phase of planetary extractivism, questions of human survivability, and the revolutionary struggle to create a society of planned degrowth and ecological civilization geared to sustainable human development. All of this, however, depends on the recovery, development, and unification in theory and praxis of the dialectical-ecological critique of capitalism, which is an indispensable and indisputable legacy of classical historical materialism.

THE DUAL NEGATION OF DIALECTICAL MATERIALISM

Soviet Marxism and the Dialectics of Nature

The reconstruction of Marxian ecology based on classical historical materialism is a very recent and still very incomplete development, largely confined to the present century and to the rise of ecosocialism. Both official Marxism associated with the Soviet Union of the late 1930s and after, which removed the *critical* element within philosophy together with Marx's ecological analysis, and the Western Marxist philosophical tradition, which rejected dialectical naturalism altogether, presented enormous obstacles to the further development of the historical-materialist ecological critique. This, then, constituted a dual negation of the dialectics of nature emanating from the Cold War antagonism between East and West. But it is one that has been increasingly transcended in recent decades as material conditions have changed.

Soviet philosophy, as originally conceived under the leadership of V. I. Lenin, Leon Trotsky, and Nikolai Bukharin on the occasion of the launching of its original flagship publication *Under the Banner of Marxism* in 1922, was intended to bring together the materialist perspectives of both Mensheviks and Bolsheviks (representing, respectively, the relatively reformist and revolutionary tendencies within Russian Marxism), mechanists and dialecticians, and philosophers and natural scientists, with the object of the concretization of a wider and internally differentiated philosophy of *dialectical materialism*—a term introduced by the working-class philosopher Joseph Dietzgen, which owed its influence mainly to the work of the founding Russian Marxist (and Menshevik) Georgi Plekhanov.[16]

Lenin set the tone in his 1922 letter to *Under the Banner of Marxism*, which was published as an article titled "On the Significance of Militant Materialism." Here, he insisted that it was necessary to bring "materialists of the non-communist camp" together with revolutionary materialists in order to promote a mutually engaged philosophical discussion. The object was to develop a fundamentally Marxist, "militant materialist" view, and at the same time guard against rigid dogmas. "One of the biggest and most dangerous mistakes made by Communists (as generally by revolutionaries who have successfully accomplished the beginning of a great revolution) is the idea that a revolution can be made by revolutionaries alone." Rather than excluding leading Menshevik philosophers such as the talented Liubov Isaakovna Akselrod (a former assistant to Plekhanov) and Abram M. Deborin from the new journal, Lenin insisted on the necessity of their inclusion. To protect against mechanistic materialism or mechanism (today more often called reductionism), he declared as essential the critical incorporation of Hegelian dialectics, despite its idealist basis, within the purview of the journal. Thus, *Under the Banner of Marxism* should, in his words, "be a kind of 'Society of Materialist Friends of Hegelian Dialectics.'" [17]

Soviet philosophy was from the start aimed at developing dialectical materialism as a general theoretical view applicable to both philosophy and science, based proximately on the work of Engels, Plekhanov, and Lenin, but rooted more fundamentally in the work of Marx, G. W. F. Hegel, and Baruch Spinoza. (Marx's philosophical discussions in his early *Economic and Philosophical Manuscripts* were at that time unknown.)

Engels's *Anti-Dühring* and the incomplete *Dialectics of Nature* provided a guiding thread that, in its most succinct expression, revolved around the three ontological principles or "laws," derived from Hegel, of the (1) transformation of quantity into quality, and vice versa; (2) the identity or unity of opposites; and (3) the negation of the negation.[18] The first of these was meant to capture what are often called in today's scientific language *phase changes* or threshold effects, in which quantitative changes lead to new qualitative realities. Through such qualitative transformations, which can be observed both in nonhuman nature and in society, a "new power," Marx and Engels observed, emerges that is "entirely different from the sum of its separate forces."[19] The second ontological principle addresses the contradictions that arise due to incompatible developments within the same relation intrinsic to all processes of motion, activity, and change. The third ontological principle of the negation of the negation refers to how the processes associated with the first two principles set the stage for dialectical negations, that is, the negation of the previous negation, and a process of *Aufhebung* (referring simultaneously to transcendence, suppression, preserving, overcoming, and superseding), giving rise to sharp reversals and transformations, establishing qualitatively new emergent realities arising at a higher level, and a complex "spiral form of development" in which negation is never mere negation, but contains within it the positive (and vice versa).[20]

"The 'dialectical moment,'" Lenin wrote in his *Philosophical Notebooks*, "demands the demonstration of '*unity*,' i.e., of the connection of negative and positive, the presence of this positive in the negative. From assertion to negation—from negation to 'unity' with the asserted—without this dialectics becomes empty negation, a game, or scepsis [skepticism]."[21] Although it has been common to reduce dialectics to the unity of opposites, such an approach would be completely barren, in Lenin's view, since it excludes *dialectical negation*.[22]

In 1924, a major debate broke out between the mechanists, who were associated with figures like Akselrod and the militant mechanist-atheist Ivan Ivanovich Skvortsov-Stepanov, and the more dialectically oriented thinkers under the leadership of Deborin and his Institute of Red Professors.[23] The mechanists were tied more directly to natural science and to such leading theorists as Bukharin, and before him Plekhanov, both of whom had displayed mechanistic tendencies, though neither were entirely averse to dialectical analysis.[24] The dialecticians, in contrast, were far more removed from natural science and focused on Hegelian idealism as critically mediated by the materialist tradition of Ludwig Feuerbach, Marx, Engels, and Lenin.[25]

The main theoretical dispute dividing the mechanists and the Deborinists revolved around the proposition of the former that both organic and inorganic nature could be reduced simply to mechanical properties. This ran counter to a dialectics predicated on the existence of irreducible organizational forms, associated in particular with Engels's analysis in *Anti-Dühring* and the *Dialectics of Nature*, the latter being published for the first time in 1925.[26] Deborin, as well as most other Soviet philosophers, argued that it was impossible to reduce in its entirety a qualitatively higher form, such as organic life, to a lower form, such as inorganic matter. Commenting on William Robert Grove's *The Correlation of Physical Forces* (1846), Engels wrote: "Chemical action is not possible without change of temperature and electric changes; organic life [is not possible] without mechanical, molecular, chemical, thermal, electric, etc. changes. But the presence of these subsidiary forms does not exhaust the essence of the main form in each case. One day we shall certainly 'reduce' thought experimentally to molecular and chemical motion in the brain; but does that exhaust the essence of thought?"[27] In this view, higher organizational levels, such as mind/thought, could not be reduced simply to lower organizational levels, even though the former were dependent on the latter. It was the distinction between different qualitative forms/levels/planes within material existence, Engels explained, that was the basis for the division of the various sciences, separating, for example, biology from chemistry and physics.

Nevertheless, the mechanists, representing the then dominant scientific outlook, challenged Engels's view that qualitative forms/levels differentiated reality, as well as thought. Thus, Skvortsov-Stepanov declared that Engels's claim that higher forms of material existence could not be explained simply by lower ones, and thus that mechanical forms of motion could not account in their entirety for the human psyche, had to be rejected outright.[28] Reductionism, in conformity with modern mechanistic science, was seen as a general principle applicable to all of existence, in line with positivism. Thus, it was often said that "the mind was a mere secretion of the brain"—a proposition first put forward by Pierre Jean Georges Cabanis in 1802 and even seemingly accepted by Charles Darwin.[29] In contrast, the Deborinist philosophers based their analysis on the dual critique of Hegelian idealism and of mechanistic materialism. On the issue of reductionism, they relied heavily on Engels's notion of quantitative change leading to qualitative transformation.

It soon became clear that neither side had the upper hand intellectually, since this was in large part a division between positivist natural science and dialectical philosophy. Yet, despite the philosophical stalemate, the Deborinists managed to triumph over their rivals through purely political means by 1929, using their superior control over the main institutions of Soviet philosophy to exclude the competing view.[30]

The Deborinist triumph proved to be short-lived since, within a year, they were placed on the defensive due to an attack from a more powerful political quarter: the Communist Party hierarchy itself. This represented the direct intervention of the so-called Bolshevizers of the party hierarchy into the struggles on the philosophical front. Although not directly defending the mechanists, considered a "right deviation," the party hierarchy decided that it was necessary to rein in the Deborinists, as a "left deviation." The Deborinists were variously accused of being Mensheviks, idealists, vitalists, and weak in their criticisms of Trotsky and other left deviationists. The crushing blow, however, was Joseph Stalin's official declaration in December 1930 that the Deborinists were "Menshevizing Idealists." Deborin himself was denounced based on his Menshevik past of some three decades prior, while the dialecticians were also charged with being associated with the brilliant Marxist economist I. I. Rubin, author of *Essays on Marx's Theory of Value*, who was executed in 1937.[31]

The suppression of Soviet philosophy in the 1930s was inscribed in stone with the publication of Stalin's "Dialectical and Historical Materialism" in 1938, as part of the official *History of the Communist Party of the USSR—Bolsheviks: Short Course* (often referred to as simply *The Short Course*).[32] In the rigid, dogmatic formulation provided in Stalin's "Dialectical and Historical Materialism," the notion of the negation of the negation, fundamental to the critical thought of Marx, Engels, and Lenin, was formally excluded. Historical materialism was reduced to a separate area subordinate to dialectical materialism. All categories were frozen. Marx's *Economic and Philosophical Manuscripts* of 1844, first published in 1932, were treated as belonging to a pre-Marxist stage in his thought and were generally ignored or downplayed.

Soviet natural science, particularly the life sciences, including ecology, suffered a similar fate to that of philosophy. Bukharin had provided a crucial link between dialectical-materialist philosophy and natural science, working with agronomist, botanist, and geneticist Nikolai Vavilov, physiologist and biologist B. Zavadovsky, and historian of science-physicist Boris Hessen. All of these thinkers, together with other leading Marxist scholars such as the philologist David Riazanov, editor of a critical edition of Marx and Engels's *Works*, were purged. Bukharin himself was executed in 1938. The revolutionary dialectical insights that had emerged in the USSR in natural science and philosophy were replaced with narrow formulas that excluded critical thought.

As a result of these developments, the official doctrine of dialectical materialism was reduced to a crude mechanistic monism

and positivism, opposed to a tendentious, if somewhat more critical, neo-Kantian dualism that was to pervade Western Marxism.[33]
Nevertheless, a genuine *dialectical materialism* continued to exist
in the recesses, refusing to be buried. As Galileo Galilei, caught up
in the Inquisition, is reported to have said of the earth, no doubt
apocryphally: "And yet it moves."[34]

Western Marxism and the Negation of Dialectical Materialism

In contrast to official Soviet Marxism, what came to be known as
Western Marxism, or the Western Marxist philosophical tradition, followed a radically different course. In this perspective, the
dialectics of nature and, with it, the notion of dialectical materialism, was invalidated on the basis that the dialectics required the
identical subject-object—that is, the notion that human beings
were both the subjects and objects of their own actions—and thus
was not applicable to external nature, where the human subject
was not present. With the exclusion of the natural realm insofar
as it was separate from and even prior to human history, Western
Marxism thus severed any direct relation of historical materialism
to natural science and the universal metabolism of nature, effectively relegating the natural world to the realm of positivism. The
result was a dualistic, two-world conception in which dialectics
related simply to human history, not natural history (the realm of
the Kantian thing-in-itself), and in which Marxism was confined
exclusively to the social.[35] Historical materialism was then robbed
of any connection to nature as a force in itself, reducing the notion
of materialism within Western Marxism simply to denaturalized
political-economic relations. Western Marxist thinkers such as
Herbert Marcuse and Theodor W. Adorno railed against the Soviet
Short Course and Stalin's "Dialectical and Historical Materialism,"
but also frequently went beyond that, as in the case of Adorno and
Lucio Colletti, to reject the transformative dialectics of Engels and
Lenin, and even in some respects that of Marx and Hegel, gravitating instead toward Immanuel Kant.[36]

Theodor Adorno's *Negative Dialectics*, often viewed today as one of the greatest contributions of the Frankfurt School within Western Marxism, had as its object the rejection of the "negation of the negation" and thus the positive moment in the dialectic. As Adorno wrote in the preface to his work: "*Negative Dialectics* is a phrase that flouts tradition. As early as Plato, dialectics meant to achieve something positive by means of negation; the thought figure of a 'negation of negation' later became the succinct term. This book seeks to free dialectics from such affirmative traits without reducing its determinacy."[37]

In Adorno's conception, "Marx was a Social Darwinist" in the sense that he saw natural history as the realm of natural necessity (also impinging on social history), to be transcended in human history by a leap to the realm of freedom. Marx's concept of nature was then, according to Adorno, ultimately the Enlightenment one, in which nature was simply there to be conquered and transcended by social praxis. For all their discussions in *Dialectic of the Enlightenment* concerning "the domination of nature," Max Horkheimer and Adorno acquiesced to the view, which they imputed to Marx himself, of the "wholesale racket in nature"—that is, a kind of Hobbesian and Darwinian state of nature or war of all against all, seen as characterizing all of Enlightenment thought. Marx himself was said to have shared these views, simply seeing freedom as the transcendence of necessity.[38] As Adorno opined: Marx "underwrote something as arch-bourgeois as the program of an absolute control of nature."[39] Moreover, by specifying at the outset of his book *Negative Dialectics* that the object of his analysis was to exclude the *negation of the negation*, and thus the positive element in the dialectic, in a manner that ironically paralleled the dogmatic elimination of the negation of the negation within Stalin's "Dialectical and Historical Materialism," Adorno cast a light on his own negativity with respect to the prospect of revolutionary change.

Alfred Schmidt—who worked under Horkheimer and Adorno in writing his thesis and *magnum opus*, published in 1962 as *The*

Concept of Nature in Marx—observed that Marx's notion of the social metabolism between nature and society raised the issue of the dialectic of nature, or "nature's self-mediation," in an entirely defensible way. Schmidt, however, later disavowed this on the grounds that Marx saw such self-mediation of nature as restricted to human action, and then only in traditional communal societies, no longer applicable to modern bourgeois society, in which *first nature*, that is, nature in and of itself, had been largely subsumed by *second nature*, the social realm. "It is only the process of knowing nature," Schmidt declared, "which can be dialectical, not nature itself."[40] This formulation retained the neo-Kantian dualism between nature and society, arguing that dialectical mediation was impossible without an active human subject, which was confined to the historical-social realm. Such views pushed dialectics, as envisioned in Western Marxism, in the direction of idealism.[41]

Given the systematic exclusion of nature/ecology from dialectical thought within Western Marxism, it was often contended, even within Marxist circles, that the philosophy of praxis had nothing to contribute to ecological analysis. This was codified in Perry Anderson's influential 1976 *Considerations on Western Marxism*, which claimed that "no major figure in the third generation of classical Marxism," which Anderson narrowly associated with Western Marxism and its rejection of the dialectics of nature, was affected by "developments in the physical sciences."[42] In his 1983 work, *In the Tracks of Historical Materialism*, Anderson declared that "problems of the interaction of the human species with its terrestrial environment [were] essentially absent from classical Marxism"—a proposition that would have been seen as absurd on its face even then, if it had not been for the fact that the entire domain of the dialectics of nature had already been systematically absented from Western Marxism, while classical Marxism's ecological critique was simply treated as nonexistent.[43]

Hence, both the Soviet conception of the "dialectics of nature" in the 1938 *Short Course*, centered on Stalin's rigid separation of dialectical materialism and historical materialism, and the Western Marxist rejection of the dialectics of nature altogether, fell prey to narrow conceptions of reality. They thus failed to embrace what Engels called the totality of bodies, from the stars to the molecules, including the human mind and human society. "In effect, the problem of the dialectics of nature," critical-realist philosopher Roy Bhaskar wrote, "reduces to a variant of the general problem of naturalism, with the way it is resolved depending on whether dialectics is conceived sufficiently broadly and society sufficiently naturalistically to make its extension to nature plausible."[44]

The Struggle for Materialist Dialectics
Dialectical Materialism Redux

Still, it would be a mistake to think that the classical Marxist notion of the "dialectical conception of nature," as Engels referred to it, was brought to a dead end, reduced to nothing without a remainder, either in the Soviet Union or in the West.[45] Rather, materialist dialectics constantly reemerged in all sorts of unexpected ways in changing historical circumstances. This can be seen most distinctly in the famous visit of Soviet natural scientists and philosophers to the Second Congress on International History of Science in London in 1931, where Bukharin, Vavilov, Zavadovsky, Hessen, and others presented the results of Soviet dialectical natural science and philosophy.

In the audience at this historic meeting were world-renowned scientists and socialist thinkers, including Joseph Needham, J. D. Bernal, Lancelot Hogben, and Hyman Levy. (J. B. S. Haldane was not present but would take up the new ideas partly under the impetus of the same event.) In the course of the Soviet presentations, Bukharin sought to generate a dialectical-humanist conception of Marxist analysis, conducive to natural science, rooted in Marx's "Notes on Adolph Wagner," where some of Marx's underlying ontological conceptions were made evident, along with the integration of biogeochemist Vladimir Vernadsky's concept of the biosphere.

Recognition of the reality in which human beings could be seen as "living and working in the biosphere" demanded, Bukharin insisted, an integrated materialist-dialectical view of process and interaction, contradiction, negation, and totality, in which both external nature and society participated. Hessen presented for the first time a sociology of science embodying materialist dialectics that explained Newton's discoveries as they related to a bourgeois mechanistic view of the world. Vavilov provided an account of the Soviet discovery, through historical and materialist investigations, of the original geographical locales (now known as the Vavilov centers) of the world's germplasm from which the major agricultural crops had arisen.[46]

For Needham, it was Zavadovsky's critique of both vitalism and mechanism from a dialectical-naturalist perspective in his article on "The 'Physical' and 'Biological' in the Process of Organic Evolution" that was to have the greatest impact in the development of his own approach to dialectical emergence in his famous theory of "integrative levels." Zavadovsky argued that "biological phenomena, [although] historically connected with physical phenomena of inorganic nature, are none the less not only not reducible to physico-chemical or mechanical laws, but within their own limits as biological processes display varied and qualitatively distinct laws," that have "*relative autonomy*" from those of inorganic, physical forms. The "dynamic connection" between the inorganic and the organic in the biological sphere was captured, he argued, by the concept of metabolism, linking higher biological forms to their physical-inorganic preconditions.[47]

It was this concept of metabolism, seen as the material phenomenon connecting the physical-chemical and the biological through exchanges within nature, that was to become the basis of ecosystem analysis. In the new ecological systems analysis, biological order as a form of emergent organization was irreducible to the various elements of which it was constituted. "Translated into terms of Marxist philosophy," Needham wrote, "it is a new dialectical level." The core idea of dialectical naturalism was "that of transformation. How do transformations occur, and how can we make them occur? Any satisfactory answer must also be a solution to the problem of the *origin of the qualitatively new*."[48]

The British Red scientists of the 1930s and '40s were themselves products of a materialist tradition that was emergentist and ecological in its orientation. Most of these figures had also embraced socialism, particularly Marxian socialism. Needham recalled the influence of the "legendary" British zoologist E. Ray Lankester, who had been Darwin's and Thomas Huxley's protégé and a close friend of Marx, as well as the foremost representative of Darwinian evolutionary theory in Britain in the generation after Darwin and Huxley.[49] Lankester developed a systematic approach to the natural world with his concept of "bionomics," which was the original term for ecology in Britain. (He also helped introduce the term *œcology* into the English language through supervising the 1876 translation of Ernst Haeckel's *History of Creation*.) He focused on the complex interrelationships between organisms and their environments and on humans as disturbers of global ecological relations, developing a critique of "the effacement of nature by man" rooted in the critique of capitalism.[50]

It was Lankester's student Arthur Tansley, the foremost plant ecologist in England in the early twentieth century, who introduced the concept of ecosystem, based in part on the wider systems theory of Levy. As depicted by Tansley, the ecosystem concept included both the inorganic and organic realms and encompassed human beings themselves as both living within and major disturbers of ecosystems. The ecosystem notion was rooted fundamentally in the concept of metabolism, which had been the basis of early ecological systems analysis, and the treatment of nutrient cycling, a subject that occupied German chemist Justus von Liebig, Marx (in his concepts of social metabolism and the metabolic rift), and Lankester.[51] Tansley's ecosystem concept was thus to play a crucial role in the development of modern systems ecology.[52] Levy developed the notion of phase changes along with a unified systems theory rooted in historical-materialist conceptions in his *The Universe of Science* (1932) and *A Philosophy for a Modern Man* (1938).

Haldane was both the co-discoverer, alongside the Soviet geneticist A. I. Oparin, of the modern materialist theory of the origins
of life on Earth, and was a major figure in the modern Darwinian
synthesis, to which he later applied Marxian conceptions. Bernal,
influenced by Engels's dialectics of nature, developed an analysis
of the negation of the negation within material processes in terms
of the action of residuals, leading to new combinations and novel
emergent developments, representing qualitatively new powers.
Hogben applied critical materialist and dialectical methods to dis-
prove the genetic theories underlying biological racism.[53] Other
closely related figures included the literary and science critic
Christopher Caudwell, who sought to bring together the dialectics of art and science (and who died fighting in the Spanish Civil
War); the historian of ancient philosophy Benjamin Farrington,
who built on Epicurean philosophy and its relation to Marxism
(inspired in part by Marx's dissertation on Epicurus); and the novelist, cultural theorist, and poet Jack Lindsay, whose 1949 *Marxism
and Contemporary Science* was an exploration of ways in which to
develop a broad dialectical and emergentist method encompassing nature and society.[54]

Despite the suppression of the mechanists and the Deborinists,
important work was still being done in Soviet philosophy in 1931,
as evidenced by *A Textbook of Marxist Philosophy*, prepared by the
Leningrad Institute of Philosophy under the direction of Mikhail
Shirokov and published in English translation in 1937.[55] This
work, which influenced Needham, was engaged in the critique of
both mechanism (reductionism) and vitalism, a view that assumes
some mysterious life force added to material reality that explains
evolution.[56] *A Textbook of Marxist Philosophy* stood out at the
time, since it relied on the conception of *emergence* as the key to
materialist dialectics. As Shirokov wrote in a passage that was later
singled out by Needham:

A living organism is something that arose out of inorganic
matter. In it there is no "vital force." If we subject it to purely

external analysis into its elements, we shall find nothing except physico-chemical processes. But this by no means denotes that life amounts to a single aggregate of these physico-chemical elements. The particular physico-chemical processes are connected in the organism by a *new form of movement*, and it is in this that the quality of the living thing lies. The new in a living organism, not being attributable to physics and chemistry, arises as the result of the new *synthesis*, of the new *connection* of physical and chemical movements. This synthetic process whereby out of the old we proceed to the *emergence* of the new is understood neither by the mechanists nor the vitalists. . . . The task of each particular science is to study the unique forms of movement characteristic of a particular degree of the development of matter.[57]

According to Shirokov, in the ancient philosophy of Epicurus, which had attracted Marx, "emergence is the uniting of atoms; disappearance their falling apart." This served to explain a process of self-generation, "the origin and development of the universe, the movement of the human soul, etc." Out of this had arisen the fundamental materialist view. In materialist dialectics, there is "ceaseless emergence and annihilation of the forms of . . . movement," which continue to reproduce themselves "in ever new movement and in ever new qualities."[58]

However, all such advancements in materialist dialectics and science were shut down completely in 1938 with the publication of Stalin's "Dialectical and Historical Materialism." What remained of Soviet philosophy consisted of a formalistic and mechanistic presentation of rigid "dialectical laws" conceived as a world outlook, rather than a critical philosophy. It was this that formed the background against which the more creative thinkers had to work. Nonetheless, in the next generation, the USSR produced major dialectical philosophers, most notably Evald Ilyenkov, whose dialectical logic was rooted not only in the Hegelian and Marxian traditions but also in the work of the pioneering psychologist Lev Vygotsky, who argued that human cognitive abilities in general were substantially the result of activity and mediation with the social and cultural environment. Ilyenkov's philosophy was directed primarily at challenging, on materialist-dialectical grounds, the dualistic "two-worlds" epistemology of British empiricism, Cartesianism, and neo-Kantianism that dominated the bourgeois philosophical outlook.[59]

Ilyenkov saw Marx's epistemology as one in which human activity or praxis creates the ideal world of thought through human production—that is, attempts to transform the world.[60] Hence, there is a *real identity* of humanity and nature at the base of human cognition that is rooted in *real activity*. The "ideal," in Ilyenkov's sense, is not properly seen as something apart, an abstract entity, but is the basis of conceptions, knowledge, and information emanating from the dialectical process of human-social encounters with the material world, of which human beings themselves are a part. Dialectics is thus itself a manifestation of this active mediation with totality, arising "out of the process of the metabolism between man and nature."[61] However, despite, or perhaps because of, the power of his analysis, Ilyenkov had trouble getting his work published. At the time of his death, half of his handwritten publications—including his much-celebrated *Dialectics of the Ideal*—remained on his desk, unpublished.[62]

Despite the purge of some of the leading figures, there continued to be remarkable developments in Soviet science based on dialectical analysis up through the 1940s. This includes, notably, Vladimir Nikolayevich Sukachev's concept of biogeocoenosis in his work on forest ecology, representing a concept parallel to ecosystems but directly integrated with biogeochemical cycles and the entire biosphere in the sense pioneered by Vernadsky, thus pointing to a dialectical Earth System analysis.[63]

Of even greater importance was the work of I. I. Schmalhausen in his *Factors of Evolution: The Theory of Stabilizing Selection*, first published in the USSR in 1947 and quickly translated into English in 1949. Theodosius Dobzhansky called Schmalhausen "perhaps the most distinguished among the living biologists in the

USSR."[64] Schmalhausen, like the Red geneticist C. H. Waddington in England, developed a theory of the triple helix of gene, organism, and environment that provided a dialectical evolutionary and ecological view, one that constituted a sophisticated alternative to Lysenkoism with its anti-geneticist (or anti-Mendelian genetics) basis. Schmalhausen's dialectical approach was particularly evident in his notion of hierarchies or integrative levels structuring biological evolution, and in his explanation that latent, assimilated genetic traits that were accumulated during long periods of stabilizing selection would come to the surface only when organisms faced severe environmental stress or certain thresholds were crossed, resulting in a process of rapid change.[65]

Following Engels, Schmalhausen saw heredity as both negative from an evolutionary standpoint, insofar as it blocked the historical evolution of organisms, and positive, in that it preserved organization and created new organizational forms.[66] The significance of what came to be known as Schmalhausen's Law of stabilizing selection, according to dialectical biologists Richard Lewontin and Richard Levins, was that it indicated that "when organisms are living within their normal range of the environment, perturbations in the conditions of life and most genetic differences between individuals have little or no effect on their manifest physiology and development, but under severe or unusual general stress conditions even small environmental and genetic differences produce major effects." The result is that normal evolution of species is characterized by stabilization punctuated by periods of rapid change, in which latent traits are mobilized in relation to environmental stress.[67] What sometimes appeared as a Lamarckian inheritance of acquired characteristics was actually a process of "genetic assimilation, the process whereby latent genetic differences within populations are revealed but not created by environmental treatment and therefore become available for selection" when certain thresholds are reached.[68]

Factors of Evolution came out, however, just prior to Trofim Lysenko's political triumph in Soviet biology/agronomy in 1948.

Soon after his book was published, Schmalhausen was denounced for promoting genetics and denying Lamarckian inheritance of acquired characteristics in his work on evolutionary ecology. As a result, Schmalhausen was dismissed from his posts as director of the Institute for Evolutionary Morphology at the Academy of Sciences and as head of the sub-department of Darwinism at Moscow University. This was only reversed around the time of Stalin's death in 1953, when Sukachev led the way in combating and defeating Lysenko. As a result, Schmalhausen was eventually able to resume his career.[69] The final decades of the Soviet Union saw important new developments in Soviet environmental thought, including the introduction of the concept of ecological civilization based on classical historical materialism, incorporating Marx's concept of social metabolism.[70]

The Struggle for a Critical Dialectics of Nature in the West

Within Marxism in the West, parallel struggles occurred, challenging the dominant Western Marxist philosophical tradition. Georg Lukács, a giant presence, was universally viewed as having generated Western Marxism as a distinct theoretical tradition, based on a brief footnote in *History and Class Consciousness* in which he had raised doubts about Engels's argument with respect to the dialectics of nature.[71] Yet, contrary to myth, Lukács did not reject the dialectics of nature altogether in *History and Class Consciousness*, since in a later chapter in that work he referred, in a manner akin to Engels, to the "merely objective dialectics of nature" of the "detached observer."[72] Moreover, several years later, in his previously unknown and only recently published *Tailism* manuscript, Lukács defended the notion of the "dialectics in nature" on the basis of Marx's concept of social metabolism, representing the dialectical mediation of nature and humanity through production.[73] Lukács worked under David Riazanov at the Marx-Lenin Institute in 1930, helping to decipher the text of Marx's *Economic and Philosophical Manuscripts* of 1844. These manuscripts greatly affected his subsequent analysis. This change in viewpoint was highlighted in his 1967 preface to *History and Class Consciousness* and in his later *Ontology of Social Being*.[74] The latter was based on Marx's social metabolism concept, seen as forming a dialectics of nature and society rather than expressly following Engels's approach to the dialectics of nature. Although examining with great depth Marx's metabolism analysis in *Capital*, Lukács failed to address Marx's notion of the metabolic rift, or ecological crisis.[75] Nevertheless, the social-metabolic ontology that he derived from Marx served to further undermine the negation of the dialectics of nature within the Western Marxist tradition that *History and Class Consciousness* had inspired. It is significant that Lukács's later work was largely disowned by the Western Marxist tradition, becoming so invisible that references to him in the West identified him almost entirely with what he had written in 1923 or before, largely excluding the almost five decades of work that were to follow.

If the dominant philosophical tradition within Marxism in the West was primarily defined by its rejection of the dialectics of nature, not all Western Marxist philosophers agreed. In 1940, the prominent French Marxist philosopher Henri Lefebvre published his *Dialectical Materialism*. In this work, Lefebvre sought to challenge the interpretation provided in Stalin's "notorious theoretical chapter in the *History of the Communist Party of the USSR*," reestablishing the dialectics of nature as a critical outlook while rejecting the simplistic view of dialectical materialism derived merely from reified "laws of Nature," viewed apart from the mediation of self-conscious thought. As Lefebvre wrote: "It is perfectly possible to accept and uphold the thesis of the dialectic in Nature; what is inadmissible is to accord it such enormous importance and make it the criterion and foundation of dialectical thought."

A crucial component of Lefebvre's argument was directed at the refusal of "institutional Marxism . . . to listen to talk of *alienation*." In Lefebvre's conception of dialectical materialism, it was necessary to integrate Marx's theory of alienation within the general conception of the metabolism of nature and society. He

drew heavily on Levy's dialectical systems theory as presented in
A Philosophy for a Modern Man in order to capture the reality of
emergence. "Man's world," Lefebvre wrote in a passage that was to
prefigure much of his later thought, "appears as made up of emer-
gences, of forms (in the plastic sense of the word) and of rhythms
which are born in Nature and consolidated there relatively, even
as they presuppose the Becoming in Nature. There is a human
space and a human time, one side of which is in Nature and the
other independent of it."[76]

Lefebvre's subsequent work proceeded in an increasingly ecological direction. In the early 1970s, he began to reflect on what is now
known as Marx's theory of metabolic rift. As he wrote in *Marxist
Thought and the City*, drawing on Marx, the growth of the capitalist urban structure "disturbs the organic exchanges between man
and nature. 'By destroying the circumstances surrounding that
metabolism, which originated in a merely natural and spontaneous
fashion, it compels its systematic restoration as a regulative law of
social production and in a form adequate to the full development
of the human race'. . . . Capitalism destroys nature and ruins its own
conditions, preparing and announcing its revolutionary disappearance." Testifying to a kind of "reciprocal degradation" of the urban
and the rural, external nature and society, he continued, "a ruined
nature collapses at the feet of this superficially satisfied society."[77]

On December 7, 1961, six thousand people crowded into a Paris
auditorium to hear a debate on the topic "Is the Dialectic Simply a
Law of History or Is It Also a Law of Nature?" On the side of those
who rejected the dialectics of nature were the existentialist Marxist
Jean-Paul Sartre and the left Hegelian philosopher Jean Hippolyte;
on the side of those defending it were the French Communist
philosopher Roger Garaudy and the prominent young physicist
Jean-Pierre Vigier. Sartre, Hippolyte, and Garaudy had all written
extensively on the issue of the dialectics of nature, while Vigier's
views on dialectical materialism were less well-known and stood
out since directly related to natural science.

Vigier argued that notions of the dialectics of nature long

preceded historical materialism and could be traced back hundreds and thousands of years. "Every day," he declared, "science further verifies the profound saying of Heraclitus which is at the root of the dialectic: everything is flux, everything is transformed, everything is in violent movement." Such dialectical movement was the product of "the assemblage of forces that necessarily evolve along opposing lines, [and] illustrate the notion of contradiction." Moreover, "the unity of opposites," at the core of most conceptions of the dialectic, has to be "understood as the unity of the elements of one level which engender the phenomena of a higher level." This was in accordance with the "abrupt rupture" of the preceding equilibrium and emergence of new integrative levels and novel forms, which constitute new "totalizations," or "partial totalities." In this sense, "qualitative leaps of the dialectic are found precisely on the borderlands where one passes from one state of matter to another, for example from the inorganic to the organic." In ecological terms, the problem, as Bernal had stated, is one of determining the "order of succession" arising from the metabolism, or material exchange, within nature (and society). "The very practice of science, its progress, the very way in which today it has passed from the static analysis of the world to the dynamic analysis of the world, is what is progressively elaborating the dialectics of nature under our eyes." In Vigier's view, "with Marx, science broke into philosophy."[78] Vigier's work reflected the rapid development of dialectical conceptions in science in the twentieth century with the rise of systems theory, often seen in dialectical terms, overtaking the contributions of dialectical social science.[79]

ECOSOCIALISM AND THE DIALECTICS OF ECOLOGY

In a dialogue with Hegel on dialectics on October 18, 1827, Johann Wolfgang von Goethe commented: "I am certain that many of those made ill by dialectics would find healing in the study of nature." Goethe's statement makes sense only if dialectics is seen as simply something apart from nature, merely "the systematized spirit of contradiction that we all have inside of us," as Hegel defined it on that occasion.[80] Yet, in the Hegelian idealist conception—as in the classical Marxian materialist one—there can be no rigid separation between a dialectics of society and a dialectics of nature. Notions of the dialectics of nature and organicist forms of materialism precede Marxism by thousands of years (not only in the work of the ancient Greeks, but also in Chinese philosophy, beginning in the Warring States Period during the Zhou Dynasty).[81] Nevertheless, Marxism has been able to bring new dialectical tools of analysis to bear on deciphering human society as an emergent form of nature, which is now, in its current alienated form, pointing toward its own annihilation.

Criticism and self-criticism are essential in the development of science. In the case of Marxism, this requires that the contradictions and divisions that arose over the dialectics of nature—contradictions and divisions that largely emanated from political realities—have to be healed in a new synthesis of theory and practice. Ecosocialism, which first emerged as a definite theoretical and political movement in the 1980s, matured in this century largely through the recovery of Marx's theory of metabolic rift, which has enabled a more complete understanding of the ecological crises of our time. But ecological materialism cannot go forward on the basis of Marx's now-famous metabolism analysis alone. It requires the recovery and reconstruction of classical Marxism's notion of *dialectical naturalism*, which constituted the second foundation of Marxism and has played a crucial role in the development of critical ecology from the late nineteenth and early twentieth centuries to the present day. This means overcoming the divisions that have developed within Marxism, in which both official Soviet Marxism and Western Marxism reduced nature to positivism while negating the negation of the negation.

Since the ecological crisis has placed the question of the dialectics of ecology front and center, it is significant that one of the bases from which today's ecosocialist/ecological Marxist critique stems is natural science. This is most clearly evident in the work of figures like Levins, Lewontin, and Stephen Jay Gould, who pushed forward a dialectical critique of reductionist science in the context of the developing catastrophic relation of capitalism and the environment. Intrinsic to this was a recognition of the weaknesses in much of Marxian theory due to the abandonment of the dialectics of nature. Levins was inspired from his youth by such figures as Marx, Engels, Lenin, Bernal, Needham, Haldane, Caudwell, Oparin, Schmalhausen, and Waddington. Levins was explicit about the failure of the Western Marxist tradition to unify its analysis with that of the Red scientists, and thus its inability on this basis to develop a meaningful analysis of the ecological crisis.[82] Writing in "A Science of Our Own" in *Monthly Review* in 1986, he stated:

> In the quest for respectability many Western European Marxists, especially among the Eurocommunists, are attempting to confine the scope of Marxism to the formulation of a progressive economic program. They therefore reject as "Stalinism" the notion that dialectical materialism has anything to say about natural science beyond a critique of its misuse and monopolization. . . . Both the Eurocommunist critics of dialectical materialism and the dogmatists [those who reduce dialectical materialism to mere formalism], accept an idealized description of science.[83]

A Marxist approach to science, Levins argued, required recognizing the importance of critical dialectical materialism in combating reductionism and positivism, as well as attention to how science itself had often been corrupted by capitalism, damaging the human relation to the earth. Levins and Lewontin published their seminal work *The Dialectical Biologist* in 1985, bringing back dialectical materialism as the basis of a critique of reductionism in biology, ecology, and society. This was followed in 2007 by *Biology Under the Influence*, which advanced a dialectical systems ecology. A key proposition was that "contradictions between forces are everywhere present in nature, not only in human social institutions."[84]

Gould, like Levins and Lewontin, consciously employed the dialectical method in all of his work on evolutionary theory, focusing in particular on (1) "*emergence*, or the entry of novel explanatory rules in complex systems, laws arising from 'nonlinear' or 'nonadaptive' interactions among constituent parts that therefore, in principle, cannot be discovered from properties of parts considered separately"; and (2) *contingency*, which meant that phenomena in nature, particularly those at higher emergent levels, had to be examined *historically*.[85] Gould warned that Earth as a place of species habitation would recover in hundreds of millions of years from the worst that humanity could deliver in terms of global thermonuclear war (or climate change)—but humanity itself would not.[86] Levins, Lewontin, and Gould all rejected the crudities of the official *diamat* in Soviet thought, while seeking to rescue the dialectics of nature as crucial not only to the Marxian critique, but also to a realist orientation to the world as a whole. Other dialectical biologists, such as John Vandermeer and Stuart A. Newman, have followed along in the same tradition.[87]

Analysis of the two most important works in Marx's hitherto unpublished intellectual corpus resulted in major developments in materialist dialectics in István Mészáros's two pathbreaking works, *Marx's Theory of Alienation* (1971) and *Beyond Capital* (1995). Mészáros was Lukács's close colleague prior to the 1956 Soviet invasion of Hungary, which compelled him to leave the country. In *Marx's Theory of Alienation*, Mészáros showed that Marx's basic ontological conception in the *Economic and Philosophical Manuscripts* embraced both the alienation of labor and the alienation of nature, tied together in Marx's ontological notion of human beings as the "self-mediating beings of nature" and their self-alienation under capitalism.[88] In *Beyond Capital*, which drew on Marx's *Grundrisse*, he argued that the planetary ecological crisis was the product of capitalism's inability to accept even the boundaries of the earth itself as a limit on uncontrolled accumulation, and that the ecological crisis was thus a core aspect of the structural crisis of capital.[89] Utilizing Marx's concept of metabolism, Mészáros presented capital as an alienated form of social metabolic reproduction based on second-order mediations of labor and nature. This analysis was to play an important role in the development of ecological Marxism, undermining narrow conceptions of Marx's dialectic and providing a systems theory rooted in Marx that bridged the ecological and social divide and helped reunify revolutionary theory and practice, impacting Hugo Chávez and the Bolivarian Revolution in Venezuela.[90]

Another key development in dialectical thought, bridging the gulf between the crude formalism of official Soviet thought and Western Marxism, was provided by the dialectical critical-realist philosophy of Bhaskar, which sought to renew ontology on materialist/realist foundations by reintegrating the question of naturalism into Marxian philosophy and ultimately developing a dialectical critical realism. It represented a full-scale attack on both neo-Kantian dualism, along with two-world dualisms in general, and on what Bhaskar called "the epistemic fallacy" that had subsumed ontology (the theory of the nature of being) within epistemology (the theory of knowledge). This went hand in hand with Bhaskar's rejection of the "anthropic fallacy," or the exclusive "definition of being in terms of human being."[91]

Bhaskar's work started from naturalist, realist, and materialist foundations and working from there systematically developed a dialectical ontology with a transformative praxis. In *Dialectic: The Pulse of Freedom*, this led to a dialectical critical realism that incorporated on multiple planes Engels's three ontological principles of the transformation of quantity into quality and vice versa, the unity of opposites, and the negation of the negation. In Bhaskar's analysis, the first of these principles was represented by the dialectics of emergence, the second by the dialectics of internal relations, and the third by what Bhaskar was to call the *absenting of absence*, incorporating the reality of past, present, and future potentials and possibilities in the understanding of the dialectic of continuity and change.[92]

Bhaskar's dialectical naturalism, like that of Marx and Engels, led him in the end to a consideration of ecological crisis. As he explained, "The limit at the plane of material transactions with nature"—Marx's social metabolism—"comes from the fact that human beings are natural beings. Nature is not apart from us, we are a part of it. The destruction of nature is not only murder but suicide and must be treated as such." From this it could be adduced that there "is a double impossibility theorem: it is not possible [at this stage] to have growth and ecological viability, and because it is not possible to have capitalism without growth, it is also not possible to have ecological viability with capitalism."[93] It followed that "at the level of material transactions with nature . . . it is absolutely unarguable that what we need is, from the point of view of the climate as a whole, less growth, that is, degrowth, and degrowth coupled with a radical redistribution of income. . . . This idea of degrowth would be associated with the idea of a simplification of social existence."[94] For Bhaskar, there was never any question about the necessity of a conception of the dialectics of nature, only about the conceptions currently held, leading him to develop his dialectical critical reason and ultimately resulting in his promoting a revolutionary praxis of degrowth.

Marx's theory of metabolic rift, or his theory of ecological crisis, was fully recovered only in the twenty-first century.[95] It derives its importance from its materialist dialectical conception of the alienated metabolism of nature and society under capitalism, a system that is now exploiting the world's population as never before while expropriating the earth on which humanity depends. This is the one critical perspective that fully encompasses both the social and extrahuman dimensions of the environmental crisis, seeing the class and ecological contradictions of capitalism as two sides of a single dynamic. The social metabolism represented by production mediates the material relation of humanity to ecological systems all the way from local ecosystems up to the Earth System.

This accords with Earth System science itself, which focuses on the disruption of the Earth System metabolism resulting in the *anthropogenic rift* in the biogeochemical cycles of the planet, creating the present habitability crisis. The result of this recovery of Marx's metabolic rift theory has been a formidable array of explorations of the social dimensions of the Earth System crisis, stretching from the metabolism of the soil to the climate to Earth System analysis.[96] Nevertheless, Marx's conception of the metabolic rift is only truly useful insofar as it provides us with a more active understanding of the social metabolism of human beings and the earth in all of its complexity as part of an overall materialist dialectics. For this, what is necessary is both a dialectics of society and a dialectics of nature, forming the basis of a new global environmental praxis.

Today, the world is faced with two opposing tendencies. One is the attempted acceleration of capital through the financialization of nature based on market forces and associated with processes of so-called decarbonization and dematerialization. The goal here is to subsume the world within the abstract logic of money as a substitute for real-world existence—an alienated logic that can only lead to total disaster, the barren negation of humanity itself. The other is the emerging struggle for planned degrowth and sustainable human development aimed at shifting power from global capital to workers on the ground and in their communities throughout the planet, representing the potential new power of an emerging environmental proletariat. This necessitates the merging of the economic and environmental struggles of the exploited and expropriated populations throughout the world in a new, broader form of cooperation. People at the grassroots are being driven to defend not just their work, but also their environments and their communities, and indeed, the habitability of the planet itself, conceived as a home for humanity and all other species. For this, however, we need a new, revolutionary dialectics of ecology.

1

The Return of the Dialectics of Nature: The Struggle for Freedom as Necessity

IT IS A FUNDAMENTAL PREMISE of Marxism that as material conditions change, so do our ideas about the world in which we live. Today we are seeing a vast transformation in the relations of human society to the natural-physical world of which it is a part, evident in the emergence of what is now referred to as the Anthropocene Epoch in geological history, during which humanity has become the major driver in Earth System change. An "anthropogenic rift" in the biogeochemical cycles of the earth, arising from the capitalist system, is now threatening to destroy the earth as a safe home for humanity and for innumerable species that live on it on a timeline not of centuries, but of decades.[1] This necessarily demands a more dialectical conception of the relation of humanity to what Marx called the "universal metabolism of nature."[2] The point today is not simply to understand the world, but to change it *before it is too late*.

Given that Marxism has been, since its conception in the mid-nineteenth century, the primary basis of the critique of capitalist society, it naturally could be expected to lead the way in the ecological critique of capitalism. But while historical materialists and socialists can be said to have played the leading, formative role in the development of the ecological critique—particularly within the sciences—the key contributions of socialist ecology, principally in Britain, took place outside the main tendencies that were to define twentieth-century Marxism as a whole. Beginning in the 1920s and '30s, a deep chasm emerged within Marxian theory, impeding the development of a coherent ecological view within the left. The dogmatism with which, on one side of this chasm, official Soviet thought by the mid-1930s approached the issue of the dialectics of nature and dialectical materialism more generally, had its counterpart, on the other side, in Western Marxism's categorical rejection of the dialectics of nature and the materialist conception of nature. To speak of "The Return of the Dialectics of Nature: The Struggle for Freedom as Necessity," in this chapter title, is thus to refer to the transcendence in our time, based on classical historical materialism and the dialectical naturalism that arose in Britain in the interwar period, of the principal contradictions hindering the development of a unified Marxian ecological critique.

POST-LUKÁCSIAN MARXISM AND THE CRITIQUE OF THE DIALECTICS OF NATURE

A major shift occurred in Marxian thought nearly a century ago following the publication in 1923 of Lukács's *History and Class Consciousness*, giving birth to what is now known as the Western Marxist philosophical tradition, but which could more accurately be referred to as "post-Lukácsian Marxism."[3] Lukács employed Hegelian dialectics to argue that the proletariat was the identical subject-object of history, giving a new philosophical coherence to Marxism and at the same time redefining dialectical thought in terms of totality and mediation.

Yet, in what was to become a defining trait of Western Marxism, Lukács, in conformity with the neo-Kantian tradition, rejected Engels's own notion of a dialectics of nature, on the alleged grounds that Engels had followed "Hegel's mistaken lead" in seeing the dialectic as fully operative in external nature.[4] Lukács applied Giambattista Vico's principle that we can understand history (the transitive realm) because we have "made it," and thus dialectical reflexivity can be said to apply in all such situations. Conversely, by the same logic, we cannot understand nature (the intransitive realm) dialectically, in the same sense, since it is devoid of a subject.[5]

At the same time, Lukács, it should be noted, did not categorically reject the dialectics of nature in *History and Class Consciousness*, subscribing rather to the notion, as Engels himself did, that there exists a "merely objective dialectics" of nature, capable of being perceived by the "detached observer."[6] This could then be seen as underlying the higher historical subject-object dialectics of human social practice. In this way, Lukács, following Engels in this respect, conceived of a *hierarchy of dialectics*, extending from merely objective dialectics, all the way up to the dialectics of the identical subject-object of history. Moreover, in his later works, beginning with his *Tailism* manuscript, written within just a few years of *History and Class Consciousness*, Lukács was to become a strong advocate of a *dialectics of nature and society* rooted in Marx's theory of social metabolism.[7]

Yet post-Lukácsian Marxists took the categorical rejection of the dialectics of nature as a defining principle of Western Marxism and even of Marx's own thought. Engels was in this way separated from Marx. As Sartre wrote: "In the historical and social world . . . there *really* is dialectical reason; by transferring it to the 'natural' world, and forcibly inscribing it there, Engels stripped it of rationality: there was no longer a dialectic which man produced by producing himself, and which, in turn, produced man; there was only a contingent law, of which nothing could be said except *it is so* and not otherwise."[8] This criticism went hand-in-hand with a hostility toward materialism and scientific realism, in the sense of the rejection of the materialist conception of nature, and a distancing from the achievements of science.[9] Serious ecological analysis was therefore missing from the Western Marxist philosophical tradition.

Although there was the famous criticism of "the domination of nature" in the work of Horkheimer and Adorno, it never got past the criticism of Enlightenment science—only to accede pessimistically in the end to its unavoidable necessity.[10] Marcuse's treatment of "The Revolt of Nature" in *Counter-Revolution and Revolt* did not go beyond the notion of the domination (and pollution) of nature's "sensuous aesthetic qualities" as a means for the domination of humanity and the need for an environmental rebellion in response.[11] There could, in fact, be no meaningful analysis of nature-society where both the materialist conception of nature and the dialectics of nature were denied, leaving Marxist theory with no dialectical critical-realist analysis on which to base an ecological critique. At most, within Western Marxist philosophical discourse, the relation of human beings to nature was reduced to instrumentalism, which was then subject to critique as the positivistic fetishism of technique, divorced from the wider question of the natural world and the human-social relation within it.

What was missing in such a one-dimensional approach was any notion of nature itself as an active power. As Bhaskar wrote in criticism of these tendencies of Western Marxism: "Marxists [meaning Western Marxist philosophers] have . . . for the most part considered only one part of the nature-social relation, that is, technology, describing the way human beings appropriate nature, effectively ignoring the ways (putatively studied in ecology, social biology, and so on) in which, so to speak, nature reappropriates human beings."[12]

Yet a powerful strain of ecological dialectics and critical, non-mechanistic materialism persisted in the natural sciences in the British Isles, evolving out of a tradition that drew on both Marx and Darwin, and that later became the heir of the early revolutionary Soviet ecology of the 1920s and early '30s. It was this "second foundation" of Marxist thought within the natural sciences which survived in the West, particularly in Britain, and that stretched back to Marx and Engels themselves, that was to play the formative role in the development of an ecological critique and constitute the main story told in *The Return of Nature*.[13]

FROM *MARX'S ECOLOGY* TO *THE RETURN OF NATURE*

The Return of Nature's central area of inquiry is the question of the organic interconnections between socialism and ecology that emerged in the century following the deaths of Darwin and Marx in 1882 and 1883, respectively, focusing in particular on developments in Britain and the United States. It follows a thread that was established in my book *Marx's Ecology* twenty years earlier. That work is best known for its explanation of Marx's theory of metabolic rift. But the real intent of the book was to explain how Marx's materialism had developed, going back to his confrontation in his doctoral thesis with Epicurus's ancient materialist philosophy. Marx's ecological perspective, it was argued, had developed as a counterpart to his understanding of the materialist conception of nature underlying the materialist conception of history.

A full materialist outlook, such as that developed by Marx, has three aspects: (1) ontological materialism, focusing on the physical basis of reality independent of human thought and existence, and out of which the human species itself emerged; (2) epistemological materialism, which is best understood as dialectical critical-realist; and (3) practical materialism, focusing on human praxis and its basis in labor. Since Marx and Engels rejected mechanical or metaphysical materialism, their materialism was necessarily dialectical in all three aspects: ontology, epistemology, and practice.[14] In Marx, materialism was closely related to mortality—"death the immortal"—applicable to all of existence, defining the material world.[15] In this perspective derived from ancient Greek materialism, nothing comes from nothing, and nothing being destroyed is reduced to nothing. In Marx's conception, the human social world was, in the sense of Epicurean materialism, an emergent form or level of organization within the natural-material universe. Energy (matter and motion), change, contingency, and the emergence of new assemblages or organizational forms all characterize the natural-physical world, which could be explained in terms *of itself*, as a process of natural history.[16] Marx's analysis was from the outset rooted in the evolutionary theory of which Darwin's theory of natural selection was the nineteenth-century culmination.

In his critique of political economy, Marx added to this overall materialist view the threefold ecological conception of: (1) the universal metabolism of nature; (2) the social metabolism (or the specifically human relation to nature through the labor and production process); and (3) the metabolic rift (representing the ecological destruction that ensues when the social metabolism comes into conflict with the universal metabolism of nature).[17] The labor and production process was thus the key not only to the mode of production in a given historical form of society, but also represented the human relation to nature, and thus social-ecological relations. Marx's theory of metabolic rift, which was first developed in the context of the rift in the soil nutrient cycle caused by the shipment of food and fiber to the new urban centers— where the essential nutrients, such as nitrogen, phosphorous, and potassium, ended up as pollution rather than returning to the soil—constituted the most advanced attempt in his day to capture the human-ecological relation. All subsequent ecological thought, up to ecosystem theory and Earth System analysis, was to be rooted in this same essential approach, focusing on metabolism.

Nevertheless, the argument of *Marx's Ecology* left the story of the formative role played by socialist thinkers after Marx in the emergence of ecology largely unaddressed. Moreover, there remained the contentious issue of the dialectics of nature, associated with Engels in particular. These issues were to be taken up in *The Return of Nature*. Although *Marx's Ecology* was a straightforward attempt to capture Marx's materialist and ecological views, the story told in *The Return of Nature* was much more complex, not least of all because it had to transgress certain divisions within Marxism itself.

Here we have to understand that the simultaneous rejection
of both the materialist conception of nature and the dialectics of
nature within Western Marxism was an inheritance of the neo-
Kantian tradition, which had its origin within German philosophy
with Friedrich Lange's 1865 work, *The History of Materialism*.
Lange attempted to use Kant's notion of the *noumenon*, or the
unknowable thing-in-itself, as the basis for demolishing materialism, a viewpoint that was carried forward in more sophisticated
ways by later neo-Kantians. It was with the rise of neo-Kantianism that epistemology came to occupy its dominant place within
philosophy, pushing aside ontology and also displacing the dialectical logic associated with Hegel. Materialist ideas and natural
science were seen as inherently positivistic. Room was made again
for religion and idealist philosophy via the Kantian *noumena* or
things-in-themselves.[18] Closely related to this, as Marx and Engels
noted, were the agnostic, dualistic views of British scientists such
as Huxley and John Tyndall.[19]

In opposition to the neo-Kantian dualism of Lange, which
rejected both materialism and Hegelian dialectics, Marx responded
by boldly stating: "Lange is naive enough to say that I 'move with
rare freedom' in empirical matter. He hasn't the least idea that
this 'free movement in matter' is nothing but a paraphrase for the
method of dealing with matter—that is, the *dialectic method*."[20]
Likewise, in *Capital*, Marx wrote: "My dialectical method is, in
its foundations, not only different from the Hegelian, but exactly
opposite to it. . . . With me . . . the ideal is nothing but the material
world reflected in the mind of man, and translated into the forms
of thought."[21]

In referring to the reflection of the "material world in the mind
of man," Marx had no simplistic notion of mirroring in mind, but
rather a dialectical conception of reflection (and reflexivity) and
a situated conception of knowledge, in which reason and both
objective and subjective agency play central roles within an ever-
changing historical reality. Marx's position was therefore a form
of "dialectical critical realism." As Bhaskar has explained, Marx's

dialectical "method, though naturalist and empirical is not positivist, but rather *realist*. . . . His epistemological dialectics [his critical
realism] commits him to a *specific* [materialist] ontological dialectics and a *conditional* [historical] relational dialectics as well."[22]

From a classical historical-materialist standpoint, the dialectics
of nature can be seen as part of a dialectical hierarchy. Thus, in
terms of what Marx in *Capital* called "its foundations," it stands for
the material world characterized by motion, contingency, change,
and evolution: the dialectic as material process. Central here is the
notion that nature (apart from human beings) in the contingent,
emergent effects of its manifold processes can be said to have a
kind of agency, even if this is unconscious agency. At a social level,
the dialectic can be seen in terms of human consciousness and
practice, the realm of the identical subject-object of the human-
historical realm, standing for human society as an emergent form
of nature. In its alienated form under capitalism, the human-social
realm often appears to be independent of the material world of
nature, or even as completely dominant over nature—though this is
a fallacy. In between these two abstract realms of the merely objective and the merely subjective dialectics lies the mediating realm
of human labor and production, the *dialectics of nature and society* (what Lukács was to call the "ontology of social being"), arising
from practice, which is, for Marx, the key to materialist dialectics.[23]

Marx gives us two basic ways of looking at this mediation of
nature and society through production—which, for him, in its
broadest sense accounts for all human appropriation of nature
and thus all material activity. In one of these pathways (most
evident in his early writings but also apparent in his later works,
such as his *Notes on Adolph Wagner*, written in 1879–80), the
human relation to the universal metabolism of nature is seen
in terms of human *sensuous* interaction with nature, which in
classical German philosophy was closely tied to aesthetics, but
which Marx linked to production as well. The second is in his
theory of the labor and production process as the social *metabolism* between human beings and nature, representing the active

relation of human beings to the earth. For Marx, we can know the world, including, to a considerable extent, the intransitive realm beyond the human subject, because we are part of it through our production and our sensuous existence. We live in a context conditioned by nature's laws, albeit in an emergent form in which historical laws, via specific modes of production, also condition human existence, mediating between nature and humanity.[24] Engels later adds to this, in line with Marx, the role of mathematics and scientific experiments as ways in which humanity connects dialectically to the wider, "merely objective" realm, employing methods of scientific inference arising originally from the human material relation to nature.[25]

In essence, whereas neo-Kantianism was rooted within a categorical division between the human subject and the objective natural world—between phenomena and noumena—that could not be transcended, Marxian materialist dialectics was grounded in human corporeal existence within the physical world, in a context of emergence, or integrated levels. Here, the dualism between humanity and nature was not a fundamental assumption but rather was seen as a result of an alienated consciousness rooted in an alienated system. We can know nature, as Engels was to write in *The Dialectics of Nature*, because "we, with flesh, blood and brain, belong to nature, and exist in its midst."[26]

THE DIALECTICS OF NATURE AND THE CREATION OF ECOLOGY

The Return of Nature, moving on from where *Marx's Ecology* left off, had a double burden. The historical narrative was concerned with explaining the various ways in which a tradition of socialist ecological analysis had arisen within art and science, in many ways dominating the ecological critique of contemporary capitalist society in the century from the deaths of Darwin and Marx up to the rise of the modern environmentalist movement. But at a deeper, more theoretical level, *The Return of Nature* was concerned as well with the ways in which a *materialist dialectics of nature*, often combined with other traditions, such as radical Romanticism and Darwinian evolutionary theory, had guided the development of modern ecology, based on the insights of social-ist thinkers. Here, the conception of the dialectics of nature, in its various forms—despite its categorical rejection by post-Lukácsian Marxists—could be perceived as playing the crucial role in a process of ecological discovery and critique.

A dialectical aesthetic as well as a dialectical conception of labor could be seen as underlying William Morris's understanding of nature-society relations. Dialectical conceptions also informed Lankester's evolutionary and ecological materialism. But the thread of the dialectics of nature only fully enters the narrative of *The Return of Nature* once the work of Engels is considered. In many ways, Engels's famous claim that "nature is the proof of dialectics" is the key, provided we understand what he meant by this in more contemporary terms by saying, "Ecology is the proof of dialectics."[27]

Although Engels has been heavily criticized by numerous thinkers for adopting a crude "reflectionist" view of knowledge, a close inspection of his work shows such claims are clearly false when placed in the context of his actual arguments.[28] Almost invariably, when Engels refers to "reflection," he immediately turns around and indicates that what we perceive as objectively conditioned by the material world around us (of which we are part) is a result not simply of conditions external to ourselves, but also a product of our active role in changing the world around us, and our understanding of it through our self-conscious reason. Our rules of scientific interference, our logic, our mathematics, our scientific experiments, our modeling, all have their roots in principles derived from human labor and production—that is, our metabolic relation to the world at large. "Reflection," as Marx and Engels use it—which invariably implies reflexivity, and is employed by them in the Hegelian, dialectical sense—is anything but positivist in character.[29]

Similarly, in attributing dialectical relations of a "merely objective" kind to nature itself, Engels emphasizes reciprocal relations, reflexivity, change, contingency, development, attraction-and-repulsion (contradiction), and emergence (or integrative levels) within nature as a totality, relying on Hegel's complex notion of "reflection determinations" from the "Doctrine of Essence" in his *Logic*.[30] The purpose is to capture the active, systemic, non-mechanistic relations that constitute the natural world, from which evolution (in the broadest sense) arises, and out of which humanity itself emerges. For Engels, as for Marx, it is our understanding of our own position within nature and our metabolism with the universal metabolism of nature that gives us the essential clues to those physical properties and principles that extend beyond ourselves. In this regard, Engels does not hesitate to attribute a kind of agency (auto-creation) to nature, the material world itself, understood in its broadest terms as in motion and constituted by the "*transformation* of energy" and the emergence of new organizational forms.[31]

Engels's well-known three "laws" of the dialectics of nature, better understood today as underlying ontological principles, perfectly manifest this outlook.[32] The first law, or the transformation of quantity into quality and vice versa, is now known in natural science as "phase transition" (or as a "threshold effect") and was explained in precisely that way by the Marxist mathematician Levy.[33] It can be seen as referring to the general phenomenon of integrative levels or the emergence of new organizational forms and assemblages within the material world, a view directly opposed to reductionist approaches to nature, and leading to a hierarchy of natural laws, the product of evolution, transformation, and change. Such an analysis is essential to all science today.

The notion of the unity/identity of opposites, or what Lukács, following Hegel, called "the identity of identity and non-identity," which has played such a large role in Marxian dialectics, was aimed at overthrowing notions of fixity, dualism, reductionism, and mechanism, focusing on the contradictions and feedback loops that induce transformative change.[34]

This then points to the third ontological principle, in which emergence now can be seen as the result of contradictions ("the incompatible development of different elements within the same relation") arising from material-historical changes, and leading to the "negation of the negation," an expression common to Hegel, Marx, and Engels. In the Marxian version, this phrase stands for the way in which the past mediates between the present and the future in material-historical development, producing a dialectic of continuity and change.[35] Engels himself referred to the "spiral form of development," which occurs when the residuals of the past and the active elements of the present coalesce to generate what Ernst Bloch was to call the "not-yet," or an altogether new reality. For Bhaskar, this takes the form of the "absenting of absence," or the transformative action directed at what has been inherited from the past in order to create a future existence.[36]

In a sense, the negation of the negation is a historical, evolutionary conception of emergence. Although emergence of new levels of organization was articulated in Engels's first "law," in terms of the transformation of quantity to quality and vice versa, now, following the generative principle of the unity of opposites (of contradiction), it takes on a developmental character: the emergence of a new form as a result of a historical process of reciprocal action or contradiction. This is what Bloch meant when he wrote that the "essential distinction between Hegel's dialectic and all previous candidates" was that "it is not stilled in the unity of contraries or contradictions."[37] In Marxian terms, the past is never simply past but rather mediates between the present (the moment of praxis) and the future.

In this way, Engels, in line with Marx, provided a dialectics of nature that was also a dialectics of emergence.[38] His analysis recognized the unity and complexity of nature, as well as the "alienated mediation" of nature and society represented by capitalism's irreversible rifts in nature's own metabolism.[39] This led to his powerful condemnation of capitalism's conquest of nature, as if of a foreign people, undermining ecological conditions. What Engels referred

to metaphorically as the "revenge" of nature was evident in deforestation, desertification, species extinctions, floods, destruction of
the soil, pollution, and the spread of disease.[40] Few other thinkers
(outside of Marx and Liebig) in the nineteenth century captured
so powerfully and succinctly the dialectic of ecological destruction
under capitalism.

Contrary to those who have argued (but without any substantive warrant) that Engels sought to subsume the dialectic of human
society in the dialectic of nature, his work *The Dialectics of Nature*,
although incomplete, was structured to move from the analysis
of the "merely objective dialectics" of nature via natural science,
to an anthropological basis in "The Part Played by Labour in the
Transition from Ape to Man." Here, the analysis was grounded in
the *dialectics of nature and society*, evolving out of human labor
and production and the human social metabolism with nature.[41]
This conformed to the structure adopted in *Anti-Dühring* in
which the argument proceeded logically from natural philosophy
to political economy and socialism, with political economy and
the mode of production seen as relatively autonomous from the
dialectics of nature as such, since conditioned by the dialectics of
human history. What in fact mediated between the two, for Engels
as for Marx, was human labor and production, that is, the social
metabolism. Herein lay the actual material realm of human beings
constituting the *dialectic of nature and society*, or what Lukács was
to later call the "ontology of social being."

Indeed, all critical-dialectical thought, encompassing both the
"merely objective dialectics of nature" and what could be called
its polar opposite, the "merely subjective dialectics of society,"
began for Engels, as for Marx, with the human social metabolism
via labor and production, constituting the objective ground of all
human existence: the *dialectic of nature and society*. Human self-
consciousness required that the objective world become its own,
but this could only be achieved on the basis of ontological principles expressing the specifically human relation to the universal
metabolism of nature.

All of our most fundamental scientific concepts regarding extra-human nature had their historical origins in human interactions with nature and the inferences drawn from them. To picture how this works, we can turn to the ancient Greeks. Empedocles in the middle of the fifth century BCE developed an experiment proving the corporeal nature of invisible and motionless air by demonstrating its resistance. This influenced Greek notions of flight. Thus, in Aeschylus's play *Agamemnon*, in which two eagles in flight (representing the two heads of the house of Atreus) are said to be rowing with "winged oars beating the waves of the wind," like the ships below, what is being presented is something more than simply a loose poetic metaphor. Rather, it was a direct application of the physical principle (the corporeal nature of air) derived from Empedocles's experiment.[42] In order to describe poetically the resistance that a bird's wings would experience in flight, Aeschylus drew on experience derived from human labor, referring to the oars of ships and the resistance that propelled the ships forward as they rowed. While such an example may seem quaint, and although we have infinitely more sophisticated explanations of a bird's flight today, what is significant is that basic scientific principles with regard to external nature arose from the earliest times through inferences from human interactions (primarily human production) with the natural world; inferences that then, in Epicurus's famous phrase, had to "await confirmation."[43] Although the scope of our experiments, our instruments, and our interactions with the universe have expanded, the fact that the basic concepts with which we approach extra-human natural phenomena arise first and foremost from our own material experience in interacting with nature remains the same.

Engels's analysis of the dialectics of nature was developed mainly in his *Anti-Dühring*—which he read to Marx as it was written in draft form (and to which Marx contributed a chapter as well as notes on the Greek atomists)—along with his unfinished *Dialectics of Nature*.[44] It was all clearly provisional, a work in progress, and incomplete. The British socialist scientists who were to be strongly influenced by Engels's materialist dialectics viewed it as a great, unfinished, and open-ended work of scientific inquiry; one far exceeding, as J. D. Bernal noted, the works in the philosophy of science in Engels's own time, represented by Herbert Spencer and William Whewell in England and Lange in Germany.[45]

For many of the leading British socialist thinkers of the early twentieth century—figures as varied as Lankester, Tansley, Farrington, George Thomson, Bernal, Needham, Hogben, and Caudwell—a key point of reference was Epicurean materialism, which was seen as offering not only a deep "materialist conception of nature," but also, via the swerve (*clinamen*, declension), the concept of contingency, understood as a movement away from a purely mechanical worldview. The Epicurean swerve was a notion stressed by Marx in his doctoral dissertation, which became available in the 1920s.[46] This was viewed by the British socialist scientists as connecting to a dialectical worldview and to Engels's dialectics of nature. Epicurus, as Needham emphasized, conceived nature as arising *of itself*, while swerving away from all rigid determinism.[47]

The result of this historical-materialist *Wissenschaft* (a term often translated as "science," but also referring to knowledge more generally when approached systematically on any topic) was a great renaissance of dialectical naturalism.[48] To point to just a few of the many pioneering developments, this included:

- Lankester's thesis that all major epidemics in animals and humans in the present age are the result of human production, and capitalism in particular[49]
- Haldane's theory (in parallel with that of the Soviet biologist A. I. Oparin) of the material origins of life—a discovery that was tied to a recognition of how life had created the earth's atmosphere, linked to the Russian biochemist V. I. Vernadsky's analysis of the biosphere[50]
- Haldane's role in the neo-Darwinian evolutionary synthesis and his integration of this with the dialectics of nature based on Engels's writings[51]

- Bernal's operationalization of the dialectics of nature and the negation of the negation in terms of a theory of the role of residuals in effecting the emergence of new forms of inorganic/organic organization[52]
- Needham's theory of integrative levels or emergence, encompassing both natural and social history[53]
- Tansley's introduction of the concept of ecosystem, in which he was influenced by Lankester's earlier ecological analysis and Marxist mathematician Levy's dialectical systems theory[54]
- Hogben's and Haldane's devastating scientific refutation of the genetic basis of race[55]
- Haldane's early empirical analysis, based on his father's research, of the buildup of carbon dioxide in the atmosphere[56]
- Bernal's leading role in the critique of the social relations of science[57]
- Caudwell's attempt to explore the interconnections in the dialectics of art and science[58]
- Farrington's and Thomson's pioneering research into Epicurean materialism and its relation to the development of Marxist thought
- Bernal's critique of nuclear-weapons development and treatment of how this threatened the end of life in its present form[59]

Collectively, this manifested itself as the detailed critique of ecological degradation and destruction integrated into the work of all these thinkers.

Not only were the scientific and cultural achievements associated with these leading figures in materialist dialectics within realms of science and art of great importance in their time (though later effaced by the Cold War), they were also connected fairly directly with the battles that occurred beginning in the 1950s, with the advent of the Anthropocene, around the sustainability of the natural environment and the rise of the environmental movement. These developments helped inspire the work of leftist scientists like Barry Commoner, Rachel Carson, and later on figures such

as Gould, Levins, Lewontin, Steven Rose, Hilary Rose, and Helena
Sheehan, and still more recent analysts such as Howard Waitzkin,
Nancy Krieger, and Rob Wallace. The reality is that there is a pow-
erful tradition of historical-materialist analysis within and related
to natural science that has often fallen outside the purview of
Western Marxism.[60]

The problem here is well illustrated by a couple of statements
by Perry Anderson, one of the premier Marxist cultural theorists and historians in Britain from the 1960s to the present day.
Writing in the *New Left Review* in 1968, Anderson referred to the
"false science . . . and the fantasies of Bernal."[61] The undeniable
fact that Bernal was one of the leading scientific figures in Britain
from the 1930s through the '60s, famous for his major discoveries,
and a Marxist, recognized as one of the great intellectual luminar-
ies of his time—even if sometimes deviating into a kind of Soviet
positivism—gets short shrift by Anderson. More significantly,
Anderson felt compelled to declare in 1983 that "problems of the
interaction of the human species with its terrestrial environment
[were] essentially absent from classical Marxism," thereby excluding Marx and Engels's contributions in this respect, suggesting
that the whole tradition of explorations of the dialectics of nature
(and of nature and society) by Marxist theorists was outside the
sphere of historical materialism properly speaking.[62] Similar posi-
tions were adopted by a host of other thinkers, such as George
Lichtheim, Leszek Kołakowski, Shlomo Avineri, David McLellan,
and Terrell Carver, all of whom sought to separate Engels from
Marx and the dialectics of nature from Marxism.[63]

Insofar as this tendency of post-Lukácsian Marxism had a
common basis, it had to do with postulations, inherited from neo-
Kantianism and deeply embedded in the dominant traditions of
philosophy, that rejected realism (critical or otherwise), and with
it any possibility of a dialectics of nature. How is it, then, that a dialectics of nature has been so powerful in unlocking the secrets of
the universe? The reason is that nature and society are not different

realities, but are co-evolving existences, in which society is asymmetrically dependent upon the larger natural world of which it is a part. Our knowledge of nature, of ourselves, and of our place in the world derives from this fact, spurred on in part by the very alienation of nature and the resulting self-consciousness that the capitalist system has generated. As Needham wrote:

> Marx and Engels were bold enough to assert that it [the dialectical process] happens actually in evolving nature itself, and that the undoubted fact that it happens in our thought about nature is because we and our thought are part of nature. We cannot consider nature otherwise than as a series of levels of organization, a series of dialectical syntheses. From the ultimate particle to atom, from atom to molecule, from molecule to colloidal aggregate, from aggregate to living cell, from cell to organ, from organ to body, from animal body to social association, the series of organizational levels is complete. Nothing but energy (as we now call matter and motion) and levels of organization (or the stabilised dialectical syntheses) at different levels have been required for the building of our world.[64]

For Caudwell, "The external world does not impose dialectic on thought, nor does thought impose it on the external world. The relation between subject and object, ego and Universe, is itself dialectic. Man, when he attempts to think metaphysically, contradicts himself, and meanwhile continues to live and experience reality *dialectically*."[65]

The French Marxist Roger Garaudy put this in more straightforwardly epistemological terms:

> To say that there is a dialectic of nature, is to say that the structure and movement of reality are such that only a dialectical thought can make phenomena intelligible and allow us to handle them.
>
> That is no more than an inference: but it is an inference

founded on the totality of human practice—an inference that is constantly subject to revision as a function of the progress of that practice. . . .

At the current stage of the development of the sciences, the representation of the real which emerges from the sum total of confirmed knowledge, is that of an organic whole in a constant process not only of development but also of auto-creation. It is this structure that we call "dialectical."[66]

Kant argued in his *Critique of Judgment* that, in dealing with the intransitive world of nature beyond our perceptions, it is necessary to conceive of it teleologically in order to say anything about it at all.[67] Science, however, has progressed far beyond this point, and though sometimes still presenting nature in teleological terms, it is more likely to resort to mechanical, systemic (systems theory), or dialectical terms.[68] The last of these most fully captures the universal metabolism of nature, encompassing its different integrative levels—including the inorganic and organic, the extra-human and human—connected with the results of human praxis.

The Dialectic of the Anthropocene

Why are these issues so important today, and why is there now a return to the dialectics of nature? This has to do with our own material conditions, which are increasingly dominated by the planetary emergency and the emergence of the Anthropocene, commencing around 1945 with the first nuclear detonation (followed by the bombings in Hiroshima and Nagasaki), representing a fundamental change in the human relation to the earth. As a result, the dialectic of nature in the twenty-first century is in many ways a dialectic of the Anthropocene. The Anthropocene Epoch is designated by science, though not yet officially, as a new epoch in the geological time scale, following the Holocene Epoch of the last 11,700 years. In the Anthropocene, humanity has arisen as the primary driver effecting changes in the Earth System. The dialectic of

nature and society has thus evolved to the point that human production is generating an anthropogenic rift in the biogeochemical
cycles of the planet, resulting in the crossing of various planetary
boundaries and representing the transgressing of critical thresholds in the Earth System that define a livable climate for humanity.

Climate change is one such threshold or planetary boundary. In
essence, the quantitative buildup of carbon dioxide in the atmosphere has resulted in a qualitative change in the climate sufficient
to threaten human existence, and even that of most life on Earth.
Other planetary boundaries that have been crossed or are in the
process of being crossed are represented by ocean acidification,
loss of biological diversity (and species extinction), the disruption
of nitrogen and phosphorus cycles, loss of ground cover (including forests), loss of fresh water sources (including desertification),
and chemical and radioactive pollution of the environment.[69]

The sources of these changes are not simply anthropogenic—
something that will not be reversed so long as industrial civilization
continues to exist—but are due more concretely to the worldwide
expansion of capitalism as an accumulative system geared to its
own internal growth *ad infinitum* and embodying in that respect
the most destructive relation to the earth conceivable. This was
captured by Marx's theory of metabolic rift, now raised to the level
of an anthropogenic rift in the Earth System.[70]

Although we have a widely accepted name for the new *geological epoch*, characterized by the human economy's current
role as the primary geological force on the level of the Earth
System itself, we still have no name for the new *geological age*,
nested within the Anthropocene Epoch that underlies the current Anthropocene crisis. Officially, in terms of geological ages,
we are still in the Meghalayan Age of the last 4,200 years, dating
from a period of climate change that was thought to have brought
down some of the early civilizations (though this is currently a
matter of dispute among scientists). But how are we to conceive
of the new *geological age* associated with the inception of the
Anthropocene Epoch?

My *Monthly Review* colleague Brett Clark and I, as professional environmental sociologists, proposed the name Capitalinian (also referred to by geologist Carles Soriano as the Capitalian) for this first geological age of the Anthropocene, standing for the fact that it is the capitalist world-system that has created the present planetary emergency.[71] The only solution—indeed, the only way of preventing the present mode of production from bringing about an Anthropocene extinction (or Quaternary Period extinction) event—is for human society to move beyond capitalism and the Capitalinian toward a future, more sustainable geological age within the Anthropocene, which we have labeled the Communian, after *community*, *commune*, and *communal*.

What is called the practical, relational dialectic—the dialectic of history—is now therefore caught up with the dialectic of nature and society reflected in Marx's theory of metabolic rift. This has now been given a wider field of operation, only truly apparent in our time, in which the metabolism of the entire planet, or the dialectic of nature, is being affected by an anthropogenic rift in the Earth System and in ways that threaten our own existence, calling to mind Engels's "revenge" of nature and Lankester's "Nature's revenges."[72]

It is important to understand that this Earth System crisis in the Capitalinian is tied to the long history of expropriation and exploitation that together constitute the foundation of capitalism's relation to the earth and humanity. *Expropriation*, in Marx's terms, meant appropriation without equivalent or reciprocity—that is, robbery. Marx thus spoke of the *robbery* of nature underlying the metabolic *rift*.[73] But he also wrote about the expropriation of the land from the population, removing the workers from the most basic means of production and thus control over their own lives. The age that Marx critically referred to as "so-called original accumulation" (so-called because it was defined not so much by accumulation as by robbery) was an age of expropriation.[74] Expropriation went beyond the theft of land to the theft of human bodies. This is associated with what Clark and I have designated as the "corporeal rift," marked by genocide, enslavement, and colonization of much of the world's population, underlying the relations of class exploitation.[75]

It is this wider logic of the expropriation of lands and bodies behind the capitalist system of exploitation that gave rise to the history of racial capitalism. This process of expropriation can also be seen in the robbing of women's household labor (which led Marx in his day to refer critically to women in capitalism as the slaves in the household) and in the continuing agribusiness expropriation of the land of subsistence workers, primarily peasants. Even people's leisure time away from work throughout the world is being expropriated in various ways in the accelerated accumulative society of digital capitalism. Today, capitalism is thus involved in myriad ways in the expropriation of the entire earth and its population: a system of robbery so extensive that the human relation to the earth, the very basis of human existence, is now in danger of being severed. In the end, the alienation of nature and the alienation of labor that characterize capitalism point only to destruction.

Our practical dialectics today thus require a knowledge of the *dialectics of nature and society*. The merely objective dialectics of nature, excluding the human subject, and the merely subjective dialectics of society, excluding natural-physical existence, are not enough. A greater critical unity of thought and action is being forced upon us. Dialectics, as Lewontin and Levins explained, focuses on "wholeness and interpenetration, the structure of process more than things, integrated levels, historicity and contradiction."[76]

In ancient Greece, the Ionian philosophers, such as Heraclitus, focused on *material processes as dialectical*. For Heraclitus, the basic metabolic process underlying life could be described thus:

> *As things change to fire,*
> *and fire exhausted*
> *falls back into things,*
> *the crops are sold*
> *for money spent on food.*[77]

In contrast to the Ionians, the Eleatics, such as Parmenides (followed by Plato and much later by Plotinus) conceived of a *dialectic of the idea*, or reason. Hegel can be seen as wedding these two vital streams together, building on all modern philosophy and the Enlightenment in his idealist philosophy, but giving precedence to dialectics as idea or reason.[78] Marx's materialist dialectics returned to material processes as underlying all reality, leading to an objective dialectic of change and emergence, of the metabolism of nature and society, and ending in a dialectics of human history and practice.

This materialist dialectical synthesis, the dialectic of nature and society, remains of great importance today. We live in a time, as Marx and Engels noted in *The German Ideology*, in which humanity must struggle in revolutionary ways not simply for the advancement of human freedom, but also to avoid destruction due to what can be called "capitalism's deadly threat" to the world and life in general. For Epicurus, Marx wrote, "the world [the earth] is our friend."[79] Materialist dialectics tells us that our goal in the present moment must be one of creating a world of ecological sustainability and substantive equality, one that promotes sustainable human development. But this starts in our time with an ecological and social revolution that is forced upon us. Today, the struggle for freedom and the struggle for necessity coincide everywhere on the planet for the first time in human history, creating a prospect of ruin or revolution: either a fall into the depths to which the Capitalinian has brought us, or the creation of a new Communian Age.[80]

—— 2 ——

Marx's Critique of Enlightenment Humanism: A Revolutionary Ecological Perspective

THAT MARX WAS THE FOREMOST revolutionary critic of Enlightenment humanism in the nineteenth century can scarcely be denied. No other thinker carried the critique of the Enlightenment's abstract, egoistic Man into so many areas—religion, philosophy, the state, law, political economy, history, anthropology, nature/ecology—nor so thoroughly exposed its brutal hypocrisy. But Marx's opposition to Enlightenment humanism can also be seen as transcending all other critical accounts down to the present day in its distinctive character as a dialectical and historical critique. His response to bourgeois humanism did not consist of a simple, one-sided negation, as in the Althusserian notion of an epistemological break separating the early and mature Marx. Instead, it took a more radical form in which the substance of his original humanist and naturalist approach was transformed into a developed materialism.[1] The result was a simultaneous deepening of his *materialist ontology*, which now took on a definite, *corporeal* emphasis focused on the conditions of human subsistence, together with the extension of this to the historical realm in the form of a *practical materialism*.

Marx's analysis was thus unique in offering a higher synthesis envisioning the reconciliation of humanism and naturalism, humanity and nature. Rather than stopping with a mere antithesis, as in most contemporary "post" conceptions, the object was the supersession of those material conditions of the capitalist mode of production that had made Enlightenment humanism the paradigmatic form of bourgeois thought. This radical rejection of bourgeois humanism was integrated with the critique of colonialism, where capitalism was seen as walking "naked" abroad, exposing its full barbarism.[2] In this regard, Marx's revolutionary response to Enlightenment humanism helped inspire the later critiques by such anticolonial thinkers as W. E. B. Du Bois, Frantz Fanon, and Aimé Césaire, all of whom called for the development of a "new humanism."[3]

Recent research into the ecological foundations of Marx's thought, particularly his conception of the metabolism of humanity and nature mediated by social production, has brought out more fully the depth and complexity of Marx's overall critique of capitalism's alienated social metabolism. This line of investigation demonstrates that, far from being anthropocentric, or succumbing to the Enlightenment notion of the conquest of nature, his vision encompassed the wider realm of what he called "the universal metabolism of nature." This included an appreciation of other life forms and his critique of environmental destruction in his famous theory of metabolic rift, giving rise to what can be called a revolutionary ecological perspective.[4]

Posthumanist, including so-called new-materialist, thinkers have recently sought to challenge Marx's metabolic vision and revolutionary ecology in general by promoting a phantom-like world of "dark ecology," hyperobjects, and vitalistic forces. However, such irrationalist views, as we shall see, invariably fail to address the fundamental criterion of the philosophy of praxis: the object is to *change* the world, not simply to *reinterpret* it.[5]

Enlightenment Humanism and Marx's Materialist Critique

For Marx, following Hegel, the Enlightenment criticism of religion led not to an all-out rejection of the Christian religious view, but rather its perpetuation through a pair of identical opposites: absolute idealism, stripped of an all-encompassing deity, on the one hand, and an equally absolute and mechanistic materialism, stripped of all sensuous qualities, on the other. Both of these mutually reinforcing opposites were evident in Cartesian rationalism, which carried over from Christian theology the dualistic distinctions between soul and body, mind and matter, and humanity and nature, and which was meant from the start to reconcile mechanistic science with religious doctrine.[6] As Engels wrote, the Enlightenment "merely posited Nature instead of the Christian God as the Absolute confronting Man."[7]

Bourgeois humanism, which arose in this bifurcated context, was characterized by Marx as the notion of abstract Man, or the isolated, spiritual, egoistic individual, "squatting outside the world," devoid of sensuous connections and material-social relations. Each atomistic individual was viewed as a "self-sufficient monad" emptied of all relations, yet endowed with innate rights, justifying a system of "mutual exploitation."[8]

Hidden within this abstract notion of bourgeois Man was not only class exploitation, but also the expropriation of human beings themselves, their very bodies, as in colonialism, genocide, and slavery. Deploring the blatantly racist content of such so-called humanism, Marx observed, quoting a public statement made at the time: "A Yankee comes to England, where he is prevented by a Justice of the Peace from flogging his slave, and he exclaims indignantly: 'Do you call this a land of liberty where a man can't larrup his n*****?'" What, Marx asked, could the "equal rights of man" possibly signify in this inhuman context?[9]

Bourgeois humanism was no less to be condemned for its inhumanity in the treatment of women. In an 1862 article titled "English Humanity and America," Marx chastised the English government and press for its effort to trade on "humanity" as an "export article" in its defense of wealthy, slave-owning women in New Orleans who were openly confronting and vilifying Union troops, and who had been told by the occupying Union general that if they acted like "street walkers" they would be treated as such. In the face of the supposedly high-minded protests in England over the gross "inhumanity" of such threats directed at upper-class, slave-owning women of the Confederacy, Marx noted that these same sanctimonious defenders of women's rights had conveniently lost sight not only of the slaves whose lives were in effect "devoured" by these New Orleans ladies, but also the English colonial abuse of Irish, Greek, and Indian women. Nor was there any consideration of the fate of proletarian women currently starving in Lancashire. The result was nothing less than a grand "humanity farce," concealing the most brutal inhumanity.[10]

Yet, despite his sharp attacks on Enlightenment humanism, Marx expounded a revolutionary humanism that came to be subsumed within his overall materialist conception of nature and history. What he characterized in the *Economic and Philosophical Manuscripts* as *positive humanism*, later termed *real humanism*, had nothing in common with the "pseudo-humanism" of bourgeois thought but rather, its negation.[11] "Communism," he wrote, "is humanism mediated with itself through the supersession of private property. Only when we have superseded this mediation will *positive* humanism, positively originating in itself, come into being." The emergence of an unalienated society would open the way to "the realized naturalism of man and the realized humanism of nature."[12] This would represent the "real emergence" of humanity, both as a "part of nature" and as the revolutionary realization of human social being.[13]

In the opening sentence of *The Holy Family*, Marx and Engels wrote: "*Real humanism* has no more dangerous enemy in Germany than *spiritualism* or *speculative idealism*, which substitutes '*self-consciousness*' or the '*spirit*' for the *real individual man*." *The Holy Family* can be seen as a work in which such speculative idealism was combated in the name of both humanism and materialism, and in which a more developed, dialectical conception of *real materialism* subsumed *real humanism* in Marx's thinking.[14] Thus, Marx writes that the speculative metaphysics arising in the seventeenth century and having its highest form in the nineteenth-century work of Hegel "will be defeated for ever by *materialism*, which . . . coincides with *humanism*. . . . French and English *socialism* and *communism* represent *materialism* coinciding with *humanism* in the practical domain."[15]

In recounting the origins of materialism in *The Holy Family*, Marx described how the resurrection of ancient Democritean and Epicurean materialism had in the seventeenth and eighteenth centuries generated a new materialism with "socialist tendencies," leading eventually to nineteenth-century socialism. Nothing was more opposed to the development of materialism in this sense than seventeenth-century speculative philosophy, particularly that of René Descartes, with its dualistic division of mind and body, soul and mechanism. Cartesian metaphysics, Marx declared, "had *materialism* as its antagonist from its very birth."[16]

Marx also opposed Hegelian idealism where it sought to reduce both humanity and nature external to humanity to pure thought, "abstracted from natural forms," creating a mystical realm of "fixed phantoms" operating on their own. Hegel, Marx wrote, saw "the history of mankind" as "the history of the *Abstract Spirit* of mankind, hence a *spirit* far removed from the real man." The human individual was reduced to a phantom-like abstraction. However, "if man is not human," since removed from material being, "the expression of his essential nature cannot be human, and therefore thought itself could not be conceived as an expression of man's being, of man as a human and natural subject, with eyes, ears, etc., living in society, in the world and in nature."[17]

The treatment of "positive humanism" in the *Economic and*

Philosophical Manuscripts of 1844 owed a great deal to Feuerbach's philosophy. However, as Marx's materialism developed, taking a more active form, he broke with Feuerbach's own abstract Man in which the human was nothing but "the true solemnization of each individual bourgeois" writ large.[18] In his *Theses on Feuerbach*, Marx rejected any essentialism or fixed conception of human nature, writing: "The essence of man is no abstraction inherent in each single individual. In its reality it is the ensemble of social relations." He added to this that, in creating such a rarefied conception of humanity, Feuerbach had been "obliged to abstract from the historical process . . . and to presuppose an abstract—*isolated*—human individual" that was unchanging.[19] All of [human] history, Marx wrote in *The Poverty of Philosophy*, "is nothing but a continuous transformation of human nature."[20] There was no sign in Marx's analysis, either before or after 1845, of what he called in *Capital* "the cult of the abstract man."[21]

Already in the *Economic and Philosophical Manuscripts*, Marx, in his comments on Hegel's *Phenomenology*, had referred to the human individual as a "*corporeal*, real, living, sensuous being" and "objective being," such that one finds one's objects and needs outside of oneself.[22] This was to form the starting point of *The German Ideology* and of Marx's historical materialism, in which he merged his early philosophical anthropology with a corporeal materialism:

> The first premise of all human history is, of course, the existence of *living* human individuals. Thus, the first fact to be established is the *corporeal organization* of these individuals and *their consequent relation to the rest of nature.* . . . Men can be distinguished from animals by consciousness, by religion or anything else you like. They themselves begin to distinguish themselves from animals as soon as they begin to *produce* their subsistence, a step which is conditioned by their corporeal organization.[23]

Here Marx both *materialized* humanity and made human beings the starting point for his philosophy of praxis. This, as Engels emphasized, was Marx's first great discovery: "the law of evolution in human history."[24]

MARX'S DIALECTICAL HUMAN ECOLOGY

Marx's materialist perspective, which owed far more to Epicurus than to Feuerbach, was ecological from his earliest writings, recognizing that the human *alienation from nature* was simply the other side of the *alienation of labor* (human self-estrangement). Hegel had defined nature as "externality," existing in "the form of the other being," and representing the realm of a distinct *other* that could only be transcended in thought. Marx retorted that this estrangement from the material world of nature should "be taken in the sense of alienation, a flaw, a weakness, something that ought not to be."[25] In this way, he declared as early as the *Economic and Philosophical Manuscripts* that the alienation of humanity from nature was the dialectical twin of the alienation of human labor, and a flaw to be historically transcended. The dual alienation of an externalized nature and of human labor could only be overcome through socialism and communism, or a new, revolutionary relation to human labor and production.

Marx has sometimes been mistakenly criticized for Prometheanism, in the contemporary sense of adherence to extreme productivism and a machine-centered technological determinism. Yet, not only are there no signs of this in his thought, but he devoted part of *The Poverty of Philosophy* to a strong condemnation of Pierre-Joseph Proudhon's very explicit, extreme mechanistic view and his myth of a "new Prometheus," which stood for the human "conquests over Nature" seen as part of a "providential aim."[26] Hence, the direct critique of mechanistic Prometheanism began with Marx himself. Marx's own identification with Prometheus was of a much earlier variety, dating back to Aeschylus's ancient Greek play, which saw Prometheus as a revolutionary figure and the bringer of light (later giving rise to the notion of Enlightenment), one who defied the gods and who was bound in chains.[27]

Nor is there any sign in Marx's work, even in his earliest writings, of a sharp separation of the human species being and the other species beings represented by nonhuman animals, except in the sense that human individuals were seen as the "self-mediating beings of nature," and thus the authors of their own self-estrangement.[28] Marx drew his understanding of psychological development of animal species from Hermann Samuel Reimarus's studies of animal *drives*, rejecting the notion of *instincts* projected by Cartesian rationalism. Instead, he identified both human and nonhuman animals as material, objective beings, motivated by inner drives, while seeking satisfaction of their needs outside of themselves, as objective beings.[29] Human beings were distinguished within this by their role as *Homo faber*, or the tool-making animal.[30] Nevertheless, as late as his *Notes on Adolph Wagner*, Marx continued to argue that not simply human beings but "animals" more generally "learn to distinguish 'theoretically' from all other things the external things which serve the satisfaction of their needs . . . and the activities by which they are satisfied."[31] Marx was a severe critic of Descartes's bourgeois reduction of nonhuman animals to machines, observing that "Descartes in defining animals as mere machines, saw with the eyes of the period of manufacture. The medieval view, on the other hand, was that animals were assistants to man."[32]

Quoting Thomas Müntzer, Marx pointed to the intolerability of the fact that in bourgeois society, "all creatures have been made into property, the fish in the water, the birds in the air, the plants on the earth—all living things must become free."[33] In his critique of early capitalist agribusiness, Marx condemned the conditions imposed on animals reduced to the state of commodity machines. In previous agricultural practices, he noted, nonhuman animals had been able to remain in the free air. Now they were confined to stalls with the accompanying box-feeding mechanisms. "In these prisons," he observed, "animals are born and remain until they are killed off," resulting in "serious deterioration of life force."

Referring to these conditions as "disgusting!" he declared that it was nothing but a "system of prison cells for the animals."[34]

Marx's wider material-ecological perspective was to manifest itself fully only in his theory of the social metabolism and the metabolic rift. What he called the "universal metabolism of nature" stood for fundamental processes underlying all existence, both inorganic and organic, in line with matter and motion (energy) and levels of organization (emergence). It thus prefigured the development of ecological theory in general, where such categories as the ecosystem, the biosphere, and the Earth System were to have the concept of metabolism as their basis. For Marx, the social metabolism was understood as the human mediation of the universal metabolism of nature via the labor and production process. The metabolic rift, or the "irreparable rift in the interdependent process of social metabolism," stood for the way in which the alienated social metabolism came into conflict with the universal metabolism of nature, generating ecological crises.[35] His analysis of the metabolic rift in the industrial capitalism of his day focused initially on the robbery of the soil of nutrients, such as nitrogen, phosphorus, and potassium, sending them hundreds and sometimes thousands of miles away in the form of food and fiber to the new urban manufacturing centers, where these "elementary constituents" of the earth ended up polluting the environment, rather than returning to the soil.[36]

On this basis, Marx developed a way of looking at how the destruction of ecological conditions, in capitalist production in particular, undermined human habitability—a viewpoint that extended beyond the issue of the soil to manifold ecological problems, including the role of the social system in spreading periodic epidemics. Marx's ecological critique, coupled with that of Engels, embraced nearly all of the ecological problems known in his time: the expropriation of the commons, soil degradation, deforestation, floods, crop failure, desertification, species destruction, cruelty to animals, food adulteration, pollution, chemical toxins, epidemics, squandering of natural resources (such as coal), regional climate change, hunger, overpopulation, and the vulnerability to extinction of the human species itself. It has now been extended by Marxian ecologists via his theory of metabolic rift to the entire set of anthropogenic rifts in the Earth System present in the twenty-first century, including the contemporary rift in the earth's carbon metabolism.[37]

POSTHUMANIST PHANTOMS VERSUS THE PHILOSOPHY OF PRAXIS

In recent years, much of Marx's critique of Enlightenment humanism has been replicated in what is called the "posthumanist turn" in philosophy, embracing a variety of attempts to deconstruct and destabilize Enlightenment humanism. These new philosophical perspectives draw principally on Nietzschean, Freudian, and, more recently, Foucauldian-Derridean-Deleuzian deconstructions of the human subject and of nature.[38] This has led to a variety of posthumanist traditions including object-oriented ontology, Latourian hybridism, new materialism, and the cyborgism of thinkers like Donna Haraway. Such views have gained considerable prominence within sectors of the left. Still, posthumanism (even when compared with the postmodernism that preceded it) has had relatively little influence thus far on Marxian theory itself, since it is radically divorced from the philosophy of praxis.

According to Marx's eleventh thesis on Feuerbach, "The philosophers have only *interpreted* the world in various ways; the point, however, is to *change* it."[39] A corollary of this is that in order to *understand* the world you have to seek to *change* it. Since posthumanism generally has been content to destabilize the human and the natural in ways that remove the theoretical bridges and ladders for changing the world—at times even seeking to undermine the notion of human praxis itself—its relation to Marxism has been quite limited. Posthumanism is caught in the world of "fixed phantoms" depicted by Marx, where the complete destabilization of the concept of the human means a disruption of the "human and natural subject, with eyes, ears, etc., living in society, in the world, and in nature." The result is a flat, monistic world of objects without subjects, populated by windowless monads, limitless assemblages (divorced from any conception of emergence), actants, hybrids, cyborgs, and enchantments—anything but a conception of material-sensuous human beings, production, and practice.[40]

This spectral world of phantoms might easily be dismissed as a pure distraction for those concerned with needed social and ecological change. However, the last decade or so has seen a shift of posthumanism (particularly in the form of so-called new materialism) into the ecological domain, where it has come into confrontation with Marxian ecology. New materialist (or new vitalist) thinkers in the humanities, such as Jane Bennett, have taken their inspiration in part from Epicurus's swerve, which was originally meant to introduce contingency into the mechanistic world of Democritean materialism. However, Bennett and other new materialists fail to note that by far the most penetrating analyst of Epicureanism in the nineteenth century, and the first to emphasize the importance of the swerve, was Marx, who deeply admired and drew upon Epicurus's nonmechanistic, nondeterministic materialism with its "immanent dialectics."[41]

New materialists, coming primarily out of the humanities, insist—as if this were a surprising new discovery—that human beings are not separated from the physical world as a whole, but instead that *becoming human* translates into "*becoming with*" nonhuman persons, who make up what was formerly called external nature.[42] Such analysts deny any special status to humanity, while embracing a flat ontology in which all life, and indeed all existence, is treated as web-like in its interconnections and fundamentally indistinguishable, even by the force of abstraction.

Replicating a tradition of thought within environmental ethics going back half a century or more, based on the notion of the intrinsic value of all things, the vitalistic new materialism places its emphasis on the moral equality of all existence (or a "democratic ontology") as the very basis of its ecological per-spective.[43] Moreover, it insists on what it calls the "vibrancy" of all nature, both organic and inorganic. Still, it does so outside of anything that could be described as a dialectical-naturalist or critical-realist perspective. Such posthumanist views are divorced from the long development of ecological theory, the critique of political economy, and the whole realm of natural science, as well as the philosophy of praxis.

In Bennett's work, nature is given a vitalist, reenchanted mean-ing, simply adding vital powers to material forms.[44] The goal, as in posthumanist thought in general, is to destabilize the concepts of both humanity and nature by creating phantom-like objects. For Timothy Morton, "dark ecology" is an approach that preserves "the dark, depressive quality of life in the shadow of ecological catastro-phe." Dominating this dark ecology are "hyperobjects," standing for spectral forces more massive than humanity and beyond its reach—as if the immensity of nature had not always been part of the materialist and dialectical conception of nature from ancient times to today.[45]

Morton, whose nihilistic dark ecology has nothing whatsoever to do with engaging with capitalism or the planetary ecological crisis (other than occasional references to the Anthropocene), nonetheless finds it necessary to enter into direct combat with Marx's ecology, given its emphasis on revolutionary praxis.[46] Marx's core concept of "social metabolism" becomes, in Morton's inven-tive rephrasing, a mere "human economic metabolism" that leaves out the rest of ecological existence. We are told that Marx adopted a "mechanical and reified" view of nature that is "frozen in the past."[47] Marx is repeatedly charged with being "anthropocentric" in introducing the notion of *human species being*—discounting that this also left room, in Marx's conception, for *nonhuman spe-cies beings* (species).[48]

All of this allows Morton to ignore or downplay the ecologi-cal analysis of classical historical materialism entirely, including Marx's notion of human society as an emergent form of nature, his broad adherence to Darwinian evolutionary theory, and his conception, along with Engels, of the dialectics of nature.

Yet, having dismissed dialectics and historical materialism, Morton's dark ecology, with its myriad phantom-like objects, cannot get "beyond antithesis," and has nothing meaningful to say about ecology itself.[49] In *Ecology without Nature, Dark Ecology*, and *Humankind*, he portrays a posthumanist, new-materialist world rife with "paranormal" spiritual phenomena, "spectral beings" and "hyperobjects." It is a *postworld* dominated by flat assemblages of humans and nonhumans, filled with "ghostly, quivering energy," and existing within the "symbiotic real." A biological species is reconceived as a "sparkling entity" beyond all rational definition. Hyperobjects become mysterious forces removed from a materialist and scientific understanding.[50]

Historical materialism is condemned by Morton for its anti-ecological perspective in excluding a conception of all objects as nonhumans to be placed on the same philosophical plane as humans. Marx's analysis is said to have come up short in its failure to recognize that oil, wind, water, and steam belong to the realm of "nonhuman people." Marxism, we are told, can only work if it becomes a new form of "animism," extending beyond the human, and even beyond living species themselves, encompassing within its conception of *persons* everything from rocks to microbes—in line with a vitalistic new imperium that embraces the "paranormal."[51]

The inner logic of this posthumanist, phantoms-of-the-opera world with its destabilizing mysticism is evident in the attacks on Marx's critique of the fetishism of commodities in the work of Bruno Latour, Bennett, and Morton. Latour famously rejected Marx's critique of commodity fetishism, along with *critique* altogether. Marx had argued that behind the fetishized forms of appearance of capitalist commodity relations lay human-productive relations. More concretely, as Lukács put it: "Fetishism signifies, in brief, that the relations between human beings which function by means of objects are reflected in human consciousness immediately as things, because of the structure of the capitalist economy. They become objects or things, fetishes in which men crystalize their social relations. . . . Human relations, as Marx says, acquire a 'spectral objectivity.'" [52]

Yet such a view of commodity fetishism, according to Latour, was too arbitrary, since it was rooted in particular conceptions of nature, humanity, production, and so on, and indeed, particular types of "facts." Having summarily dispatched in this way the critique of fetishism, Latour himself was then free to present the world of appearances as one of infinite things, commodities, objects, hybrids, and "actants," existing within a "flat ontology," with no up and down or inside and outside, blurring all distinctions. Reification in this world of "imbroglios" was no longer the subject of critique, which had thus "run out of steam." [53] Rather, the goal was to universalize the reification of human-social relations such that commodity fetishism became the model for analyzing an infinity of assembled things, forming an object-oriented ontology.

Such a total destabilization of the concept of humanity also requires the total destabilization of any concept of nature itself, of which humanity is an emergent part. So integral to Latour's theory was the negation of nature as a concept standing for the whole of material reality that, when he belatedly recognized the existence of the earth crisis, whereby humanity was destroying its own planetary habitat, he sought to replace the notions of *nature* and *ecology* with the *earth*, the *terrestrial*, and *Gaia*—a discursive change that constituted his entire contribution to the ecological discussion. For Latour, the posthumanist rejection of Marx's critique of the capitalist fetishism of the commodity had to be preserved, even to the point of claiming together with the capitalist ecomodernists of the Breakthrough Institute that we should uncritically "love" our technological Frankenstein monsters—disregarding the fact that adopting such a position would ensure a total incapacity to address the human-social dimensions of the planetary ecological emergency. [54]

Following in the footsteps of Latour, Bennett and Morton both explicitly reject Marx's critique of commodity fetishism (and of reification), insisting that instead of the "demystification" of things/objects/commodities, the goal should rather be one of their *reenchantment*, even *remystification*. Bennett thus seeks to speak on behalf of the inner "force of things" as nonhuman actants, both living and nonliving, organic and inorganic. She characterizes Marx's critique of the fetishism of commodities in *Capital* as inherently anthropocentric, since "what demystification uncovers is always something human," thereby screening out nonhumans. Adopting Spinoza's seventeenth-century metaphysical doctrine of *conatus*—or the inner impetus to be found within all physical entities aimed at preserving themselves and their motions—Bennett insists that "there is a power in *every* body." Quoting Spinoza, she pronounces: "In this respect all things [objects] are equal." In a questionable interpretation of Spinoza, she suggests that even stones have "thing-power." As Engels observed, "The notion of a 'vital force' latent in all things has been the last refuge of all supernaturalists."[55]

Morton similarly argues that human-centered demystification and defetishization, aimed at the world of commodities/things, should be rejected and replaced by a kind of remystification, thereby opening up space for nonhumans. By nonhumans, Morton, like Bennett, is not simply concerned with real, material, living species, but extends this to the realm of objects generally, embracing a flat ontology that puts Theodor Adorno's collection of plastic dinosaurs, a chocolate bar, and a microbe on the same physical and moral plane as a human individual living in society.[56] Marx's critique of commodity fetishism is thus rejected by posthumanist object-oriented ontology and by what has been called the "vitalistic new materialism," in the name of a phantom-like world akin to the mystical realm of religion, where objects of all kinds take on the role of spectral beings.[57]

For Morton, the issue is not that capitalism fashions a mystical veil associated with commodity fetishism, but rather that "capitalism is not spectral enough," and hence needs to become more so.

"The realm of the 'object' (the nonhuman in its most basic guise),"
he writes, "is precisely the realm in which commodity fetishism is
happening." But what is fetishistic, in his view, inverting Marx, is
not the failure to perceive the underlying human-social relations,
but rather the failure to give full spectral identity to the object.
Thus, defetishization or "demystification, rudely stripping the
appearance from things, is the capitalist operation par excellence,"
and needs to be reversed by privileging the mystical, the spectral,
and the paranormal. Only by means of animating commodities/
objects, no longer seeing them as mere things, will "solidarity with
nonhuman beings"—encompassing everything from microbes
to clouds—become possible.[58] In line with object-oriented ontology, we are told that "all beings [both organic and inorganic] have
agency, even mind."[59]

Posthumanist ecology, along with posthumanism more generally, thus closes off the philosophy of praxis in the name of the
leveling of all things within its flat ontology. Here there is no
room left for the consideration of the long history of capitalism,
colonialism, racism, imperialism, or ecological destruction, only
infinite webs of vital assemblages and hyperobjects, all circulating
nomadically on the same ontological plane without essential order
or meaning.[60]

The sharp contrast with historical materialism can be illustrated
by the way in which Morton selects for criticism a passage from
Marx's technical description of how raw materials are absorbed
in the process of production (in the account of constant capital
in the first volume of *Capital*). Quoting a sentence in which Marx
says, "The coal burnt under the boiler vanishes without leaving a
trace; so too the oil with which the axles of the wheels are greased,"
Morton pronounces that Marx here adopts the "anti-ecological
concept of '*away*'" toward such "nonhumans" (that is, the coal, the
oil, and the grease) denying that "objects have agency."[61] However,
what Morton, caught up in his posthumanist/postnaturalist conceptions, fails to comprehend is that coal, oil, and grease do not
themselves have historical agency—though, like everything else

in existence, they are in perpetual flux—and cannot usefully be treated as "nonhuman persons," comparable to human beings. Coal burned under the boiler is not its own self-mediating being of nature any more than a lump of coal could willfully decide to combust itself and distribute the resulting carbon dioxide molecules into the atmosphere, contributing to climate change.[62]

Here, a turn from posthumanism to reality is necessary. The current planetary ecological emergency is the greatest environmental threat that the human species has ever encountered, endangering the lives of billions of people along with the majority of known species on Earth. As Kate Soper said, in responding to the posthumanist destabilization of the concepts of humanity and nature, it needs to be remembered that "it is human ways of living," and, more specifically, capitalist ways of producing, "that are wrecking the planet, and [it is] humans alone who can do something about it."[63] In the struggle before us, focusing on phantoms, spectral beings, and cyborgs will not help. Everything in existence is not on the same plane and the world will not be rescued by the actions of objects.[64] What is needed instead is a revolutionary humanity inspired by reason and dedicated to the struggle to create what Marx called "the perfected unity in essence of man with nature." This can only be achieved through the transcendence of the capitalist order and the rational regulation of "the interdependent process of social metabolism" by the associated producers.[65] There is no other way.

3

Engels and the Second Foundation of Marxism

ON THE OPENING PAGE of *The Return of Nature*, I referred to the "second foundation" of socialist thought as follows:

> For socialist theory as for liberal analysis—and for Western science and culture in general—the notion of the conquest of nature and of human exemption from natural laws has for centuries been a major trope, reflecting the systematic alienation of nature. Society and nature were often treated dualistically as two entirely distinct realms, justifying the expropriation of nature, and with it the exploitation of the larger human population. However, various left thinkers, many of them within the natural sciences, constituting a kind of *second foundation* of critical thought, and others in the arts rebelled against this narrow conception of human progress, and in the process generated a wider dialectic of ecology and a deeper materialism that questioned the environmental as well as social depredations of capitalist society.[1]

The origins and development of this *second foundation* of critical thought in materialist philosophy and the natural sciences, and how this affected the development of socialism and ecology, constituted the central story in *The Return of Nature*. The initial challenge confronting such an analysis was to explain how historical materialism, in the dominant twentieth-century conception in the West, had come to be understood as strictly confined to the social sciences and humanities, where it was divorced from any genuine materialist dialectic, cut off from natural science and the natural-physical world as a whole.

Explorations of the dialectics of nature by Engels along with Marxian contributions to natural science were commonly treated in the Western Marxist philosophical tradition as if they simply did not exist. The natural-physical world was seen within the dominant view of Marxism in the West as outside the domain of historical materialism. The realm of biophysical existence was thus ceded to a natural science that was viewed as inherently positivist in orientation. This was so much the case that, with the rise of the environmental movement in the 1960s, it never occurred to those on the left who wrongly charged that Marxism had contributed little or nothing to the development of ecological analysis to look beyond the social sciences to socialist contributions in the natural sciences, out of which today's systems ecology arose. The irony was that not only had socialism engaged with the natural environment, but it had from the very beginning played a pivotal role in the development of a critical ecology within science and materialist philosophy.

Part of the problem was that the entire tradition of "dialectical materialism," associated with Soviet Marxism in particular, was declared by the Western Marxist philosophical tradition to be erected on false foundations. The *dialectics of nature*, as opposed to the *dialectics of society*, it was claimed, needed to be rejected since it lacked an identical subject-object and thus absolute reflexivity. But in rejecting the dialectics of nature, Western Marxism was compelled to absent itself from the natural world almost entirely, except insofar as it could be said to impinge on human psychology or human nature or to have an indirect impact via technology. This then encouraged a shift toward a more idealist interpretation of Marxism.[2]

To be sure, the classical Marxism of Marx and Engels in the mid-nineteenth century had its origin in the critique of social science. As Engels wrote, "Classical political economy" was "the social science of the bourgeoisie" and, as such, the enemy of socialism.[3] Marx's critique of classical political economy was aimed at uncovering the "hidden abode" of class-based exploitation and expropriation on which the capitalist mode of production was based.[4] It was this critique, therefore, that constituted the initial foundation of Marxism. But from the beginning, the materialist conception of history in critical social science was inextricably tied to the materialist conception of nature in natural science. No coherent critique of political economy was possible without exploring the actual biophysical conditions of production associated with what Marx called the "universal metabolism of nature."[5]

Human beings themselves were seen by Marx as *corporeal beings*, and thus *objective beings*, with their objects outside of themselves. There was, in the end, only a "single science" looked at "from two sides," those of natural history and human history.[6] It was necessary, therefore, to go beyond philosophy and social science to engage in the critique of bourgeois natural science as well. Indeed, as a theoretical method, the philosophy of praxis could not be confined to the realm of social sciences and humanities—that is, it could not be divorced from natural science, without undermining its overall critique.

The fact that natural science and social science, nature and society, are inextricably bound together in any attempt to confront the current mode of production and its consequences is dramatically demonstrated to us today by the current Anthropocene Epoch of geological history, in which capitalism is generating an "anthropogenic rift" in the biogeochemical cycles of the Earth System, endangering humanity along with innumerable other species.[7] In these circumstances, the role of Marxian ecology in understanding our current environmental predicament is of crucial importance. It is here that the second foundation of Marxian theory within materialist philosophy and natural science proves to be indispensable to the development of a revolutionary praxis.

THE SECOND FOUNDATION

Marx and Engels did not see science, or what they called "scientific socialism," in terms of the narrow conceptions of science that prevail in our day, but rather in the broader sense of *Wissenschaft*, which brought together all rational inquiries founded on reason.[8] Reason as science had its highest manifestation in the application of dialectics, which Engels defined in the *Dialectics of Nature* as "the science of the general laws of *all* motion," contending that "its laws must be valid just as much for motion in nature and human history as for the motion of thought."[9] Indeed, a consistent materialist dialectic was not possible on the basis of social science alone, since human production and human action occurred "in society, in the world and in nature."[10]

Engagement with natural science became a more urgent necessity for Marx and Engels as their work proceeded. Darwin's evolutionary theory, in Marx's words, was "the basis in natural science for our view." Engels depicted Darwin as the leading "dialectical" thinker within natural history.[11] Revolutions in natural science, such as Liebig's soil chemistry, allowed Marx to develop his theory of metabolic rift. The emergence of anthropology as a result of the revolution in ethnological time pulled Marx and Engels into this new realm having to do with prehistory.[12] They incorporated the new revolution in thermodynamics within physics into their political-economic critique.

However, there were also negative developments that compelled the founders of historical materialism, beginning in the 1860s, to shift their research more in the direction of natural science and the second foundation of Marxist theory. The defeat of the 1848 revolutions in Germany in particular had encouraged the growth

of a mechanistic philosophy of science in a line extending from
the post-1840s writings of Feuerbach to thinkers such as Ludwig
Büchner, Carl Vogt, and Jacob Moleschott. At the same time,
Lange had introduced neo-Kantianism as a dualist philosophical perspective aimed at circumscribing a one-sided mechanical
materialism, which was then separated from an equally one-sided
social/ideal realm. Coupled with this was the spread in Germany
of irrationalism in the philosophies of Arthur Schopenhauer and
Eduard von Hartmann, who saw materialism and dialectics, principally Hegel and Marx, as the enemy.[13] Eugen Dühring entered
into all of this with an eclectic mix of neo-Kantian, pseudosci-
entific, and positivistic ideas that targeted Marx. Agnosticism in
Britain, in the work of figures like Huxley and Tyndall, was closely
identified with neo-Kantianism. Social Darwinism first arose in
this period principally as an attack on historical materialism in
the work of the German zoologist Oscar Schmidt. As a result of
these various attacks on materialism and dialectics, both Marx
and Engels were pulled into the task of articulating a dialectics
of nature consistent with a socialist conception of the metabolism
of humanity and nature, in what was later variously referred to
as dialectical materialism, dialectical naturalism, and "dialectical
organicism."[14]

Engels's dialectical naturalism was first advanced in a comprehensive form in his influential work *Herr Eugen Dühring's
Revolution in Science* (better known as *Anti-Dühring*), completed
in 1878. His wider, unfinished work, written in the 1870s and '80s,
Dialectics of Nature, was not published in German and Russian
until 1925, and had to wait another decade and a half before it
was to appear in English translation. Nevertheless, Engels's central argument, that "Nature is the proof of dialectics," was clear
from the start. Translated into today's terms, it meant: *Ecology is
the proof of dialectics.*[15]

In Engels's words, "dialectics," in its materialist form, was "a
method found of explaining . . . 'knowing' by . . . 'being,'" rather
than "'being' by . . . 'knowing.'" It "interprets things and concepts

in their interdependence, in their interaction and the consequent changes, in their emergence, development, and demise." Viewed in this way, "nature," he wrote, "does not move in the eternal oneness of a perpetually recurring circle, but [goes] through a real evolution." Thus, "the whole of nature accessible to us forms a system, an interconnected totality of bodies, and by bodies here we understand all material existences extending from stars to atoms. . . . It is precisely [their] mutual reaction that creates motion."[16] Within the course of natural history, nature as matter and motion (transformed energy) generates new, emergent forms or integrated levels of material existence that arise out of, and yet remain dependent on, the physical world as a whole. Human society is, in this sense, an emergent form of the universal metabolism of nature with its own specific laws.[17]

As we noted in chapter 1, Engels has often been criticized on the left for his three dialectical "laws," more properly referred to today as general ontological principles, that he presented in his works on the dialectics of nature: (1) the law of the transformation of quantity into quality, and vice versa; (2) the law of the identity or unity of opposites; and (3) the law of the negation of the negation. However, the first of these ontological principles has been long recognized within science through the concept of phase change, while the second is the main way in which dialectics is commonly approached in philosophy and social science through the concept of contradiction, or "the incompatible development of different elements within the same relation."[18] Most criticisms thus focus on the third of these laws, the negation of the negation, which is often simply dismissed.[19]

Nevertheless, it is important to understand these three laws or ontological principles in terms of a dialectics of *emergence*. For Engels, everything is motion—attraction and repulsion, contingency, and development—leading to new forms or levels of organization in nature and human history. The law of the transformation of quantity into quality and vice versa refers to material transformation and transcendence at the most general level. Given such tendencies, arising out of the transformation of matter and motion (or energy) in organic and inorganic processes, contradictions or incompatible elements naturally ensue, leading to change as development, evolution, or emergence: the *negation of the negation.*

We can see the significance of this in Engels's approach to geology. He treated geology and paleontology as "the history of the development of the organic world as a whole," which practically came into being as a developed field of scientific research only in the late eighteenth century. The world that geology describes exists even "in the absence of human beings."[20] Nonetheless, geological history can be approached dialectically, since "the whole of geology is a set of negated negations" resulting in massive transformations on the surface of the planet that can be discerned by means of careful scientific investigation. Engels questioned Georges Cuvier's crucial emphasis on geological "revolutions" or catastrophes as contaminated by religious dogma, and argued that Charles Lyell, with his gradualism, had introduced a more scientific approach to geology. But Lyell himself had made the error of "conceiving the forces at work on the earth as constant, both in quantity and quality," so that "the cooling of the earth" associated with ice ages "does not exist for him." In this view, there are no "negated negations" and no major, permanent changes.[21]

There was, for Engels, no constant, noncontingent, inconsequential process of earth surface formation in line with Lyell's uniformitarianism. Massive transformations of the earth at certain intervals in its history, as emphasized by Cuvier, were not to be denied. Some of these criticisms (and appreciations) of both Cuvier and Lyell, advanced by Engels, were later developed in the twentieth century by Gould, who used precisely these antinomies to explain the origins of the theory of punctuated equilibrium within the evolutionary process.[22]

Anti-Dühring, because of its sheer range—addressing philosophy, natural science, and social science—became one of the most influential works of its time. It helped spark the development of left materialism in science, which was later given a further boost by the publication of *Dialectics of Nature*. This facilitated major ecological discoveries, especially in the Soviet Union in the first two decades after the revolution, and in the British Isles, where a tradition emerged drawing on both Darwin and Marx. Among the major figures in Britain were Marx's friend, and Darwin and Huxley's protégé, Lankester, and later leading Red scientists and related cultural figures such as Bernal, Haldane, Needham, Hogben, Levy, Caudwell, V. Gordon Childe, Farrington, Thomson, and Lindsay.[23] Along with Engels's works on science, the British Red scientists drew heavily on Lenin's *Materialism and Empirio-Criticism*.[24] Although frequently overlooked in treatments of Marxism, this tradition included the most prominent Marxist thinkers of the day, in Britain, all of whom were connected with materialist philosophy and natural science. Their work sunk deep roots in natural science, the influence of which has extended to our own time.

Marxist scientists and materialist philosophers were the target of purges in the Soviet Union in the 1930s and in the anticommunist attacks in Britain and the United States in the 1950s. The suppression of Red science, which seemed almost to disappear for a time, had deep ramifications for Marxism as a whole. Since the leading representatives of the Western Marxist philosophical tradition rejected outright materialism apart from economic/class relations—a position closely associated with their rejection of the dialectics of nature—they had almost nothing of substance to contribute to the ecological critique. This led to the myth that socialism as a whole had failed in this area.[25] To be sure, critical theorists such as Horkheimer and Adorno referred to the "domination of nature," by which they chiefly meant the role played by instrumental rationality and technology in contemporary capitalist society, as well as the repressive effects of this on human nature. However, the material-ecological world itself was characteristically absent from their analysis. Hence, the dialectical connections associated with human social production and its metabolism with the larger environment were also absent.[26]

What has become clear with the growth of Marxian ecology since the 1980s is the close connection between the critique of *economic alienation* and *ecological alienation* under capitalism. Recognition that these constitute the two sides of the historical-materialist critique has become increasingly pronounced in the context of the planetary ecological crisis. All of this calls for the reunification of Marxian theory, symbolized by *the return of Engels*, and an attempt to grapple with the universal metabolism of nature. There is an urgent necessity to transcend the current alien-ated form of the capitalist social metabolism with its destructive mediation of the human relation to nature through generalized commodity production.

Engels and the Roots of the Anthropocene

In the twenty-first century, we live in an age of planetary eco-logical peril, represented by the anthropogenic rift in the Earth System. This is associated with the advent, around 1950, of the Anthropocene Epoch in the geological time scale, which suc-ceeded the Holocene Epoch of the last 11,700 years. Capitalism is presently in the process of crossing planetary boundaries that have defined the earth as a safe place for humanity. If all geological history, as Engels said, is the history of "negated negations," today the Holocene—the geological epoch in which human civilization arose and prospered—is being negated by the system of capital accumulation, leading to the Anthropocene crisis of today.

If we were to look back to the earliest overarching recognition of the ecological predicament imposed by capitalist society, we could not do better than to turn to Engels's famous treatment of this in "The Part Played by Labour in the Transition from Ape to Man" in the *Dialectics of Nature*. Here, Engels declared that human beings, as social beings, do not "rule over nature like a conqueror over a foreign people, like someone standing outside nature—but that we, with flesh, blood and brain, belong to nature, and exist in its midst, and that all our mastery of it consists in the fact that we have the advantage over all other creatures of being able to learn its laws and apply them correctly." Thus, for each presumed "victory" of humanity over the natural world of which we are a part, "*nature takes its revenge on us*," leading to widespread natural/ecological devastations—not simply in the ancient and medieval worlds, but increasingly, and on a far larger scale, in the world wrought by capitalism and colonialism.[27]

Failure to understand what Engels called "*our oneness with nature*" and the need to conform to its laws is itself a product of our historical class relations. Here, the capitalist domination of nature becomes a means of dominating human beings. The result is that history moves in a spiral, exhibiting both progress and retrogression.[28] Accumulation of capital is accompanied by the accumulation of catastrophe. Moreover, under such an anarchic system—as opposed to a socialist and planned society controlled by the associated producers—a fully rational pursuit of science becomes impossible, and substantive irrationalism prevails even in the midst of the advance of formal technological rationality. Pointing to soil degradation, deforestation, floods, desertification, species extinction, epidemics, and the squandering of natural resources, Marx and Engels indicated that the current mode of production was generating widening Earth catastrophes associated with the uncontrolled "interference with the traditional course of nature."[29] Engels's global analysis of nature's "revenge" was thus at one with Marx's theory of metabolic rift.

"The Part Played by Labour in the Transition from Ape to Man" was first published in 1896 in the German Social Democratic journal *Die Neue Zeit* shortly after Engels's death. Although it is difficult to chart its influence outside of Marxism, it is remarkable how close Engels's analysis was to the ideas put forward not long afterward by Lankester in 1905 in his Romanes Lecture at Oxford, "Nature and Man" (later retitled "Nature's Insurgent Son"), and his related 1904 article "Nature's Revenges: The Sleeping Sickness," both of which were reprinted in his 1911 *The Kingdom of Man*.[30] We do not know if Lankester read Engels's article, though he was fluent in German, communicated with social democratic circles, and would have been deeply interested in Engels's analysis in this respect, which overlapped in many ways with his own.[31] As a close friend of Marx and an acquaintance of Engels, a strong materialist, and a critic of capitalism (who had read Marx's *Capital*), as well as the leading figure in British zoology at the time, Lankester's radical ecological critique was necessarily related to historical materialism. In referring to the *Kingdom of Man*, Lankester sought to describe a new period in Earth history in which human beings were now the main force affecting the natural world, with the result that they increasingly must take responsibility for it. He presciently highlighted the ecological consequences of a capitalist economic system engaged in the unheeding destruction of nature, ultimately undermining humanity itself.

In "Nature's Revenges," Lankester referred to the human-social being as "the disturber of Nature," including being the instigator through world capitalism and finance of all epidemics in animals (including humans) and probably plants as well, which could be traced largely to social, and primarily commercial, causes, including the "mixing up of incompatibles from all parts of the globe."[32] Under these circumstances, humanity had no choice but to control its production and its relation to nature, relying on science and superseding the narrow dictates of capital accumulation, thus ushering in a coevolutionary development. Human society was on a permanent ecological knife-edge in its relation to the natural world, which Lankester described somewhat ironically as the "Kingdom of Man." Such "effacement of nature by man" not only undermined living species, but also threatened civilization and human existence itself.[33] The only answer was for social humanity to take responsibility for its relations to the natural world, in conformity with natural laws and principles of sustainability, in opposition to the capitalist mode.

Today, resistance to the notion of the Anthropocene Epoch is evident in many of those on the left, who, while largely oblivious of the scientific discussion, are horrified by the implications of a dominant

Anthropos. This seems, in their minds, to point to an exaggerated humanism or anthropocentrism in the understanding of the physical world, and to a downplaying of the social causes of the geological climacteric that we are now witnessing. Yet, from a geological and Earth System perspective, the issues are clear. By crossing certain critical thresholds or planetary boundaries, the global system of capital accumulation has generated quantitative changes that represent a qualitative transformation in the Earth System, shifting it from the Holocene Epoch in the Geologic Time Scale to the Anthropocene Epoch, where *anthropogenic* rather than *nonanthropogenic* factors are for the first time the major drivers of Earth System change and are imperiling human civilization and existence.[34]

From a historical and dialectical perspective, the planetary ecological contradictions that we are now witnessing have been long coming. The issue of a new "Kingdom of Man," which was at the same time subject to the *revenge of nature* or *nature's revenges*, can be traced back to Engels and Lankester. Such views were related to the conception of nature as a dialectical totality mediated by processes of evolutionary change, in which humanity was increasingly playing a dominant role. The notion of what was called the Anthropogene Period in geological history, connected to the disruption of the biosphere as defined by Vernadsky, was introduced in the Soviet Union during the 1920s by the geologist Aleksei Pavlov. The word *Anthropocene* itself, as an alternative to *Anthropogene*, first appeared in English in the early 1970s in the *Great Soviet Encyclopedia.*[35] It was by uniting the awareness of ecological destruction with the concept of ecosystem, the theory of the origins of life, and the analysis of the biosphere—all products of dialectical science—that Rachel Carson was able to warn the world population of the full scale of the planetary peril confronting them in her lecture introducing the concept of ecology to the general public. Moreover, it was socialist scientists who pointed to a decisive change in the human relation to the entire Earth System, or "ecosphere," beginning around 1945.[36]

More recently, we can point to the breakthrough in the treatment of the Anthropocene Epoch in Earth history represented by the geologist Soriano. The conception of the Anthropocene Epoch in the Geologic Time Scale derives from the recognition that for the first time in the more than two billion years of Earth history after the oxygenation of the oceans and atmosphere by cyanobacteria, a living species, *Homo sapiens*, is the primary driver of Earth System change. This revelation of the human role in geological change was thus the product of both the emergence of Earth System science and the growing perception of an "anthropogenic rift," undermining the earth as a safe home for humanity. It has its theoretical roots in the concept of metabolism, which formed the basis for the notion of ecosystem (first introduced by Lankester's student, the British ecologist Tansley, a Fabian-style socialist) and the later concept of the Earth System metabolism.[37]

Once human society has emerged as the primary force in Earth System change due to the scale of production, inaugurating the Anthropocene Epoch, this becomes unalterable—barring the collapse of industrial civilization in an Anthropocene extinction event. Like it or not, industrial humanity is now permanently responsible, on pain of its own extinction, for limiting and controlling its effects on the Earth System. Nevertheless, if capitalism by the mid-twentieth century has ushered in a planetary ecological rift, the possibility still remains of the transformation of the human metabolism with nature in conformity with natural laws in a society devoted to substantive equality and ecological sustainability.

Rooting his analysis in materialist dialectics, Soriano, as previously noted, proposed in *Geologica Acta* in 2020 that the first geological age of the Anthropocene, following the current geological age of the Meghalayan (the last age of the Holocene Epoch), be designated as the *Capitalian*, in recognition of the destructive relation that capitalism is now playing with respect to the entire Earth System, creating a habitability crisis for humanity.[38] The Capitalian Age stands for the fact that behind the current Anthropocene crisis lies the capitalist mode of production. Clark and I, as environmental sociologists, independently issued a similar proposal shortly after, suggesting that the new geological age associated with the advent of the Anthropocene Epoch should be called the *Capitalinian*, and that the future geological age toward which humanity must now necessarily strive—introducing a new climacteric surmounting the planetary emergency—should be named the *Communian*, after *community, communal,* and *commons.*[39] If all of geological history, according to Engels, is one of "negating negations," leading to the Earth System crisis of today, we are now presented with the choice between the negation of the material conditions of human society to which capitalism is leading us, or else the negation of the capitalist mode of production (and thus of the present Capitalian/Capitalinian Age). What is essential in these circumstances is the creation of a new, socially mediated geological age of the Communian (the negation of the negation), embodying a restored, developed, and sustainable metabolism of humanity and the earth.

Dialectics, Engels argued, encompassed interaction, contradiction, and emergence, and was a general expression of the evolving totality of material things and of motion (matter and energy), applicable to all of existence. From this standpoint, it was possible to understand more fully the material world around us, providing the basis of a grounded scientific socialism. In the past, Marxist scholarship with respect to Engels's forays into the dialectics of nature has focused simply on the question of the rejection or acceptance of his general views, leaving out the more positive challenge of exploring their significance for the philosophy of praxis. Today, we need to go beyond this stale debate to recognize, in line with the neglected second foundation of Marxism within science and materialist philosophy, that the dialectics of nature offers new insights and methods for the understanding of our time, precisely because its approach is a unified one, bridging the great gulf that has emerged in the ecology of praxis.

As Soriano explains, "most natural sciences" today—if "spontaneously" and without full awareness—take "a dialectic and materialist epistemic view in understanding the natural side of the Earth System and of the Anthropocene crisis. From the social side of the problem, however, the epistemic view adopted by most natural scientists turns into a positivist and idealist one," deferring to mainstream liberal social science and philosophy.[40] Meanwhile, the so-called Western Marxist tradition, while holding on to the notion of dialectics, has applied this only in ways related to the identical subject-object of the human historical realm. The tendency here has been to portray natural science as primarily positivistic, while seeing no relation between nature and dialectics. In this way, the two realms of dialectical thought in the natural sciences and the social sciences have remained separate, making a unified praxis based on reason as science impossible. This can only be overcome by reunifying Marxism's *first foundation* in the critique of bourgeois political economy with its *second foundation* in the critique of mechanistic science.

Writing in the tradition of Engels, Soriano states: "Nature is dialectical too, and the dialectics of Nature is not merely a theoretical construct but a construct that is only possible because Nature is inherently so. Otherwise, how is it possible to 'construct' dialectics if it is not yet in the studied object, which is the ultimate source of any empirical perception?"[41] Today, the dialectics of nature must be reunited with the dialectics of society, the critique of political economy with the ecological critique of capitalism. This requires that the second foundation of Marxism be accorded a central place in the philosophy of praxis. The human relation to the earth lies in the balance.

POSTSCRIPT: DID ENGELS BREAK WITH MARX ON METABOLISM?

Kohei Saito's important work *Marx in the Anthropocene: Toward the Idea of Degrowth Communism*, published by Cambridge University Press in 2023, has raised the critical question of whether Engels departed fundamentally from Marx's analysis of social metabolism.[42] Saito charges that Engels, in editing the third volume of *Capital*, removed the adjective *natural* from the original draft in Marx's *Economic Manuscript of 1864–1865*, and thus in effect removed the term *natural metabolism* from Marx's passage on the "irreparable rift."[43] This is then backed up by a criticism of Engels for allegedly "rejecting Liebig's concept of metabolism." On these bases, Saito argues that Engels was largely responsible for the suppression of Marx's social metabolism/metabolic rift argument, helping "to make Marx's ecology invisible," with disastrous effects for later Marxist theory. The reason given for Engels's alleged transgression in this respect is that his notion of the dialectics of nature represented an approach to nature/natural science that was in direct conflict with Marx's social-metabolic analysis. "It was precisely due to this difference" between Marx's and Engels's approaches to dialectics and ecology, we are told, that "the concept of metabolism and its ecological implication were marginalized throughout the 20th century."[44]

It is true that the term *natural metabolism* was missing from the passage on the "irreparable rift" in Engels's edition of the third volume of *Capital*. (This same term is also absent in Ben Fowkes's recent English-language translation of Marx's original manuscript for the third volume in the *Economic Manuscript of 1864–1865*.) Hence, instead of capitalism leading to "an irreparable rift in the interdependent process of social metabolism, a metabolism prescribed by the natural laws of life itself," as conveyed in Engels's edition of the third volume, in Saito's rendering the same passage should read: "an irreparable rift in the interdependent process between social metabolism and natural metabolism prescribed by the natural laws of the soil." (An even more literal translation would be "an irreparable rift in the context of the social and natural metabolism prescribed by the natural laws of the soil.") Engels, in editing the third volume of *Capital*, thus removed the term *natural metabolism*, though *natural* still remains in the rest of the sentence. In Saito's view, this omission reflected a "profound methodological difference" between Marx and Engels on the concept of metabolism.[45]

Yet, examined closely, it is debatable that the removal of *natural metabolism* substantially changed the meaning of Marx's original passage—certainly not enough to raise a significant issue in that regard. Although Marx referred in his original incomplete draft to the "social and natural metabolism," definitely including the term *natural metabolism*, there is a certain redundancy here. The notion of natural metabolism is basic to Marx's entire materialist approach and is already assumed in the very concept of "social metabolism" itself, which *mediates* the relation of humanity with what Marx called the "universal metabolism of nature."[46] The social metabolism for Marx is nothing but the specifically human relation (via the labor and production process) to the universal metabolism of nature. Moreover, even without the words *natural metabolism*, the passage indicates that the "irreparable rift in the interdependent process of social metabolism" violates "the natural laws of life [soil]," which itself refers to a break with the universal metabolism of nature. The omission of the word *natural*, and thus the term *natural metabolism*, does nothing to alter the fundamental point being made. Saito declares that what is lost in Engels's version is Marx's second-order mediation, or alienated mediation.[47] But that too is problematic, since the very context of the passage, as it appears in the third volume of *Capital*, is a rift in the social metabolism—that is, a disruption of the social-metabolic *mediation* of humanity and nature as a result of *alienated* capitalist production.

Saito supplements his philological argument on the missing term in Engels's editing of Marx's "irreparable rift" passage with the additional charge that Engels developed a "critique of Liebig's theory of metabolism."[48] However, evidence of this "critique" is nowhere to be found in Engels's writings. In fact, Saito himself is unable to offer a single sentence indicating such a critique of Liebig on metabolism issued from Engels's pen. Instead, he resorts to highlighting Engels's quite different criticisms in *Dialectics of Nature* of Liebig's vitalism, including his rejection of Darwin's theory of evolution and his hypothesis that life had existed eternally. Saito illogically infers from Engels's criticisms of Liebig in this regard that since Engels objected to Liebig's vitalistic and anti-evolutionary notions in biology, he must also have objected to Liebig's use of the metabolism concept in his chemistry. However, Liebig was a "dilettante" in biology and at the same time a leading scientist in chemistry, a distinction that Engels stressed. What makes Saito's criticism here even more problematic is that Engels repeatedly utilized Liebig's analysis of the rift in soil metabolism in his own writings—even if he did not choose, as Marx did, to use the word *Stoffwechsel* (metabolism) in this context.[49]

But the deeper theoretical problem confronting Saito in his attempt to find evidence of Engels's supposed "rejection" of Liebig's concept of metabolism is that Liebig, in utilizing the notion of metabolism, was referring to the *natural-science concept of metabolism*. Liebig did not, as in the case of Marx, develop the category of *social metabolism*. Saying that Engels rejected Liebig's concept in this regard then amounts to charging that he rejected the notion of natural metabolism, of which Engels was a major nineteenth-century proponent. The concept of metabolism originated in German cell biology early in the nineteenth century and was applied broadly in Liebig's midcentury writings in agricultural chemistry.[50] Metabolism in this sense was a concept that Engels employed many times, including in his famous analysis of metabolism (and proteins) as the key to the origins of life.[51] Indeed, the notion of *Stoffwechsel* was central to the development of the first law of thermodynamics in Julius Robert Mayer's "The Motions of Organisms and their Relation to Metabolism" (1845), which strongly influenced Engels (as well as Liebig and Marx).[52]

All of this throws into further disarray the contention that Engels, supposedly encumbered by his dialectics of nature perspective, failed to appreciate the significance of Marx's inclusion of "natural metabolism" in the "irreparable rift" passage. It was due to this failing, Saito tells us, that Engels "intentionally" deleted the term *natural metabolism*, effectively "marginalizing" and making "invisible" Marx's core ecological critique, which was thereby

"suppressed."[53] Yet, here Saito is confronted with the inconvenient
fact that Engels, who was certainly one of the most erudite figures of his day, wrote again and again on the subject of nature's
metabolism, a concept for which he demonstrated a very deep
appreciation.[54] Moreover, Engels's edition of the third volume of
Capital, far from suppressing the conception of "natural metabolism," includes it in other places where Marx employed it in his
original text.[55]

Behind Saito's entire argument is an attempt to reinforce the
notion within the Western Marxist philosophical tradition that
Engels's dialectics of nature, with its wider materialism, was antithetical to Marx's own historical materialism. Thus, rather than
looking at how Marx's and Engels's ecological analyses are complementary and reinforce each other, we are presented with the notion
of a theoretical break between the two that is rooted in Engels's
dialectics of nature, which supposedly led Engels to distance himself from Marx's ecology. Yet, in the course of his argument, Saito
is unable to find any satisfactory way of demonstrating that the
dialectics of nature as developed by Engels is actually at odds with
Marx's ecology. He merely contends that Engels's approach to
Earth history was "transhistorical" in that it transcended human
history in the manner of positivistic natural science when addressing nonhuman nature.[56] Yet, one wonders what kind of natural
science there would be if it were to restrict its analysis simply to
human history, that is, if it were not transhistorical in the sense of
superseding the human world. Clearly, our social being influences
our understanding of nature, something that Engels emphasized
as well as Marx. But science is necessarily concerned with domains
beyond the human.[57] Surely, an analysis of Earth history extending
beyond human history did not contradict Marx's own thinking,
since he exhibited a deep fascination with paleontological developments within geological time *prior to human existence*.[58]

Engels is also criticized by Saito for developing a more "apocalyptic" theory of ecological crisis than Marx through his use of the
metaphor of the "revenge" of nature and the notion that human

beings are capable of undermining the conditions of their existence on a planetary scale.[59] Engels even contemplates human extinction in the distant future. Saito attributes such views to Engels's "apocalyptic" conception of the dialectics of nature as opposed to Marx's non-apocalyptic ecological conceptions in his theory of metabolic rift. But surely Engels, from the standpoint of the twenty-first century, is to be commended for conceiving of the reality of human-generated ecological crisis throughout the globe! Nor does this in any way contradict Marx's theory of metabolic rift, the contemporary relevance of which has mainly to do with the Earth System crisis.[60]

The full extent of Saito's adherence to the notion of a break between Marx and Engels on the dialectics of nature, depicting a deep ecological split between the two thinkers, can be seen in his direct support for Terrell Carver's position that Engels most likely *lied* in his 1885 preface to *Anti-Dühring* when he indicated that he had read the various parts of that work to Marx prior to their publication in serial form. In Saito's own words, Engels's statement here was "not necessarily credible."[61] Engels, it is insinuated, might very well have lied about his interactions with Marx in this respect. The fact that there is absolutely no basis for believing that Engels would have lied on such an important point, which does not at all fit with his character or his lifelong loyalty to Marx, does not seem to deter those sowing such doubts. Indeed, the nature of this argument is that Engels *must have lied*, because otherwise, Marx, who had contributed a chapter to *Anti-Dühring*, could be assumed to have been entirely familiar with that work prior to its publication and presumably broadly agreed with its contents. This would then undermine the notion of a fundamental break between Marx and Engels.[62]

Saito's attempt to establish a methodological break between Marx and Engels with respect to the concept of metabolism adopts a similar form for essentially the same reasons. Engels *must be responsible for intentionally suppressing* the term *natural metabolism* (and with it the significance of the metabolic rift) in editing the third volume of *Capital*, since otherwise notions of the complementarity of Marx's and Engels's writings on ecology might carry the day, contradicting Saito's contention that "Marx never really adopted the project of materialist dialectics that Engels was pursuing."[63]

Yet, the fact that Saito's whole supposed proof of a methodological break between Marx and Engels depends on the absence of a single term—the word *natural* preceding *metabolism*—in a single passage, constituting a small change of highly debatable significance, points to the total absence of any substantive evidence of such a break. To rend asunder Marx and Engels on metabolism and ecology on such a basis is unwarrantable. The truth is, while Engels did not directly employ Marx's notion of "social metabolism," except in his 1868 *Synopsis of Capital*, nor develop Marx's analysis in this regard, there is no indication that his outlook contradicted that of Marx in this area.[64]

If Marx's theory of metabolic rift was not better known among Marxists prior to this century, this had nothing to do with Engels's alleged suppression of Marx's ideas, a claim for which there is no concrete basis. Rather, it had to do with the reality that the metabolism concept was embedded in the deep structure of Marx's work and thus was often overlooked, while a great deal of what he wrote on this was incomplete, developed only in his later years. More important, much of Marx's science, as Rosa Luxemburg emphasized, was well ahead of the socialist movement itself and would only be taken up as new problems presented themselves.[65] It was the development of ecosocialism a century after Marx's death that led to the rediscovery and reconstruction of Marx's theory of metabolic rift, rather than the reverse. This unearthing of Marx's ecological argument was partially enabled by the substantial (if somewhat indirect) influence that it had exerted, along with the work of Engels, on subsequent socialist ecological analyses within natural science and materialist philosophy.[66]

Rather than perpetuating old divisions within the left, it is necessary today to bring Marx's social-metabolism argument together with Engels's dialectics of nature, seeing these analyses as integrally related. The object should be to unite the first and second foundations of Marxist thought, providing a broader material basis for the critique of the capitalist mode of production as the essential ground for a revolutionary ecosocialist praxis in the twenty-first century.

4

Nature as a Mode of Accumulation: Capitalism and the Financialization of the Earth

THE EXPROPRIATION OF THE COMMONS, its simplification, division, violent seizure, and transformation into private property constituted the fundamental precondition for the historical origin of industrial capitalism. What Marx referred to as the *original expropriation* of the commons in England and in much of the world (often involving the expropriation of the laborers themselves in various forms of slavery and forced labor) generated the concentrations in wealth and power that propelled the late eighteenth- and early nineteenth-century Industrial Revolution.[1] In the process, the entire human relation to nature was alienated and upended. As Karl Polanyi wrote in *The Great Transformation*, "What we call land is an element of nature inextricably interwoven with man's institutions. To isolate it and form a market for it was perhaps the weirdest of all the undertakings of our ancestors."[2]

It is hardly surprising in this context that the first references to "natural capital" and to the "earth's capital stock" arose in this same period in the work of radical and socialist political economists, who sought to defend nature and the commons against the intrusions of the market. Here, the notion of *natural capital* was viewed in terms of the stock of physical properties and natural-material use-values constituting *real wealth*, and was seen as opposed to the growing "sense of capitalism" as a system of mere exchange-value or cash nexus.[3]

This nineteenth-century notion of "natural capital," conceived in physical, use-value terms, was to be revived in the 1970s and '80s as part of an emerging ecological critique. In more recent decades, however, mainstream neoclassical economics (sometimes with the help of ecological economists), together with corporate finance, have completely separated the concept of natural capital from its original use-value–based critique, the memory of which has long receded, instead conceiving natural capital entirely in exchange-value terms, as just another form of financialized capital. This is then used to reinforce the view that the solution to the current ecological crisis of the planet is to make a market out of it.

A turning point in the financial expropriation of the earth occurred from September to November 2021, overlapping with the 2021 UN Climate Change Conference negotiations in Glasgow. Three major interrelated developments occurred at this time: (1) the creation of the Glasgow Financial Alliance for Net Zero embracing most of global capitalist finance; (2) the approval of key elements of Article 6 of the Paris Agreement, creating the unified financial rules for global carbon trading markets; and (3) the announcement that the New York Stock Exchange together with the Intrinsic Exchange Group (IEG)—whose investors include the Inter-American Development Bank and the Rockefeller Foundation—was launching a new class of securities associated with natural asset companies (NACs). As the IEG now tells investors, while the asset value of the world economy is $1,540 trillion, the asset value of the earth's natural capital is estimated at $5 quadrillion ($5,000 trillion), all potentially there for the taking.[4]

Together, these developments represent a sea change in the capitalization of nature, such that all natural processes that involve ecosystem services to the economy are now increasingly seen to be subject to exchange on the market for profit—all in the name of conservation and climate change. This represents the culmination of a theoretical shift in the dominant economic paradigm aimed at the unlimited accumulation of total capital, now seen as including "natural capital." The result is to reinforce the Great Expropriation occurring in this century aimed at what Charles Darwin called the earth's "web of complex relations."[5]

In order to develop a critical analysis of the current capitalist expropriation of world ecology, it is necessary to explore the concept of natural capital in the work of Marx and other early radical critics within classical political economy. It will then be possible to contrast this to current approaches in neoclassical economics, which views natural capital in purely exchange-value terms, offering this as a solution to the environmental problem. If, in Marx's analysis, the human economy existed within what he called "the universal metabolism of nature," in today's dominant neoclassical economics, according to Dieter Helm, chairman of the UK Natural Capital Committee, "The environment is part of the economy and needs to be properly integrated into it so that growth opportunities will not be missed. Integrating the environment into the economy is hampered by the almost complete absence of proper accounting for natural assets."[6] Here, the whole of the Earth System is conceived as a largely unincorporated "part" of the capitalist economy. In Helm's conception, the capitalist economy faces no outer boundaries but is capable of subsuming all of nature, which then simply becomes part of the overall capitalist system.

CLASSICAL POLITICAL ECONOMY AND NATURAL CAPITAL AS USE-VALUE

Most accounts of the origin of the term *natural capital* trace it to economist E. F. Schumacher's 1973 book *Small Is Beautiful*.[7] However, the notion of natural capital and the related concept of the earth's capital stock were widely used in nineteenth-century

classical political economy, particularly among radical and social-
ist critics, appearing in the works of thinkers as various as Victor
P. Considerant, Marx, Engels, Ebenezer Jones, George Waring,
Henry Carey, and Justus von Liebig.[8]

Considerant was a utopian socialist, Charles Fourier's leading
disciple, who did much to establish the Fourierist tradition. In
his *Theory of the Right to Property and the Right to Work* (1840),
Considerant insisted that there were two forms of capital: (1) land,
which in classical political economy stood for all forms of nature,
and which he referred to as *natural capital*; and (2) *created capital*,
produced by human labor (utilizing natural capital).[9] According
to Considerant, property rights to nature and natural resources
are mere rights to *usufruct*, or the temporary use of that which
belongs to the chain of human generations. Thus, natural capital was to be redistributed to each generation on an equal basis.
However, under bourgeois civilization, natural capital had been
usurped by a minority of private landholders, who had established
land monopolies violating the principles of usufruct applying to
all of humanity.[10]

Later in the same decade, the British poet and radical political
economist Ebenezer Jones provided a similar argument in *The
Land Monopoly*. For Jones, the principal evil affecting the welfare
of the population of England and Ireland was the land monopoly exercised by landlords, who appropriated "natural capital,
God's gift to all men." In the next century (the twentieth), Jones
indicated, the inhabitants of the land may have difficulty understanding "how the land they have come to live on [and its natural
capital] could have been thus sold, not only (to use an expressive
phrase) over their heads, but actually over their cradles, or even
before they were born." In these terms, natural capital was treated
as the annual "produce of the land" (nature), or, in today's terms,
ecosystem services. Jones provided estimates of what the land was
capable of generating in terms of the number of people it could
support.[11] He punctuated his argument on the land monopoly
by pointing to the English colonial exportation of the proceeds

of the land from Ireland during the Great Famine of only a few
years before, amounting to sufficient food to have fed half the
Irish people.[12] With great acuity, he queried: "Suppose a body of
men should consider the air of London to be in need of cultivation, and should unsolicitedly establish round the metropolis a
circle of aerial purification—what would be conceived of their
sanity, if they should in consequence consider themselves air-
lords, with the air of London for their private property, for them
to do what they like with, even to the exclusion of people from
the use of it . . . ?"[13]

Marx studied Considerant's political-economic work in October
1842.[14] In *The German Ideology* of 1845, Marx and Engels employed
the term *natural capital* to refer to natural resources, such as the
cotton and wool fibers used, for example, in textile production, and
to capital as it emerged in the towns of the Middle Ages and then
in the Mercantilist "putting out" system, tied to estates. The growth
of textile production, they wrote, required the "mobilization of
natural capital through accelerated circulation." They contrasted
"natural capital," rooted in the land, estates, and concrete use-values, to "movable capital" associated with the "beginning of money
trade, banks, national debts, paper money, speculation in stock and
shares, stockjobbing in all articles and the development of finance
in general," resulting in capital losing "a great part of the natural
character that still clung to it."[15]

The natural capital concept, as used by Marx and Engels in *The
German Ideology*, was thus tied to the natural-material use-value
structure of the economy and to landed capital and estates, as
opposed to the greater mobility and fungibility of capital as pure
exchange-value or finance, which evolved under mercantilism and
became dominant in industrial capitalism. If capital could originally be seen primarily in physical terms, it increasingly became
measured in exchange-value forms. Marx and Engels's overall
emphasis here corresponded to the classical political-economic
conception that real wealth consisted of natural-material use-
values while private riches were based on exchange-value, that is,

purely monetary claims to wealth. Yet, since reference to natural capital seemed to naturalize capital, Marx was to drop all direct reference to the term in his subsequent work.[16] Nevertheless, the basic distinction was reflected in his contrast between the "natural form" of the commodity, related to natural-material use-values, and the "value form" associated with exchange-value, as well as his distinction, as we shall see, between *earth matter* and *earth capital*.[17]

For classical political economists in general, including such figures as Adam Smith, Thomas Malthus, David Ricardo, and John Stuart Mill, nature, as distinct from labor, created no value, and was treated as a "free gift" to capital—long before Marx pointed to the ecological contradictions that this entailed for the capitalist economy.[18] As the Ricardian John Ramsay McCulloch put it, "In its natural state, matter is *always destitute of [exchange] value*."[19] Or, as Marx wrote, "Value is labour, so surplus-value cannot be earth."[20]

Nevertheless, the notion of natural-material use-values, if no longer referred to as natural capital, remained integral to Marx's conception of the capitalist economy and its ecological basis, including conceptions of the expropriation of nature and of natural processes turned into capital. The decisive shift in his analysis, in this respect, was already evident in *The Poverty of Philosophy* in 1846. In his critique of Pierre-Joseph Proudhon's *System of Economic Contradictions: Or the Philosophy of Misery*, written earlier that same year, Marx, as he later recounted in the third volume of *Capital*, introduced "the distinction between *terre-matière* and *terre-capital*," or between *earth matter* and *earth capital*:[21]

Land, so long as it is not exploited as a means of production, is not capital. Land as capital [*terre-capital*] can be increased just as much as all the other instruments of production. Nothing is added to its matter, to use M. Proudhon's language, but the lands which serve as the instruments of production are multiplied. The very fact of applying further outlays of capital to land already transformed into means of production increases land as capital without adding anything to land as matter [*terre-matière*], that is, to the extent of the land. M. Proudhon's land as matter is the earth in its limitation. As for the eternity he attributes to land, we grant readily it has this virtue as matter. Land as capital is no more eternal than any other capital.[22]

In this passage, Marx draws a distinction between land, viewed on the one hand as eternal earth matter (*terre-matière*, or mere matter) and, on the other, as historically generated earth capital (*terre-capital*). He is already pointing to the contradiction between capitalism and its natural conditions of production, a historical and materialist view that will govern his developing ecological critique, leading eventually to his metabolic rift concept. Although natural capital, now called *earth capital*, exists, it is seen as an alienated product of capitalism and by no means eternal. In *Capital*, Marx writes: "Capital may be fixed in the earth, incorporated into it, both in a more transient way, as is the case with improvements of a chemical kind, application of fertilizer, etc., and more permanently, as with drainage ditches, the provision of irrigation, leveling of land, farm buildings, etc." This is connected to "ground-rent . . . paid for agricultural land, building land, mines, fisheries, forests, etc. . . . Ground rent is . . . the form in which landed property is economically realized, valorized."[23] By incorporating capital into the earth, Marx explained, capitalists "transform the earth from mere matter into earth-capital."[24] In this conception, the earth as matter (*terre-matière*) remained the basis of all life and production, while the valorization of portions of the earth as earth capital (*terre-capital*) represented a fundamental contradiction between the eternal laws of nature and the law of value of capitalism.

In some cases, Marx noted, the monopolization of a "force of Nature" could be enormously profitable, as in the case of ownership of a waterfall, providing waterpower to industry. Here, "a monopolisable force of Nature, which, like the waterfall, is only at the command of those who have at their disposal particular portions of the earth and its appurtenances," generates surplus profit potential. This then allows those who own the waterfall or other forces of nature to impose rents on their use. The rent is not a product of the waterfall itself—that is, does not derive from its "natural value"—nor does it derive directly from labor, but rather emanates from the owner's *private monopoly* of a limited natural force (with the rent ultimately coming out of total surplus value).[25] Marx argued that it was only the title to a particular natural resource that allowed monopoly rent to be applied, despite the fact that owners believed they were entitled to rent simply by purchasing the land or natural resource, particularly as the price of the land contained this capitalized tribute. It was not the purchase or transfer of title that created the rent, but rather the title itself, which was a product of social relations that created the monopoly position and the power to enact rent—whether it was the title to a waterfall, a coal deposit, or other natural resources, the common inheritance of all humanity. Such rents, he argued, were being imposed "in ever greater measure" as capitalism developed.[26]

It is worth noting that the works of classical political economics in general, and Marx's analysis of production in particular, were permeated with the treatment of environmental services, or what in ecosocialist theory are known as the *eco-regulatory* aspects, which supersede human labor. Such a view was inherent in Marx's conception of the universal metabolism of nature as underwriting the social metabolism of the labor and production process. Thus, we find innumerable discussions in his work of the soil metabolism and of other "physical, chemical, and physiological processes" and "organic laws" associated with natural reproduction, operating on different time scales from human production. "The economic process of reproduction, whatever may be its specific social character," he writes, "is in this area (agriculture) always intertwined . . . with a process of natural reproduction."[27]

In 1855, a twenty-two-year-old George Waring—already recognized as an eminent agriculturalist in the United States, later to be seen as one of the great ecological figures in U.S. history for his contributions to fighting urban waste and disease—presented an extensive address titled "Agricultural Features of the Census of the United States for 1850" and subsequently published in the *Bulletin of the American Geographical Society* in 1857, to a meeting of that society in New York. Waring, who like other progressive agriculturalists had been influenced by Liebig's *Organic Chemistry in Its Application to Agriculture and Physiology* (1840, better known as *Agricultural Chemistry*), used census figures for agriculture to estimate the loss of fertilizer agents within the U.S. economy. This was at a time when the capital invested in agriculture in the U.S. economy was seven times the amount invested in manufacturing, mining, the mechanic arts, and fisheries. In depicting the enormous losses of nutrients to the soil, he wrote:

> What with our earth-butchery and prodigality, we are losing the intrinsic essence of our vitality. . . . The question of economy should be, not how much do we annually produce, but how much of our annual production is saved to the soil. Labor employed in robbing the earth of its capital stock of fertilizing matter, is worse than labor thrown away. In the latter case it is a loss to the present generation; in the former it becomes an inheritance of poverty for our successors. Man is but a tenant of the soil, and he is guilty of a crime when he reduces its value for other tenants who are to come after him.[28]

Waring's statement was taken up by Henry Carey, the foremost U.S. economist of the day, who had previously sent Marx *The Slave Trade, Domestic and Foreign*, a work that at one point characterized "man as a mere borrower from the earth."[29] Carey quoted extensively from Waring on "the robbing of the earth of its capital stock" in both his *Letters to the President: On the Foreign and Domestic Policy of the Union* (1858) and *Principles of Social Science* (1858). This was, in turn, to influence Liebig, who drew on Waring via Carey in his own *Letters on Modern Agriculture* (1859), which marked the beginning of his major attack on industrialized capitalist agriculture as a "robbery system." Liebig's critique in this respect was to culminate in the famous introduction to the 1862 edition of his *Agricultural Chemistry* that inspired Marx's theory of metabolic rift. Significantly, in the same paragraph in which Marx made the crucial distinction between land as earth matter and as earth capital in the third volume of *Capital*, he also referred to the classic criticisms of the degradation of the soil by James Anderson and Carey, pointing to the ecological contradictions of capital.[30]

In classical political economy—the logic of which in this respect was brought out most fully by Marx—nature and labor (itself a natural force) were the sources of *real wealth* as use-values, while exploited labor power under capitalist production was the source of (commodity) value.[31] The conflict that this set up between natural-material use-values (treated as free gifts to be expropriated by capital) and the system of exchange-value generated the fundamental ecological contradiction of capitalist production, associated with the robbing of nature.[32] As the eighth Earl of Lauderdale, James Maitland, declared in *An Inquiry into the Nature and Origin of Public Wealth and into the Means and Causes of Its Increase* (1804), the system of commodity production destroyed *public wealth* (natural-material use-values), generating scarcity and monopoly, thereby enhancing *private riches* (exchange-value), with negative consequences for human society as a whole.[33]

NEOCLASSICAL ENVIRONMENTAL ECONOMICS AND THE VALORIZATION OF NATURAL CAPITAL

In sharp contrast to classical political economy, neoclassical economics beginning in the late nineteenth and early twentieth centuries has sought to exclude nature and use-value altogether from its analysis, reducing everything to exchange-value and denying the distinctiveness of the natural world (as well as of human labor). It has defined capital in non-social, transhistorical terms, as any asset of any kind producing a stream of income over time—a definition that leads to an endless series of contradictions, derived from the fact that it sees capital as a kind of "social black box."[34] Nature and land were thus lumped together with other forms of "capital" and were, in effect, eliminated from the analysis, with the neoclassical production function reduced to two abstract factors of production: capital and labor. Inherent in this view was the postulate that natural resources were entirely reproducible or substitutable by human-made capital. A "weak-sustainability" postulate, representing the dominant neoclassical view, contends that all natural resources can be economically substituted by human-made or renewable resources—that is, there are no irreplaceable natural resources or processes that must be maintained. This is counterposed by a "strong-sustainability" postulate, associated with ecological economics, arguing that certain "critical natural capitals" are irreplaceable and cannot be substituted for by human-manufactured capital.[35]

The dominant weak-sustainability conception is well captured by economic growth theorist Robert Solow's claim: "If it is very easy to substitute other factors for natural resources, then there is in principle no 'problem.' The world can, in effect, get along without natural resources, so exhaustion is just an event, not a catastrophe. . . . At some finite cost, production can be freed of dependence on exhaustible resources altogether."[36] Based on such assumptions, the liquidation of natural assets with the development of capitalism is not "an obstacle to further progress," since such natural resources and processes are simply substituted by the human economy with a zero net loss of capital overall.

The concept of natural capital was reintroduced into the economic discussion in the 1970s and '80s, beginning with Schumacher's *Small Is Beautiful*, to highlight the "liquidation" of "natural capital" stock as a failure of the first order of the modern economic system, representing the view of ecological economics.[37] Thus, the use of the concept up through the 1980s was directed mainly at the idea of maintaining a constant biophysical stock of natural capital. It was at this point that the notion of weak sustainability was formally introduced by some of the same figures, such as British economist David W. Pearce, who had first insisted on maintaining a constant stock of natural capital, but then argued, in line with neoclassical economics generally, that such natural capital could be easily replaced in the human economy and thus no strict natural constraints on the economy existed. According to the weak-sustainability postulate, the notion of natural capital became largely indistinguishable from the neoclassical category of capital in general, insofar as it could be viewed as constituting productive assets providing an income stream.[38]

In response to the neoclassical weak-sustainability argument, ecological economists—initially inspired by Nicholas Georgescu-Roegen's *The Entropy Law and the Economic Process* (1971), which emphasized the importance of the second law of thermodynamics in any realistic economics—embraced the notion of natural capital as a key concept, while wedding it to the notion of "critical natural capital" in conformity with the strong-sustainability postulate.[39] Critical to the notion of strong sustainability were the three principles of sustainability introduced by Herman Daly: (1): "For a *renewable* source—soil, water, forest, fish—the sustainable rate of use can be no greater than the rate of regeneration." (2) "For a *nonrenewable resource*—fossil fuel, high-grade mineral ore, fossil groundwater—the sustainable rate of use can be no greater than the rate at which a renewable resource, used sustainably, can substitute for it." (3) "For a *pollutant*, the sustainable rate of use can be no greater than the rate at which the pollutant can be recycled, absorbed, and rendered harmless by the environment."[40] This approach established limits to growth and determined sustainability in biophysical/use-value terms, rather than in terms of exchange-value. The whole issue of natural capital, from the standpoint of the strong-sustainability postulate, thus became one of maintaining a *net zero decrease* in natural capital, viewed in biophysical terms, in which reductions in the stock of nonrenewable forms of natural capital, like fossil fuels, were offset by corresponding increases in renewable natural capital, such as the harnessing of solar energy and biomass.[41]

Ironically, economists associated with the International Society of Ecological Economics and the journal *Ecological Economics* were to do the most to expand the notion of natural capital as a monetized economic category. Although ecological economists defended the notion of strong sustainability and some, such as Daly, continued to insist on treating natural capital simply in use-value terms, the majority yielded to the temptation of putting a price on the world's ecosystem services—if only for pedagogical purposes—with the intent of establishing their importance from the standpoint of the economy. From there, it was a slippery slope toward the actual financialization of the world ecology. Moreover, the conception of what constituted critical natural capital was often watered down, while the principles of sustainability came to include the substitutability of human-made products for nature. Hence, the distinction between the weak- and strong-sustainability approaches tended to fade.

In this general slippage within ecological economics, in which much of the tradition was brought back into the dominant neoclassical fold, natural capitals/ecosystem services were increasingly reduced to a strictly economic or imputed "commodity" value basis, to the point that there emerged what Marxian ecological economist Paul Burkett called an "artificial ecumenicism" between ecological economics and the hegemonic neoclassical economic tradition.[42] Outside the relative few who stuck to the thermodynamic-based analysis of Georgescu-Roegen, or who were associated with the Marxist tradition, ecological economists found it difficult to resist the almost total dominance of the neoclassical tradition and the closely aligned corporate world.[43]

Once the natural capital concept was generally affixed to neoclassical economics—on the basis of the recognition in some way of weak/strong sustainability, with critical natural capital representing an exception and subject to change under the force of technology—it was quite possible to water down the environmental analysis altogether, to the point that the potential threat such ideas posed to capitalist accumulation could be downplayed. In practice, this meant reducing the conception of strong sustainability to the extent that it simply constituted a footnote to weak sustainability. Here, the treatment of natural capital was no longer seen as an actual limit on the expansion of the system. Thus, as the World Bank stated in its 2003 *World Development Report*:

> Limits-to-growth type arguments focus on strong sustainability, while arguments in favour of indefinite growth focus on weak sustainability. So far the former arguments have not been very convincing because the substitutability among assets has been high for most inputs used in production at a small scale. There is now, however, a growing recognition that different thresholds apply at different scales—local to global. Technology can be expected to continue to increase the potential substitutability among assets over time, but for many essential environmental services—especially global life support systems—there are no alternatives now, and potential technological solutions cannot be taken for granted.[44]

The World Bank statement subtly suggested that substitutability was high for all natural-resource inputs, except in the case of production at higher thresholds, particularly where this affected "global life support systems" (downplaying that this was precisely the issue in a globalizing economy within a limited planetary environment), while technological solutions to such scale, if not available *now*, were seen as *potentially* available in the future. The relation of the economy to natural resources should thus be one of promoting the "mix of assets that supports improvements in human well-being," which was expected to change over time, thereby posing no clear limits to "indefinite growth." The notion of critical natural capital—that is, a strong-sustainability argument—was thus carefully discounted. Entirely ignored was any consideration of the specific socioeconomic conditions governing capitalist production and the contradictions these inherently pose for the Earth System metabolism.

In 1992, the International Society of Ecological Economics held a conference in Stockholm dedicated to the full operationalization of natural capital as a concept of ecological economics. In 2003, *Ecological Economics* published an introduction to a special issue that stated: "Natural capital is a key concept in ecological economics."[45] This shift coincided with a struggle within the journal, in which Robert Costanza, the chief editor and leading proponent of the hybrid neoclassical/ecological-economic notion of natural capital, managed to remove leading systems ecologist Howard Odum and a number of other natural scientists from the editorial board. In opposition to the natural-capital concept with its attempted valuing of nature on capitalist terms, Odum had promoted a way of accounting for the embodied energy inputs in the natural economy using the notion of *emergy* (spelled with an *m*), directly related to the use-value category of classical economics. This was aimed at challenging attempts to play down the opposition between the capitalist economy and natural systems, and providing a comprehensive theory of ecological imperialism. Following Odum's ouster from the journal, the concept of emergy was effectively banned from the publication.[46]

These shifts in ecological economics opened the way to measuring the "natural income" or "welfare" flows to the human economy from natural capital stock in the form of ecosystem goods and services (shortened for convenience simply to services), thus providing putative market values for nature's contribution to economic growth.[47] Natural capital was, in effect, redefined in market terms as the natural resource stock that provided ecosystem services to the human economy. Ecosystem services did not refer to ecosystem processes as a whole, but only to those services that could be seen as subsidizing the human economy, and thus could be separated in this way from the rest of nature.[48] The implicit goal was accounting for and eventually, to some extent, "internalizing" discernible free gifts to the capitalist market economy on the basis of imputed consumer preferences. Nature, where such benefits to the capitalist economy were absent, in effect remained devoid of imputed economic value and external to this wider natural-capital conception, as if it could be sliced and diced in economic asset terms. In this respect, ecosystem services as a *natural-income* category displaced the category of natural capital itself.[49]

Costanza, who did the most to expand the notion of ecosystem services, proceeded to lead a study titled "The Value of the World's Ecosystem Services and Natural Capital," published in *Nature* in 1997, that provided estimates of seventeen ecosystem services across sixteen biomes based on a "simple benefit transfer [or value transfer] method." The study assumed a constant per unit dollar value per hectare of a given ecosystem type, which was then multiplied by the total area of each type to obtain aggregate values.[50] Values were obtained by relating benefits in the human economy to analogous benefits provided by ecosystem services. This constituted, in effect, a system of "shadow prices" based on an economist's best estimate of what price a function or thing would obtain in the capitalist market economy, rooted in what were assumed to be individual preferences.[51] Carrying out such an analysis requires, as does capitalist expropriation as a whole, what has been called "the division of nature," that is, its simplification into putatively commodifiable elements.[52] Natural, heterogeneous, and qualitatively distinct processes are "disaggregated into discrete and homogeneous value units," reducing widely incommensurable entities and processes—Darwin's "complex web of relations"—to monetary terms, allowing them to be aggregated to stand for global ecosystem services as a whole, while valued/priced in terms of capitalist commodity relations.[53]

The 1997 Costanza study was widely acclaimed among environmentalists, if only because it gave what seemed to be hard numbers to the notion that the world economy was dependent on the world ecology—now itself reduced in terms of ecosystem services to dollars. In that study, Costanza and his coauthors depicted the value of annual world ecosystem services in 1995 as $33 trillion in current dollars, slightly less than double the $18 trillion world GDP.[54] The notion of natural capital valuation was further advanced in the Millennium Economic Assessment in 2005, which took as its main message the dangers of the "running down of natural capital assets" and neglect of environmental services across the globe. The United Nations was to launch a System of Environmental-Economic Accounting, utilizing the natural capital/ecosystems services approach.[55] In 2014, in an updated analysis, "Changes in the Value of Global Ecosystem Values," Costanza and his colleagues estimated that world ecosystem services in 2011 were equal to $145 trillion annually (in 2007 dollars), compared to a world GDP of approximately $73.6 trillion.[56]

Yet, while current attempts to place values on nature can serve useful pedagogical roles and help enhance strategic planning, they are increasingly being integrated with goals of capital accumulation. As Friends of the Earth noted in *The Financialization of Nature*, "Promoting ecosystem markets involves the same methodologies and institutions for pricing and trading which were developed for economic evaluation."[57] Thus, over the last three decades, "the history of ecosystems services research" has been accompanied by "a parallel history of ecosystem function commodification," operating through universities, governments, and businesses, using the same language and methods of ecosystems services accounting, but further extending the analysis to the creation of actual natural-capital markets. This occurs through three steps: (1) designating an ecological process as an ecosystem service to the human economy, (2) imputing to it a single "exchange-value," and (3) establishing ownership and managerial rights so as to link users and providers of the service in a market exchange, permitting financial investment and accumulation.[58]

For the IEG (now teamed up with the New York Stock Exchange, a minority investor), the significance of the 2014 Costanza-led study of global ecosystem values is that it shows that ecosystem services have a value far exceeding that of world GDP—one that, in the context of environmental concerns, can be opened to accumulation and financial exploitation via ecosystem function commodification.[59] "Nature's economy is larger than our current industrial economy and we can tap this store of wealth" based "on natural assets and the mechanism to convert them into financial assets," thereby transforming the economy into "one that is more equitable, resilient and sustainable." In this perspective, "intrinsic value" is used as the umbrella term for potential economic values of the natural environment that have "not yet been identified or quantified," representing vast new openings for financial investment and wealth as the boundaries between the capitalist economy and unpriced nature erode.[60]

Accumulation of Natural Capital and the Financialization of Nature

The last decade has seen an explosion of natural capital initiatives aimed at the accumulation and financialization of nature as a means of addressing environmental constraints. In 2011, the UK Environment Bank, a private institution devoted to the financialization of nature, received £175,000 from the Shell Foundation to aid it in the development of markets for ecosystem services.[61] Since 2012, the Natural Capital Committee of the UK government and the UK Department for Environment, Food, and Rural Affairs have been promoting a natural capital "aggregate rule" based on the notion of net-zero losses in natural capital in economic value terms. This has involved the development of mechanisms for treating various elements of nature as commensurate not only with each other, but also with commodity markets. A methodology for managing natural capital has been introduced in which the destruction of biodiversity or the climate would be balanced by offsets that increase (or protect) natural assets by an equal value amount elsewhere. This has required the reduction of nature/natural capital to monetary units that can then be integrated into consolidated national accounts, incorporating changes in UK natural capital, valued in 2015 at £1.6 trillion. This process has been facilitated internationally by the formation of a host of entities dedicated to natural capital accounting, including the World Forum for Natural Capital, the Natural Capital

Declaration, and the Natural Capital Financing Facility of the European Investment Bank and European Commission.[62]

Although carbon trading markets were behind much of this, of near-equal importance have been initiatives associated with biodiversity and conservation. In September 2016, the World Conservation Congress of the International Union for Conservation of Nature introduced its "natural capital charter" (Motion 63) as a framework for treating all biodiversity as natural capital values. This was preceded by the global Natural Capital Protocol of multinational corporate business initiated in July 2016 by the Natural Capital Coalition (now renamed the Capitals Coalition).[63] *The Economics of Ecosystems and Biodiversity*, published in 2010 and 2011, initiated under the auspices of the Natural Capital Coalition with the support of the UN Environment Programme and the European Commission, was to be a heavy promoter of the valuation of natural capital.[64]

A watershed initiative with respect to the accumulation of nature was launched by the Swiss-based global investment bank Credit Suisse, which in 2016 introduced a report on *Conservation Finance: Moving Beyond Donor Funding to an Investor-Driven Approach*, followed by a report that same year on *Levering Ecosystems: A Business-Focused Perspective on How Debt Supports Investment in Ecosystems Services*. The Credit Suisse scheme is to move beyond donor capital in conservation to construct a "conservation finance space." The key goal here is to reorganize conservation finance to create in each case a definite "financial vehicle" or company, controlling the natural capital/ecosystem services, which would generate major financial returns to investors. This will turn ecosystem services into "an asset treasured by the mainstream investment market."[65] This was the basis for listing NACs on the New York Stock Exchange, which used the same methodology of creating a "financial vehicle" or "natural assets company" as an intermediary in the conversion of a "natural asset" into "financial capital" consecrated by the launch of an Initial Public Offering of the natural asset company.[66]

Various means would be developed in this respect for the Payments for Ecosystem Services and trading of natural capital, involving non-financial corporations, banks, governments, and non-governmental organizations. Government-owned natural capital assets, often expropriated from Indigenous populations and subsistence farmers, could be sold in the form of debt for nature swaps or leveraged via international financial capital. More important, however, is the role envisioned by the IEG in which NACs would operate essentially like businesses that have acquired "mining rights," thus allowing them to exploit the resources and accumulate monetized assets—in this case in the name of sustaining nature.[67] Although a given state would normally continue to have sovereign ownership of the land, the financial vehicle managing and disposing of the ecosystem services would profit directly off the income streams associated with these "tradable" assets. According to the Credit Suisse *Conservation Finance* report, in order for firms to profit through investment in natural capital, it will be necessary to combine "*heterogeneous*" natural assets, "*bundling* them into a single product with a tailored risk and return sharing vehicle." In this way, it is possible to "provide a market-rate return and leverage multiple sources of finance to reduce risk," thereby maximizing value for investors.[68]

Carbon trading, which is now being fully globalized through Article 6 of the 2021 UN Climate Change Conference, is designed to promote a world market in offsets, allowing a firm to avoid actual carbon emission reductions by financing (and frequently capitalizing) an offset, usually in the Global South, involving carbon sequestration. The $100 billion that the developed capitalist countries have promised to direct at the Global South for climate finance is seen as subject to debt leverage by multinational monopoly-finance capital. This lies behind the 2021 Glasgow Financial Alliance for Net Zero initiative of global finance, which has declared at the outset that carbon-mitigation financing to developing countries will be dependent on whether they fully open up their economies to global capital. Credit Suisse sees "ecological footprints" as moving "closer to being recognized as assets and liabilities by companies allowing debt to fund natural capital investment and the creation of new profitable markets with "net-positive financial outcomes" in the Global South.[69] In general, the accumulation and financialization of nature involves the creation of titles to environmental services of various kinds, previously within the commons as the inheritance of the world's people, after which these titles can be traded and leveraged.

In the case of valorized natural capital, monopoly rights to environmental services can be established with the cooperation of governments through the creation of NACs, which then will be free to accumulate based on the "management" of this service, including trading in all sorts of offsets. As the New York Stock Exchange indicated, NACs would "hold the [economic] rights to ecosystem services produced on a given chunk of land."[70] The logic, as far as capital and finance is concerned, is not that far removed from how extractive industries themselves developed, but in this case it is putatively about sustaining natural assets by maintaining net-zero losses. In analogy with *standing timber* as a concept in forestry, these assets are now referred to as *standing natural capitals*.[71] Profiting off the extraction of environmental services is conflated with the notion of sustainable forestry, marketing the service while maintaining the overall asset. It, however, runs into the same contradictions.[72]

Governments, intergovernmental organizations, financial institutions, non-financial corporations, and non-governmental organizations, in introducing the notion of natural capital in their various reports, often begin by referring to it in broad material use-value terms as consisting of nature's resource stock—a view of natural capital that goes back to the nineteenth century. Yet, the fine print soon makes it clear that natural capital is primarily viewed today in exchange-value, not use-value, terms. One such market is the global voluntary carbon market, which is projected to reach $180 billion by the end of this decade. Only "a tiny fraction" of these carbon offsets, according to *Bloomberg* in January

2022, actually remove carbon from the air, while 90 percent of firms employing certified carbon offsets were found in a survey to have inflated their claims on carbon savings. In line with this, the term *carbon neutral* is now being used as a marketing tool with no basis in net-zero carbon accounting, in much the same way as the term *natural*, lacking any clear designation, is adopted in place of *organic* in marketing to fool the unwary consumer.[73] In this context, the Reducing Emissions from Deforestation and Forest Degradation market has become the leading vehicle for voluntary carbon offsets. Such projects, however, have been associated with the expropriation of Indigenous lands and the removal of Indigenous peoples.[74] It is significant in this respect that the Terra Bella Fund of Terra Global Capital, which is a private investment fund specializing in environmental assets, is specifically directed at "voluntary markets where regulations are uncertain or nonexistent" in emerging and developing economies and is focused on buying up "undervalued derivative instruments on environmental assets."[75]

According to Kanyinke Sena, director of the Indigenous Peoples of Africa Coordinating Committee, Indigenous people constitute less than 5 percent of the world's population but protect 80 percent of the world's biodiversity.[76] The world's peasantry also plays a vital ecosystem role, employing traditional practices. Ironically, in the name of ecology and combating the capitalist destruction of the earth as a safe home for humanity and innumerable other species, we are seeing an enormous expansion of the domain of what Marx called earth capital. This is occurring by means of the expropriation of Indigenous and peasant populations, along with the expropriation of the human natural inheritance altogether, including that of future generations. This constitutes the great tragedy of the commodification of the commons, a new Great Expropriation, pointing to the destruction of the earth, involving vast land (and ocean) grabs, particularly in the Global South.[77]

The famous Lauderdale Paradox, the destruction of *public wealth* (principally the commons) in order to generate *private riches*, introduced by the Earl of Lauderdale at the beginning of the nineteenth century, has a direct application in our time. The expropriation and degradation of the ecological commons is generating the conditions of scarcity crucial to the creation of exchange-value, private property monopolies, and monopoly rents. It is hardly surprising, therefore, that multinational capital is playing both sides of this game of the destruction and accumulation of nature. According to Portfolio Earth, the world's fifty largest banks provided $2.6 trillion in 2019 to companies linked to deforestation and biodiversity destruction, especially in Southeast Asia and the Amazon. The top three offenders are Bank of America, Citigroup, and JPMorgan Chase.[78] The *Financial Times* carried a report in October 2021 indicating that global banks and asset managers had extended $119 billion since 2016 to agribusiness companies involved in deforestation.[79] Over 70 percent of global carbon emissions can be traced to just one hundred corporations (military emissions excluded).[80] The same capitalist firms that are destroying the Earth System as a home for humanity are now supporting the financialization of the world's natural capital/ecosystem services, aimed at profiting off attempts to safeguard the earth from their own continuing destruction of it. In this conception, profits can be made on both sides of the ledger, by contributing to the creative destruction of nature as part of the accumulation of capital and by profitably investing so as to ensure a zero net loss in total human and natural assets. It would be an understatement to refer to this as a planetary-level protection racket raised to the level of the capitalist economic system as a whole.[81]

AGAINST THE ACCUMULATION OF NATURE

The concept of natural capital, including the earth as a capital stock, was introduced in nineteenth-century political economy and environmental discussions, primarily within the socialist and radical traditions, as a way of emphasizing that real wealth consisted of natural-material use-values as opposed to the commodified exchange-values of the capitalist economy. Those figures within classical political economy who initially focused on the conservation and common human ownership of material use-values as constituting real wealth, opposed land monopolies and the confiscation, commodification, and destruction of nature in the interest of capital accumulation. Such arguments with regard to natural capital could already be seen in the writings of Considerant, Jones, Marx, Waring, Carey, and Liebig, among others.

When Schumacher revived the concept of natural capital in 1973 in *Small Is Beautiful*, he was operating, as he was well aware, in this same basic tradition, seeing natural capital as constituting use-values or natural resources that could not be quantified, but represented a stock of real wealth that was being liquified by capitalist production. As he wrote: "To measure the immeasurable is absurd and constitutes [on the part of the economist] but an elaborate method of moving from preconceived notions to foregone conclusions: all that one has to do to obtain the desired results is to impute suitable values to the immeasurable costs and benefits" of nature. The only real result of such an endeavor was to perpetuate the myth that "everything has a price, or, in other words, that money is the highest of all values."[82]

As we have noted, Marx and Engels in *The German Ideology* initially used the concept of natural capital to refer to the "natural form" of the commodity tied to use-value and its concrete, physical form. In its initial development, coming out of the Middle Ages, they argued that capital was tied to physical space, in the sense of land/space, involving definite material inputs, and in this sense could be regarded as a form of "natural capital." This was contrasted to the subsequent development of "mobile capital" based on an exchange-value and the circulation of financial claims to wealth. However, the term *natural capital* was dropped by Marx by the time he wrote *The Poverty of Philosophy* only a year later, given his critique of the naturalization of capitalism. In its place, he introduced a more ecological distinction between the earth or land as a natural-material entity or *earth matter* versus the category of *earth capital*, the latter representing nature (for example, the soil or a waterfall) turned into capital.[83] The accumulation of earth capital, though indispensable to capital accumulation, led in Marx's view to the disruption of the universal metabolism of nature in favor of capitalism's alienated social metabolism, thus developing an "irreparable rift" in the metabolism of nature and society (or metabolic rift).[84]

Here, Marx's analysis was much influenced by the work of Waring, Carey, and Liebig, who wrote of the *robbing* of the earth's capital stock, a notion that Marx was to make central to his notion of metabolic rift. In Marx's own terms, what was being "robbed" through the accumulation of "earth capital" was the material metabolism and reproductive basis of the earth as matter (material nature) itself. Capitalism was to be conceived as a form of creative destruction in which the destructiveness of the system would overwhelm its creative side. As he observed, "Capital . . . is in practice moved as much and as little by the sight of the coming degradation and final depopulation of the human race, as by the probable fall of the earth into the sun."[85] A rational, sustainable relation to the earth was impossible under the regime of capital, since it saw the earth either as a mere free gift to capital accumulation or as transformed into *earth capital*. In either case, the ecological system was robbed. There was nothing eternal about *terre-capital*, which existed on the basis of the capitalization of nature; only *terre-matière*, constituting the realm of natural-material existence, the universal metabolism of nature, was eternal.

"Natural capital," Daly insists, should be seen in use-value terms, "based on the relations of physical stocks and flows, not prices and monetary valuation."[86] Yet, the notion of natural capital has to be seen as a dangerous one altogether in a capitalist society. Rather than embodying a distinction, as in Marx's analysis, between earth matter and earth capital, it is easily incorporated into an all-inclusive, ahistorical notion of capital, which is treated as homogeneous and to be measured in terms of the single yardstick of exchange-value. In this respect, it is crucial to remember that capitalism is a system of accumulation geared to exponential expansion, hence leading to the drawing down of natural resources. It represents the very opposite of conservation. It therefore cannot accept material limits or boundaries, which are viewed simply as barriers to be surmounted.[87] Faced with environmental constraints, the dominant economic approach is, therefore, to incorporate ecosystem services into the economy by placing capital values on it and selectively integrating it with capital accumulation itself—a process made easier by the fact that capital makes nature scarcer and more marketable by destroying it. Valuing nature simply by its ecosystem services to a capitalist economy is inevitably destructive of nature, with the concept of ecosystem services inviting the extreme division of nature in capitalist terms, since it has as its initial basis the "cutting" of nature into discrete pieces to be valorized.[88]

In the context of the overall financialization of the world economy, vast amounts of surplus "free cash," the growth of financial bubbles, and the promotion of debt peonage in the Global South, the financialization of nature is likely to intensify the volatility of the capitalist economy itself.[89] Nevertheless, it is the environmental bubble generated by the financialization of nature that is most dangerous.[90]

In what amounts to a victory of notions of weak sustainability, it is often contended that the continual destruction of nature required by capital accumulation can be offset by the valorization of nature and its internalization within the logic of capital itself, so that there is no net loss of natural capital in economic value terms and the exponential increase of capital accumulation in a limited environment is allowed to proceed. New financialized ecosystems can help support the entire system. If nature is itself capital, the argument goes, there is simply no problem. The destruction of one species or of a whole ecosystem can be compensated for by natural capital that provides ecosystem services for the economy elsewhere. In the words of Solow, representing the neoclassical view of sustainability:

History tells us an important fact, namely that goods and services can be substituted for one another. If you don't eat one
species of fish, you can eat another species of fish. Resources are,
to use a favorite word of economists, fungible in a certain sense.
They can take the place of each other. That is extremely important because it suggests that we do not owe to the future any
particular thing. There is no specific object that the goal of sustainability, the obligation of sustainability, requires us to leave
untouched. . . . Sustainability doesn't require that any *particular*
species of fish or any *particular* tract of forest be preserved.[91]

Like most capitalist economists, Solow fails to recognize that
each species and each ecosystem is unique, and that extinction is
irreversible, affecting the whole complex evolution of the Earth
System. For Credit Suisse, conservation finance is about turning nature into "fungible" cash flow and products in precisely
Solow's sense.[92] Species and ecosystems may be treated as commensurable and substitutable in the economic value terms of the
capitalist economy, but in reality they are incommensurable and
irreplaceable. Their individual demise represents real ecological
consequences. To think otherwise is to fall prey to what Marxist
geographer David Harvey called "the madness of economic
reason," in which there are no limits—quantitative or qualitative—
to the valorization and financialization of capital, conceived as
value in motion, absorbing all of reality, including nature itself.[93]

As ecological economist John Gowdy declared, the concept of
natural capital as it is now employed "contains two contradictory
concepts: 'natural' indicating a world governed by biophysical laws
and 'capital' indicating a world governed by the laws of market
capitalism."[94] Attempts to overcome this contradiction by subsuming material nature within capital run into the contradiction that
Marx expressed between the earth as natural-material and the earth
as capital. For Marx, human production and extra-human nature
had to be seen as complementary and co-evolutionary, requiring that natural systems be maintained in terms of their material

flows and complex web of relations, preserving the metabolism of humanity and nature for the entire chain of human generations and for the sake of life on Earth itself, in accord with the principle of acting as good heads of the household.[95] In the classical Marxian view, as emphasized by Ernst Bloch in *The Principle of Hope*, nature and humanity are "co-productive," in the sense that "the creations slumbering in the womb of nature" are the material basis of all human productivity.[96]

What this means is that other, wider ecological principles, applicable to both natural and human systems, need to displace current attempts to solve the planetary crisis generated by capitalism by simply absorbing the earth itself within the logic of the system, extending commodity fetishism to the realm of nature.[97] Ecology has generated new bases for promoting sustainable human development and the overcoming of economic and ecological imperialism.[98] Within Marxism, there is a long, if disputed, tradition of the dialectics of nature, which stands strongly opposed to reductionist approaches to nature and its evolution, exposing the dangers of all attempts to commodify the natural world and insisting that human beings "belong to nature and exist in its midst, and . . . all our mastery of it consists in the fact that we have the advantage of all other creatures of being able to learn its laws and apply them correctly."[99]

Such a critical, dialectical, and materialist perspective requires the abandonment of both the naturalization of capital and the capitalization of nature, as well as the recognition of the inescapable social character of capital, associated with a particular historical system: capitalism. Only an ecological and social revolution that would allow humanity as a whole, the associated producers, to regulate the human social metabolism with the earth in a rational and sustainable way, in accord with a broad scientific understanding and with the aim of promoting genuine, free human development, can offer a way out of the current planetary crisis.[100]

5

The Defense of Nature: Resisting the Financializaton of the Earth

ON OCTOBER 28, 2021, POLITICAL leaders in the Malaysian state of Sabah on the island of Borneo signed an agreement with the Singapore shell company Hoch Standard without the knowledge of Indigenous communities, giving the company title to the management and marketing of "natural capital/ecosystem services" on two million hectares of a forest ecosystem for one hundred to two hundred years. Although the full nature of the agreement has not been disclosed, journalistic investigations and a lawsuit filed by Adrian Lasimbang, an Indigenous leader in Malaysian Borneo, have revealed that the Nature Conservation Agreement allowed Hoch Standard—a holding company with two officers and paid-up capital of a mere 1,000 U.S. dollars provided by shareholders, but backed by undisclosed multibillion-dollar private-equity investors—to acquire commercial rights to the natural capital in Sabah's forest ecosystem. The revenue from the rights to ecosystem services, such as water provisioning, carbon sequestration, sustainable forestry, and biodiversity conservation, over the next century was estimated at some $80 billion, with 30 percent, or $24 billion, to go to Hoch Standard. It was stipulated that the Sabah government could not withdraw from the agreement, while Hoch Standard could sell its rights to the natural capital in the Sabah Forest to other investors without government consent. Singaporean Ho Choon Hou, who allegedly misrepresented himself as director of Hoch Standard (he was not listed among its officers but is said to be Hoch Standard's project director and strategic funder), is the managing director of the private equity firm Southern Capital Group, which focuses on corporate buyouts. Financial documents revealed that, as a shell company, Hoch Standard lists a single shareholder, Lionsgate Ltd., registered in the British Virgin Islands, a tax haven and financial base for "dark money" where it is illegal to disclose the name of company shareholders.[1]

The Natural Conservation Agreement between the Sabah government and Hoch Standard was brokered by the Australian consulting firm Tierra Australia, specializing in the financialization of natural capital. Peter Burgess, CEO of Tierra Australia, has defended the exclusion of Indigenous peoples from the agreement on the neocolonial, racist basis that if it were necessary to "sit around every campfire" talking to Indigenous peoples about the "jungles" they happen to live in, nothing at all would be accomplished. According to Burgess, the Indigenous communities—there are thirty-nine Indigenous ethnic groups in the forest reserves in Sabah, making up a population of more than 25,000—"actually don't know that their jungles . . . are going to be conserved for 200 years" by the agreement, which is aimed at "restoring [their] jungles," providing benefits so as to "uplift" them, "bringing them back into normal society." Tierra Australia is closely connected to major multinational banks in the capitalist core, such as Credit Suisse and HSBC, along with major Singapore banks, all of which have been heavily involved in investments in natural capital. It has partnered with Hoch Standard, along with Harvard, the Massachusetts Institute of Technology, and Cornell, in devising natural capital platforms for private investment.[2]

The two chief promoters of the Sabah-Hoch Standard deal

are Stan Lassa Golokin, signing for Hoch Standard, and Jeffrey Kitingan, representing the Sabah government. Golokin is a business partner of Burgess at Tierra Australia and is linked to eleven companies registered in the British Virgin Islands. He was listed as an associate of four companies included in the Panama Papers, a leaked database on global elite financial dealings. Kitingan is second deputy chief minister and state agricultural and fisheries minister in Sabah and was a witness to the signing of the agreement by Frederick Kugan, Sabah's chief conservator of forests. Kitingan has emerged as the main defender of the deal within the Sabah government. In the 1980s and '90s, both Kitingan and Golokin were involved in the Sabah Foundation, which was given a century-long concession to a million hectares of forest, to be managed on a sustained yield basis. Kitingan was director of the Sabah Foundation while Golokin was group general manager of a holding company for the Sabah Foundation's commercial assets. As evidence of the extraordinary corruption at the time, some $1.6 billion in timber rent went missing under their management, while Kitingan's personal wealth during his nine years as director of the foundation rose suddenly to $1 billion. During the same period, Kitingan's brother was chief minister of Sabah.[3]

As of February 2022, the Nature Conservation Agreement between the Sabah government and Hoch Standard is in a kind of legal limbo, according to Sabah's attorney general, since key aspects of the agreement are not binding or enforceable.[4] Yet, while the Sabah-Hoch Standard Nature Conservation Agreement is presently on hold, it can be seen as part of the massive "gold rush" to secure rights to the world's "natural capital" that is now taking place globally.[5] It is no mere accident that the October 28 signing of the Sabah multibillion dollar natural capital deal occurred only a month after the New York Stock Exchange and the Intrinsic Exchange Group announced the creation of a new asset category in the form of natural asset companies (NACs), stipulated as financial vehicles for the ownership, management, and control of the world's natural capital assets.

Only three days after the Hoch Standard agreement was made, the Glasgow United Nations Conference on climate negotiations began. This coincided with the consolidation and rise to global prominence of the Glasgow Financial Alliance for Net Zero, advertising itself as representing multinational banking and money managements adding up to $130 trillion in financial assets, and led by some of the very same multinational banks, such as Credit Suisse and HSBC, with which Tierra Australia was connected.[6] Golokin was present at the Glasgow UN climate negotiations seeking to drum up global finance for the Hoch Standard-Sabah Nature Conservation Agreement, which he claims is designed to draw out the potential of the Sabah forest's "lazy assets," a term referring to ecosystem services not incorporated into the market. Burgess gave a presentation at the International Heart of Borneo Conference in November 2021, aimed at attracting investment in natural capital to Borneo. He depicted Borneo's natural environment as a prime target for the global movement directed at the "monetisation of the world's Natural Capital Assets."[7]

It is impossible to exaggerate the extent of this natural-capital rush, now being promoted by global speculative finance, which since the Global Financial Crisis of 2007–10 has sought to acquire real assets in the physical environment to underpin continuing debt expansion.[8] The transmutation of so-called natural capital into tradable exchange value over the last decade is seen as opening up almost unlimited opportunities for corporations and money managers. In 2012, the Corporate EcoForum, a group of twenty-four multinational corporations including Alcoa, Coca-Cola, Dell, Disney, Dow, Duke Energy, Nike, Unilever, and Weyerhaeuser, published *The New Business Imperative: Valuing Natural Capital* in conjunction with the Nature Conservancy, insisting that the then "estimated $72 trillion of 'free' goods and services" associated with global natural capital and ecosystem services be monetized for the purpose of more sustainable growth." The report emphasized the enormous debt "leverage" opportunities represented by "emerging natural capital markets such as water-quality trading, wetland banking and threatened species banking, and natural carbon sequestration." As a result, it was imperative to "put a price on nature's value," or, stated differently, "a monetary value on what nature does for . . . businesses." The future of the capitalist economy lay in ensuring that the market pay "for once-free ecosystem services," which could thereby generate new economic value for those corporations able to convert titles to natural capital into financial assets.[9]

In 2016, more than fifty multinationals, led by firms such as Dow, Coca-Cola, Nestlé, and Shell, joined with Conservation International in the Natural Capital Coalition (now known as the Capitals Coalition) to develop the Natural Capital Protocol. This was directed at designing a framework for the monetization of the world ecology, using fabricated shadow pricing systems based on the capitalist market system.[10] The Natural Capital Protocol was soon accompanied by other initiatives like the Natural Capital Charter, introduced the same year by the International Union for the Conservation of Nature.[11]

Costanza and his associates valued the world's "seventeen" ecosystem services associated with sixteen biomes in 2011 at $145 trillion annually (in 2007 dollars). The net present value of these ecosystem services, discounted at 1 percent over the remainder of this century, was estimated to be worth over $4 quadrillion ($4,000 trillion).[12] The Intrinsic Exchange Group, now updating the value of world ecosystem services to $5 quadrillion, sees this as representing a virtually unlimited set of metaphorical gold mines for the taking by natural asset corporations. Economist Wilhelm Buiter of Citigroup expects "to see a globally integrated market for fresh water within 25 to 30 years. Once the spot markets for water are integrated, future markets and other derivative water-based financial instruments . . . will follow. . . . Water as [a natural] asset class will . . . become eventually the single most important physical commodity-based asset class, dwarfing oil, copper, agricultural commodities and precious metals." In this perspective, the world's sources of fresh water, representing one of the planetary boundaries designated by natural science, will be monopolized as natural capital by relatively few companies who will be able to charge market rents for the ecosystem services.[13]

Plans for the expropriation and accumulation of natural capital by global finance are primarily directed today at the Global South. According to the UN Environment Programme, spatial mapping of natural capital indicates there is "a high concentration of terrestrial ecosystem assets in the equatorial regions, particularly in the Brazilian Amazon and the Congo Basin." Marine ecosystem assets are highest in Southeast Asia (the South China Sea) and along coastlines.[14] Indigenous territories cover some 24 percent of the earth's land surface and "contain 80% of the earth's remaining healthy ecosystems and global biodiversity priority areas," making these primary targets for expropriation and conversion into marketable natural capital. Sub-Saharan Africa is a target since "it's estimated that around 90% of land is untitled," with the result that many Indigenous communities that have lived in these areas for untold years lack official land titles, and their land is therefore open to land grabbing.[15] The African Forum on Green Economy, working with the Natural Capital Coalition and the World Wildlife Fund, stated in 2020 that "natural capital is part of a wider economic system," implying that Africa's ecosystems can be completely subsumed within the capitalist economy.[16]

The implications of this rapid financialization of nature, which is promoting a Great Expropriation of the global commons and the dispossession of humanity on a scale exceeding all previous human history, are vast. This Great Expropriation is being justified on the grounds of saving nature by turning it into a market, thereby replacing the laws of nature with the laws of commodity value. Not only is the logic behind this fallacious, but it is also likely to widen the associated colossal financial bubbles, while accelerating destruction of planetary ecosystems and of the earth as a safe home for humanity.

In order to understand the monumental folly of the monetization of the earth, it is necessary to take a theoretical excursion into the classical critique of the "fetish character of capital" and the confusion of real wealth and debt as developed in the work of thinkers such as Karl Marx and the Nobel-Prize-winning chemist and unorthodox economic critic Frederick Soddy.[17] This will allow us to comprehend the necessary conditions for the defense of the earth in the face of the current financialization juggernaut, requiring the greatest alliance of workers, peoples, and movements in the history of humanity.

The Myth of the Innate Power of Capital: Marx and Soddy

In his critique of "the fetish character of capital," in the *Grundrisse* and *Capital*, Marx highlighted the views—far exceeding "the fantasies of the alchemists"—of the late eighteenth-century British political-economic writer and nonconformist minister Richard Price, a friend of Benjamin Franklin and Joseph Priestley. Price claimed that through the magic of compound interest a universe of riches could be obtained. In his 1772 *Appeal to the Public on the Subject of Public Debt*, Price went so far as to state: "ONE PENNY, put out at our Saviour's birth to 5 *per cent. compound* interest, would, before this time, have increased to a greater sum, than would be contained in A HUNDRED AND FIFTY MILLIONS OF EARTHS, all solid gold."[18]

For Marx, Price's "150 millions of earths all solid gold" was a cosmic fantasy of "the innate power capital," in which capital becomes "a self-reproducing being . . . a value perennating and increasing by virtue of an innate quality" without any reference to real material and historical conditions. "The good Price was simply dazzled by the enormous quantities resulting from geometrical progression of numbers. . . . He regards capital as a self-acting thing, without any regard to the conditions of reproduction or of labour," or—as Marx was also to insist—the material conditions and limits imposed by the earth itself. With capital thus conceived, in Marx's words, "as a mere self-increasing number," Price "was able to believe that he had found the laws of growth in that formula." Indeed, for Price, according to Marx, capital was "a self-acting automaton," embodying "an innate property as ever persisting and growing value." How capital accumulation actually occurred, together with its limits and contradictions, was "quite immaterial to him," since all of this was superseded by "the innate quality of interest-bearing capital." Hence, for Price and those he influenced, Marx wrote, "Adam Smith's theory of accumulation" as the basis of the wealth of nations is turned "into the enrichment of a nation by accumulation of debts." It is here that "the fetish character of capital" is complete.[19]

In Marx's critique of political economy, all human production has a *real basis* in a "material substratum . . . furnished by nature without human intervention," while the labor process "mediates the metabolism between man and nature."[20] A commodity has a dual aspect as both a natural-material use-value, meeting social needs, and as exchange-value, generating surplus value for capitalists. Use-values, constituting real wealth, are the product of both nature and human labor. A particular use-value "does not dangle in mid-air. It is conditioned by the physical properties of the commodity and has no existence apart from the latter." Human labor has a dual character as both a material-biophysical force, transforming natural-material use-values through production, and as a generator of exchange-value/value under capitalism. The conflict between the production of commodities as use-values, on the one hand, and exchange-value, on the other, lies at the core of all capitalist contradictions.[21] What nature itself provides, apart from labor time, is in the capitalist system a mere "free gift . . . to capital" and not incorporated directly in its accounting of value production, where it is treated as a mere externality.[22] Nevertheless, the monopolization of elements of scarce land/nature gives rise to monopoly rents, which are withdrawn from total surplus value, feeding the coffers of owners of natural resources.

Capitalism's exclusive focus on production for exchange-value rather than use-value, and its treatment of nature as a free gift, led in Marx's analysis to the robbing of nature of the elementary constituents of production, and thus the creation of the metabolic rift between nature and society, exemplified by the nineteenth-century soil crisis in which essential soil nutrients were shipped to the new urban centers of industrial production, where they contributed to pollution and were lost to the soil.[23] In Marx and Engels's political-economic critique, the material conditions of production were integrated with the developing science of thermodynamics of their time, which emphasized the environmental/energetic limits on production.[24] In accordance with ancient Epicurean materialism, nothing came from nothing, and nothing being destroyed was reduced to nothing.[25] Marx quoted Pietro Verri's statement that "all the phenomena of the universe, whether produced by the hand of man or indeed the universal laws of physics, are not to be conceived as acts of creation but solely as a reordering of matter."[26]

In neoclassical economics, as it emerged in the late nineteenth and twentieth centuries, distinct from classical political economy, the concept of natural-material use-values was removed from the fundamental framework in economics, leaving only exchange-value in the conception of wealth. Land as a factor of production, since it was assumed that human-made capital could substitute for it, was eventually excluded altogether from the neoclassical production function, consisting of simply labor and capital. Hence, all necessary relations of capital to nature were extinguished, together with any conception of material production being dependent on the laws of thermodynamics. The idea that the growth of capital was in any way limited by the natural environment was also eliminated.[27]

All of this fed the myth of the innate power of capital. As Daly has written, "Perhaps the standard example of misplaced concreteness [reification] in economics is 'money fetishism,' applying the characteristics of money, the token and measure of wealth, to concrete wealth itself. Thus, if money can grow forever at compound interest, then presumably, so can [real] wealth," as if there were no physical limitations.[28]

The ecological/energetic critique of the innate power of money, introduced by Marx, was taken still further a century ago by Soddy, the winner of the Nobel Prize in Chemistry in 1921 and a pioneer in ecological economics, beginning with the 1922 publication of his *Cartesian Economics: The Bearing of Physical Science upon State Stewardship*. Soddy was among the first in the study of radiation, introducing the concept of isotopes.[29] He was concerned early on about the destructive potential of harnessing atomic energy, indicating in 1926 in his *Wealth, Virtual Wealth, and Debt*: "If the discovery [of how to release atomic energy] were made tomorrow, there is not a nation that would not throw itself heart and soul into the task of applying it to war, just as they are now doing in the case of the newly developed chemical weapons of poison-gas warfare. . . . If [atomic energy] were to come under existing economic conditions, it would mean the *reductio ad absurdum* of scientific civilization, a swift annihilation instead of a none too lingering collapse."[30]

Soddy saw the capitalist economic system, and particularly the debt economy it fostered, as the greatest danger to world stability. In the early twentieth century, during his most productive period as a chemist in Glasgow, he became acquainted with socialist ideas, principally the Romantic radical tradition, in which the main sources of inspiration at the time were figures such as Percy Bysshe Shelley, Thomas Carlyle, John Ruskin, Walt Whitman, and William Morris. This was a critical milieu that had been influenced by Morris's Socialist League and by the development of municipal socialism. The miners' strike in 1911–12 paralyzed British industry and highlighted the dependence of production on fossil fuel energy, with Soddy pointing out at the time that the contemporary economic world found its basis in this particular form of low-entropy matter/energy.[31]

Soddy was associated for a number of years with the Independent Labour Party. In 1918, he joined the newly created National Union of Scientific Workers, through which he became closely acquainted with the zoologist, Marxist, ultra-materialist, and author of *An Outline of Psychology* Henry Lyster Jameson, with whom Soddy carried out an extensive correspondence.[32] In the context of his correspondence with Jameson, Soddy entered into studies of Marx and Ruskin, as well as the work of the late-nineteenth-century theorist of banking and credit Henry Dunning Macleod.[33] The result of these studies was *Cartesian Economics* (originally two lectures presented to the Student Unions of Birkbeck College and the London School of Economics), in which Soddy challenged the innate power of money. *Cartesian Economics* was published the same year as the presentation of Soddy's 1922 Nobel Lecture and marked a decisive shift in his work from research in chemistry to the criticism of economics and the role of money emanating from the energetic standpoint of thermodynamics.

Soddy, like Ruskin, entered the economic discussion as an outsider with only a cursory knowledge of economics, coupled with a radical perspective. Hence, his views have generally been ignored by the economics profession. In approaching economics from the standpoint of natural science, he brought back the notion of real wealth as the useful embodiment of matter/energy, thereby questioning the exchange-value orientation of the capitalist economy. Like Ruskin, he saw wealth as life, or as metabolism, associated with the rational utilization of energy flows, ultimately derived from the sun. Wealth was "the humanly useful forms of matter and energy."[34] All human production was rooted in energy flows, and it was of this that real wealth was composed.

In this context, Soddy resurrected the use-value perspective of classical political economy, seeing real wealth as consisting of natural-material use-values and distinguishing this from exchange value and mere financial claims to wealth. Via John Stuart Mill, Soddy highlighted the Lauderdale Paradox, whereby the destruction of public wealth increased private riches. In illustrating the Lauderdale Paradox, Mill had pointed to the calamity represented by a situation in which clean air became so scarce and monopolizable that it could be turned into a commodity, thereby enhancing private riches at the expense of the community through the monetization of the "free gifts" of nature.[35]

The main error of capitalist economics, for Soddy, was the confusion of real wealth, which was governed by physics, with money/debt, which was a mathematical quantity. Money itself was primarily to be viewed as a lien on future production and thus a debt of the public (the issuer of currency) to the holder of money.[36] All "debts" in a commodity economy, he argued, "are subject to the laws of mathematics rather than physics," and thus are divorced from physical processes and limits. In the case of money/debt, the entropy law—the tendency of physical systems to greater disorder—did not apply, replaced by the magic of compound interest.[37] Real wealth, in contrast, emanated from solar energy and photosynthesis, and was inherently limited and subject to the entropy law—if nonetheless capable of further development in terms of utilization of energy flows. Following Aristotle and Ruskin, Soddy argued that economics as practiced by capitalism had taken the form of *chrematistics* or the mere art of acquisition, rather than *oikonomia*, or household management (from which the words *economy* and *ecology* were derived).[38] The economic successes of Britain and the other developed economies, he contended, mainly emanated from the harnessing of fossil fuel energy and exercise of contemporary imperialism, as opposed to the fantasy of the innate power of capital.[39]

Soddy highlighted a number of times Marx's insistence that real wealth in the form of use-values was rooted in both nature and material labor (the latter a force of nature). If, in Marx's critique of political economy, as Soddy explained, the exploitation of socially necessary labor power was the sole source of "exchange-value or money-price" under capitalism, this was to be distinguished from real wealth, where *nature and labor* together constituted the fundamental bases—something that many of Marx's own followers had failed to understand. Marx therefore had underscored the natural-physical basis of wealth.[40] Yet, while indicating his admiration at various points for Marx's analysis and learning from Marxist thinkers such as his friend Jameson, Soddy was himself far from being a Marxist. Moreover, by the time he wrote *The Role of Money* in the 1930s, he had moved away altogether from the socialist critique of capitalism and toward schemes for radical monetary reform.[41] As distinguished from Marx, Soddy was not at all interested in the social basis of value and capital—partly because, from a physicist's perspective, he took the view that plants engaged in photosynthesis were the ultimate sources of wealth—but rather in the narrower issue of the conflict between the monetary world and the world of physics.[42]

Marx had sharply criticized Macleod for his supposed "discovery" in *The Theory and Practice of Banking* that "currency . . . is capital," discounting the question of value.[43] Soddy, likewise, was to see Macleod as standing for the fetish of money capital in his advancement of the argument that debt should not be treated as a "negative" quantity, but rather as a positive economic value in itself. Indeed, for Macleod, "The great modern discovery is to make the debts themselves saleable commodities" and to build up a whole supreme credit and financial system based on that, which would increasingly rule the capitalist world.[44] Banks, in Macleod's terms, were "shops for the express purpose of buying and selling debts" or for "the Manufactory of Credit."[45] Macleod's emphasis on how banks under capitalism internally (or endogenously) created credit money out of nothing, coupled with the explosive character of compound interest divorced from all relations to the physical world, expressed for Soddy the modern money fetish deeply embodied in the capitalist economy, which, in its irrational financial explosion, was imperiling all of existence.[46]

Indeed, the extreme fantasies of capital, money, and finance were, in Soddy's view, pointing the world toward final catastrophe. The illusory pursuit of a perpetual motion machine was propelling the globe toward another world war, as country after country sought unlimited competitive expansion and devil take the hindmost. Moreover, the mythological view that compound interest had a real basis in material reality, in defiance of the entropy law, was generating a set of unstable economic relations that further threatened human self-sufficiency. If economics were not put on a solid, physical basis, the growth of the debt economy would propel humankind to disaster.[47] In his 1935 foreword to *The Frustration of Science*, a work whose contributors included leading British left scientists such as Bernal and Patrick M. S. Blackett, Soddy referred to the loss of productivity of the soil and the general waste in the economy, arguing that society should be ruled by the productive elements of society concerned with "the creation of its wealth rather than of its debts," and who retained a connection to the earth. Science "should speak the truth though the heavens fall."[48]

As Daly, commenting on Soddy, explains in "Capital, Debt, and Alchemy," capital, when defined in financial terms, is an expected "perennial net revenue stream" derived from an underlying asset "divided by the *assumed* rate of interest and multiplied by 100." It becomes in the current money form the calculation of a "permanent lien on the future real production of the economy." Hence, the capitalist growth economy, while continuing to profit in the course of its creative destruction, is ultimately faced with physical limits of an Earth System, which does not, like compound interest, increase exponentially. Real physical wealth emanating from nature and ultimately derived from solar energy is subject to the entropy law and cannot generate endless rapid growth as in the case of "symbolic monetary debt." The conflict between finance-based economic expansion and the ecological basis of society is thus inevitable.[49]

THE FINANCIALIZATION OF NATURE AS A NEW ECOLOGICAL REGIME

The year 2009 will be remembered in world history for two globally destabilizing events, each of which represented a major turning point. Not only did 2009 constitute the peak of the Global Financial Crisis, which began in 2007 in the United States, but it also marked the extraordinary failure of the climate negotiations in Copenhagen. Perversely, when the financial explosion that has characterized modern monopoly-finance capital resumed soon after, it was to be coupled with a search for new real asset bases from which to further leverage global finance. This search immediately came to focus on the financialization of ecosystem services, not previously incorporated within the economy, building on global carbon markets and conservation finance, offering as the solution to the global ecological crisis the monetization of earth, constituting a new financialized ecological regime.[50]

The concept of *natural capital*, it is well to remember, was introduced in the early nineteenth century, prior to the term *capitalism*, in an attempt to defend land and natural resources from the developing logic of industrial capitalism and the dominance of exchange value. In this original context, it was argued that natural capital or the *earth's capital stock*—a term that arose at the same time—needed to be defended against the artificial capital being generated by the system of cash nexus.[51] This usage of the concept of natural capital, as embodying natural-material use-values underlying production, persisted into the twentieth century, but in the last three decades has given way to a notion of natural capital in exchange-value terms, and thus a tradable asset that can be internalized within the capitalist economy.[52] This is what George Monbiot, in a 2014 talk to the Sheffield Political Economy Research Institute, termed "the Natural Capital Agenda: the pricing, valuation, monetisation, financialisation of nature in the name of saving it."[53]

A turning point in this respect was the initial 1997 article on "The Value of the World's Ecosystem Services and Natural Capital" by Costanza and his associates, aimed at pricing the planet. This relied on a reductive approach that applied a system of artificially fabricated prices derived from capitalist market relations to significant parts of a given "ecosystem service" or function, such as the production of atmospheric oxygen or the synthesis of carbohydrates by plants. Each ecosystem service was then given a single dollar price, followed by the aggregation of all seventeen of the world's ecosystem services.[54]

Such a system of imposed tradable values is based on treating incommensurable natural processes as commensurable. In such

costing of the earth, demand curves are constructed by determining the consumers' willingness to pay. However, since actual
markets for ecosystem services do not exist—that is, they are not
commodity products that are actually bought—consumers' willingness to pay is imputed by various methods, known as hedonic
pricing and contingent valuation. In hedonic pricing, valuation is
made by drawing parallels with closely associated marketed services. Thus, in the United States, a category known as *wildlife fish
user days* has been utilized to calculate the worth of various species
of wildlife in cost-benefit analyses—for example, in determining
whether it is economical to eliminate wildlife by constructing a
dam. In the wildlife fish user days calculus, the various forms of
wildlife are valued by the average amount of money an individual sportsperson is willing to pay in pursuit of a particular type
of wildlife with the expectation of killing it, thus establishing its
value. Similarly, the worth of a particular wilderness area is determined in hedonic pricing by the willingness of consumers to pay
for parking to visit it. Contingent valuation, in contrast, takes the
form of the creation of hypothetical markets on the basis of which
consumers are asked to determine what they would hypothetically
pay for a particular environmental service and what compensation
they would have to receive for losing it.[55]

Based on such studies of inferred consumer preferences,
Costanza and his associates apply a "benefit transfer" (or value
transfer) method, extrapolating the imputed value for a particular ecosystem service in one localized context, such as the water
purifying role of a particular river system, where consumer preferences have been established, and then extending that to entirely
different ecological contexts on which studies have not taken
place. The results are then aggregated to determine the price/value
of the ecosystem service on a planetary basis. This same method
is applied to all seventeen of the designated global ecosystem services in order to price the planet as a whole.[56]

The object of these elaborate exercises is to impute a value to
ecosystem services or natural assets that currently lie outside the

market. The justification offered for this is that, unless an economic value is placed on nature's services, they will continue to be treated as a free gift or externality to be robbed.[57] Yet, in the words of heterodox economist Guy Standing, though it is claimed that "unless a price is placed on every bit of nature, it will not be treated as having value," it is nonetheless true that "a price only comes when something is for sale, when it becomes a commodity." The UK government is now arguing that landowners by virtue of simply owning and monopolizing land are "providers of ecosystem services" who deserve to be paid financial compensation for offering these "services" associated with the land, previously viewed as free gifts of nature, such as ecosystem services of water purification, pollination of crops, biodiversity, and carbon sequestration.[58] (Of course, in many situations, especially on conventional farms, current practices commonly generate ecosystem "disservices" such as water pollution and loss of biodiversity.) The monetization of the environment thus allows for an enormous expansion of the circuit of exchange value and monopoly rent in the name of ecological sustainability. In Monbiot's words, it means that "you are effectively pushing the natural world even further into the system that is eating it alive. . . . All the things which have been so damaging to the living planet are now being sold us as its salvation; commodification, economic growth, financialisation, abstraction. Now, we are told, these devastating processes will protect it."[59]

The laws of motion of capital are governed by the accumulation process. To monetize the environment is ultimately to draw it into the market and to subject it to the uncontrollable dynamic of accumulation, for which a rational, sustainable relation to the environment is by definition impossible. For example, according to the standard principles of forest management under capitalism, a forest consists of so many millions of board feet of standing timber. Such timber services, according to the market rule, should be "harvested" whenever the interest rate exceeds the rate of growth in value of timber, determined by the natural growth rate of the trees. Since an old growth forest, in which trees are sometimes a century or more old, means that the growth rate of the mature trees is much reduced, falling below the rate of interest, the market demands that such old growth be liquidated on the spot, to be replaced by younger, faster-growing trees. These are to be harvested within twenty to thirty years, with chemicals increasingly applied during the processing of the wood into lumber in order to make up for the lower quality.[60]

In general, the monetization of the earth's complex biological-physical-chemical web, even in the name of conservation, will tend to replace systems of natural reproduction and evolution with reductionist, market-based criteria, for which profitable expansion is the goal. Following market rules, ecosystem services are analytically embedded within commodity markets dominated by a given accumulation of private riches. Yet, this goes against the sustainable requirements of ecosystems, and indeed the Earth System. In the process of being capitalized, the global commons will be cut up and monopolized by a few private interests, who will turn them into revenue streams to be bundled together as financial assets, including various kinds of derivatives.

Where actual conservation of natural assets is concerned, a "blended" financial arrangement is typically adopted in which governments take on most of the costs, owning and investing in the forests, and private firms reap the benefits, receiving a disproportionate share of the resulting revenue. Today, alternative sources of finance like carbon credits and debt financing, piled onto the already excessive debt loads of developing countries, are making investments in forests, tradable on the market, more profitable for international capital. In the voluntary carbon market, carbon credits, offered for already standing forests (taken from Indigenous inhabitants) can be purchased or financially managed to constitute supposed offsets for carbon emissions elsewhere in the global economy, thereby making real emission reductions unnecessary within a net-zero scheme. Carbon credits can be received by simply liquidating a natural capital asset less rapidly than would presumably have been the case otherwise, relying on fabricated baselines.[61] However, some of the problems associated with using carbon offsets—aside from not actually requiring polluters to reduce their level of pollution—can be seen in those cases where the very forests that were traded as offsets for emissions elsewhere have already burned up in the massive global forest fires induced by climate change, thus increasing carbon dioxide emissions.[62]

In 2012, the UK Ecosystems Market Task Force referred to the need for "harnessing City financial expertise to assess the ways these blended revenue streams and securitisations [of natural capital assets] enhance the return on investment of an environmental bond."[63] Commenting on this and on the overall logic of the Natural Capital Agenda, Monbiot wrote:

> What we are talking about is giving the natural world to the City of London, the financial centre, to look after. What could possibly go wrong? Here we have a sector whose wealth is built on the creation of debt. That's how it works, on stacking up future liabilities. Shafting the future in order to serve the present: that is the model. And then that debt is sliced up into collateralised debt obligations and all the other marvelous devices that worked so well last time around. Now nature is to be captured and placed in the care of the financial sector. . . . The same Task Force says we need to "unbundle" ecosystem services [from the rest of the Earth System] so they can be individually traded.[64]

Once unbundled from the rest of nature, these ecosystem services can then be rebundled as financial assets to promote financial gains. In today's carbon market, focusing on offsets, financial interests purchase credits in large numbers from suppliers so as to "bundle" them, combining various tranches of derivatives and gathering these together in portfolios, consisting of carbon offsets associated with widely different forms of natural capital.[65] The financialization of biodiversity within conservation finance now involves mechanisms for "stacking and bundling," referring to the "different ways of packaging multiple ecosystem goods and services including biodiversity for sale in environmental compensation schemes or to attract [monetized] incentive-based conservation funding."[66]

As indicated in Credit Suisse's 2016 report, *Levering Ecosystems*, conservation finance is becoming increasingly reliant on debt financing, based on expectations of rapidly growing revenue from natural capital.[67] Such approaches rely, in the first place, on the notion of the "innate power of capital" (what Marx called "the fetish character of capital"), coupled with a recognition of the increasing scarcity of natural capital, allowing for the widening of the circuit of exchange-value to all ecosystem services. The financial goal in these circumstances is to "monetise ecological credits," since a "blended return" from natural capital management "can be astronomical."[68] The ultimate result, however, is to impose a system geared to economic growth and debt expansion on top of natural systems, which are physically limited, and where the crucial conditions are those of reproduction and sustainability. In a paper on capitalizing the world's ecosystems, beginning with the natural capital of Indigenous populations in Australia and Malaysia's Sabah state in Borneo, Burgess of Tierra Australia argues that the monetization of the world's ecosystem services can underwrite a whole new global financial system, providing through "its productive value . . . the underlying asset for a stable universal medium of exchange."[69] In reality, what is meant is the leveraging up of the credit/debt system worldwide through the financialization of the earth, with the expropriation of Indigenous people and lands as its basis.

The negative consequences to be expected from extending the capital fetish to nature as a whole are planetary in scale. According to one critical study by ecological economists,

> High-debt dependent production systems exert negative effects on the capacity of the economic system to enhance the sustainable use of natural resource stocks. . . . The model of debt-fuelled

growth requires ever-faster growth rates to allow the repayment of the ever-increasing debt. . . . Thus, the profit-seeking behavior of firms and speculative agents . . . drives the inappropriate use of credits (debt), which consequently brings about systemic instability. . . . Debt-bearing economic systems can result in a complete collapse of both natural and economic systems.[70]

As John Maynard Keynes observed in *The General Theory of Employment, Interest, and Money* in 1936, in the midst of the Great Depression: "Speculators may do no harm as bubbles on a steady stream of enterprise. But the position is serious when enterprise becomes the bubble on a whirlpool of speculation."[71] Today, this has become more serious still, at a time when the "enterprise" that is being turned into a "bubble on a whirlpool of speculation" is the metabolism of the Earth System itself.

The first published estimates of the global value of natural capital/ecosystem services led to a celebration in financial circles of this new "asset class" and the huge market it portended, consisting of hundreds if not thousands of trillions of dollars, now potentially open to expropriation and exploitation by capital. In this view, the pricing of the planet had resulted in a huge increase in global wealth. Yet, operating on the principle of thinkers such as Marx, Ruskin, Soddy, and Daly, in which real wealth consists of natural-material use values, and indeed the earth itself, what was being measured in the pricing of ecosystem services was not real wealth, but rather the increased drain on the world's resources, their growing scarcity.[72] Based on this, the realm of commodity exchange was being enhanced—not for purposes of conservation, but as a further basis of capital accumulation, representing the acceleration of processes that had created metabolic rifts in nature's ecosystem processes in the first place. The trajectory, at present, unless stopped through global collective action, is toward a world of widening *catastrophe capitalism* marked by interconnected financial and ecological crises, based on the myth that nature can be transformed into a new speculative asset class.

Ecological Capital and
the Environmental Proletariat

Beginning with *The Poverty of Philosophy* in 1846, Marx—who like other social and radical critics had at first referred to "natural capital" in use-value terms, counterpoising this to exchange-value and artificial human-made capital—was to abandon this approach since it tended to naturalize capital. Instead, he drew a distinction between *earth matter*, that is, material existence, and *earth capital*; between natural-material conditions and processes and the capitalization of the earth. Nature, or earth matter, was eternal (in the sense of the first and second laws of thermodynamics), while earth capital was not.[73] The creation of earth capital, as a distinct social form, required the creation of private-property titles, and thus original expropriation of the land/earth, transforming what was previously the commons into a realm of private commodity value.[74] Monopolization of the land gave rise to a system of rents, imposed by the landlords on society as a whole, paid out of the total surplus product.

Ralph Waldo Emerson observed that "nature is inexhaustibly significant," since as material beings we must return to it again in every action we take. Historical materialists have traditionally referred to the "indissoluble unity" of humanity with the "universal metabolism of nature."[75] Today, however, nature is alienated along with labor, forming the basis of the capitalist system of exploitation. The concept of natural capital as it is employed today is nothing other than an attempt to extend this alienation to nature and humanity as a whole, monetizing ecosystem services so as to generate a new financial ecological regime: a social and historical relation in which the entire earth is for sale. For Paul Hawken, Amory Lovins, and L. Hunter Lovins, capitalism cannot be said to exist unless it is "natural capitalism" bringing the entirety of nature within its logic.[76]

The playing out of the logic of the expropriation of the earth can be seen in the attempts of neoclassical environmental economist

Edward Barbier to promote the idea that ecosystems, extending to the Earth System itself, are nothing but capital, conceived in exchange-value terms. All of existence is thus capital. "If ecosystems are . . . considered capital assets," then they are by definition, he tells us, "ecological capital" to be conceived in exchange-value terms. Ecological capital as a whole thus stands for the totality of the world's ecosystems, seen as constituting mere "forms of capital." All ecological problems for Barbier have a single solution: "capitalizing on nature."[77] In this view, nature, the earth, the basis of all life and existence, presumably stretching to the universe, is capital, measured in money. This dwarfs even Price's notion of compound interest leading to wealth equal to "150 millions of earths all solid gold," since Price was referring to a mathematical process of compound interest—not to the notion that the earth and the universe was nothing but *solid capital*.[78] Here, we see the capital fetish highlighted by Marx and Soddy in its most extreme form. Not only is capital seen as an innate power; it has now, in the fantasies of contemporary economists, effectively replaced matter itself, generating what Marx called a "cosmic confusion."[79]

The historical reality of capital as a system of *social relations* is hidden behind this fetishized notion of natural capital as an innate power with a potential cash value, stemming from the earth, even replacing the earth/nature/matter as the most fundamental element of existence. The monetization and financialization of the earth's ecosystems, reenvisioned as "ecological capital" without limit, is at the same time a Great Expropriation, leading to a wider environmental proletariat (and ecological peasantry).[80] The system of original expropriation, which was the basis of the creation of the industrial proletariat and the modern system of labor exploitation, has metamorphosed into a planetary juggernaut, a *robbery system* encompassing the entire earth, leading to a more universal dispossession and destruction.[81] The result is the creation of a *global environmental reserve army of the dispossessed*, the product of capital's drive to monopolize the biogeochemical processes of the planet, at the expense of humanity as a whole.[82]

The effects of this rift in the earth's metabolism, and in humanity's social metabolism with the earth, are to be seen everywhere, including in the most developed capitalist states, as witnessed by carbon markets and water privatization. Yet, the onslaught on nature/natural capital today is principally directed at the Global South, where the financial gains from the expropriation of the earth in the name of the management of natural capital and offsets are the greatest. And it is here too that an increasingly dispossessed environmental proletariat is most in evidence. Everywhere, the class struggle of production is converging with class-based environmental justice struggles over food, air, water, and the conditions of social and ecological reproduction.

The global resistance of Indigenous communities, together with peasant subsistence producers, to increasing land grabs associated with the accelerating capitalization of nature is one of the most important developments of our time. In the case of the attempt of Hoch Standard and the Sabah government to seize the natural capital of Malaysia's Borneo forests, it is the Indigenous communities, threatened with expropriation and removal, who are at the forefront of the ecological and cultural resistance movement, defending the indissoluble unity with nature. This struggle is occurring on all three continents of the Global South and in regions of the Global North, an indication of how close the ties are between neocolonialism and the natural capital juggernaut. In Kenya, for example, members of the Sengwer community—who over the last decade and a half have faced forced mass evictions at gunpoint and the burning down and destruction of their villages by the Kenyan Forest Service in alignment with international capital—are waging a struggle to defend the forest and water towers (rainfall in mountains and highlands becomes water sources for lowland irrigation and human consumption).

Many African states inherited a dual land system from the earlier colonial era, which has continued in the postcolonial period. In Zambia until early this century, for instance, 94 percent of the land was held on the basis of customary rights, while all land was also formally held by the state. Now, with corporate-induced land grabs, frequently supported by governments, Indigenous and peasant communities are having their lands seized by large private, foreign-based interests. In Zambia, peasants have been fighting a battle against the financial expropriation of their land by Agrivision Africa, which has as one of its investors the World Bank's International Finance Corporation.[83] Some countries, such as Ghana and Botswana, have promoted laws that give customarily held lands the legal clout of private property.[84] But in most of Sub-Saharan Africa, Indigenous land rights are tenuous in private property terms. Given growing scarcity of resources and the incessant drive for natural capital, Indigenous people and smallholders are fighting to defend their lives, communities, and lands. In this context, the fact that such populations are generally the best stewards of the earth are frequently shunted aside by corporations in the drive for turning nature into gold.

A key basis for resistance to natural-capital colonialism is agroecology, presented as a more rational ecological alternative. La Via Campesina initiated its Global Campaign for Agrarian Reform in 1999.[85] The Landless Workers' Movement in Brazil has also played a leading role in fighting against the capitalization of nature. In the words of João Pedro Stedile, the national coordinator of the Landless Workers' Movement, "When you set up a car factory, you expect to obtain a 13% per year profit. When you take control of a natural resource and turn it into a product, like water, for example, you have profits of over 700%. That's what they are after."[86] India's massive farmers' movement in 2020–21 represented an enormous mobilization of small farmers against the growing agribusiness domination of Indian agriculture and the attempts to turn the earth and food into capital.[87] In the United States, the massive 2020 solidarity/George Floyd protests emanating largely from the working class and youth in support of a Black-led movement can be seen as an indication of the level of resistance to racial capitalism only waiting to burst out as material conditions, particularly in urban built environments, polarize.

In all of these struggles and numerous others, the goal is ultimately one of sustainable human development, necessarily coupled with resistance to capitalism, racism, colonialism, imperialism, and ecological devastation. Within this wider collective perspective, in agreement with natural science, human production is properly viewed as *complementary* with natural-material systems and cannot be reduced to a universal system of commodity value—based on the fallacious notion that all of existence is commensurate with and can be measured in terms of money. The goals governing the struggle for a viable future are necessarily those of substantive equality and ecological sustainability, together defining socialism in our time. Scientific and human development criteria are complementary elements in creating an integrated path to an ecological future. Behind this lies the recognition that an exploitative system that puts its faith in "the fetish character of capital" at the expense of all human existence and life on the planet can only lead, if not checked, to ultimate catastrophe.

As the North American based Indigenous (and socialist) organization Red Nation declared in *The Red Deal*, it is the *philosophy of money* that drives contemporary society and "its primary method of relationality is destruction. There is another word for a money-driven system that expresses its existence through destruction: capitalism. Capitalism destroys life. It pollutes the rivers. It scars mountains. It starves moose, wolves, and salmon. It alienates our bonds with each other and with the Earth. Its very existence demands our disappearance." The only response to such a destructive system raised to a planetary level is a universal struggle for nature and humanity, demanding a peoples' sovereignty of the earth and of production. "The global ecological revolution still to come" means "returning to our humanity and our origins as good relatives" of the earth. It means rationally regulating the metabolism of human society with the universal nature of which we are inextricably a part.[88]

6

Ecological Civilization, Ecological Revolution: An Ecological Marxist Perspective

TODAY, IT IS CRUCIAL to address the connections between *ecological civilization*, *ecological Marxism*, and *ecological revolution*, and the ways in which these three concepts, when taken together dialectically, can be seen as pointing to a new revolutionary praxis for the twenty-first century. More concretely, it is important to ask: How are we to understand the origins and historic significance of the concept of ecological civilization? What is its relation to ecological Marxism? And how is all of this connected to the worldwide revolutionary struggle aimed at transcending our current planetary emergency and protecting what Marx called "the chain of human generations," together with life in general?[1]

In 2018, cultural theorist Jeremy Lent, author of *The Patterning Instinct: A Cultural History of Humanity's Search for Meaning* (2017), wrote an article for the online site Ecowatch, titled "What Does China's 'Ecological Civilization' Mean for Humanity's Future?" This article exhibits a peculiarly Western view, which, while recognizing the distinctiveness of the notion of ecological civilization in China, nevertheless attempts to separate China's core conception in this regard from ecological Marxism and the critique of capitalism. In opening his article, Lent writes:

> Imagine a newly elected president of the United States calling in his inaugural speech for an "ecological civilization" that ensures "harmony between humanity and nature." Now imagine he goes on to declare that "we, as human beings, must respect nature, follow its ways, and protect it" and that his administration will "encourage simple, moderate, green, and low-carbon ways of life, and oppose extravagance and excessive consumption." Dream on, you might say. Even in the more progressive Western European nations, it's hard to find a political leader who would make such a stand.
>
> And yet, the leader of the world's second largest economy, Xi Jinping of China, made these statements and more in his address to the National Congress of the Communist Party in Beijing last October [2017]. He went on to specify in more detail his plans to "step up efforts to establish a legal and policy framework . . . that facilitates green, low-carbon, and circular development," to "promote afforestation," "strengthen wetland conservation and restoration" and "take tough steps to stop and punish all activities that damage the environment." Closing his theme with a flourish, he proclaimed that "what we are doing today" is "to build an ecological civilization that will benefit generations to come." Transcending parochial boundaries, he declared that his Party's abiding mission was to "make new and greater contributions to mankind . . . for both the well-being of the Chinese people and human progress."[2]

Why is it that the category of ecological civilization, which is so central for China today, is largely inconceivable even as a talking point within the imperial core of the capitalist world, lying entirely outside its ideological sphere? Lent argues that such a principle is diametrically opposed to traditional Western culture, from Plato to the present day, with its alienated view of nature, in which the environment is viewed simply as something to be conquered. This stands in sharp contrast, he argues, to the more ecological culture embedded in China's 5,000-year-old civilization—though China too has experienced thousands of years of ecological destruction.[3] He quotes the early Neo-Confucian philosopher Zhang Zai, who wrote a thousand years ago:

> Heaven is my father and earth is my mother, and I, a small child, find myself placed intimately between them.
>
> What fills the universe I regard as my body; what directs the universe I regard as my nature.
>
> All people are my brothers and sisters; all things are my companions.[4]

For Lent, China's view of ecological civilization—though laudable—has nothing really to do with the political economy of present-day China or Marxism.[5] Rather, he associates it with the "regeneration" of traditional Chinese values. Here, the fact that the Chinese Communist Party has adopted the notion of ecological civilization, while such a forward-looking view is generally incomprehensible in the West, is simply interpreted in terms of the very different cultural heritages of China and Europe. In this way, the divergence between Asia and the Western world regarding ecological civilization is largely divorced from material foundations and from such issues as capitalism and socialism. Hence, in Lent's perspective, China's emphasis on ecological civilization has nothing whatsoever—except in a negative sense—to do with ecological Marxism. Rather, the People's Republic of China is characterized as an authoritarian state that is the very symbol of unfreedom. He points to contemporary China's "hyper-industrial" economy as somehow worse than what prevails in the West, leading it down the road toward the pollution of the entire earth, and opposed to its claim to be building an ecological civilization.[6]

Lent's argument seems to be that while Europe and North America have superior political and economic foundations, their environmental progress is hindered by their more destructive traditional ecological culture. China, in comparison, has a more harmonious ecological culture extending back millennia, but it is hindered by its "hyper-industrial," authoritarian political-economic regime from bringing this to fruition, thus endangering the entire earth and all humanity—unless, of course, China's traditional ecological culture triumphs over its present Marxian-inspired political-economic goals.

This attempt, in the name of traditional Chinese values, to sever the notion of ecological civilization from ecological Marxism and the question of revolutionary-scale ecological change is ultimately aimed at disconnecting the idea of ecological progress from a socialist praxis of sustainable human development. In contrast, I contend that the concept of ecological civilization is in fact a historical product of the development of ecological Marxism. Any attempt to separate the two, notwithstanding the importance of traditional Chinese values, is to deny the historical significance of the ecological civilization concept, and its importance in conceiving the necessary worldwide ecological revolution.

ECOLOGICAL MARXISM AND THE ORIGINS OF THE ECOLOGICAL CIVILIZATION CONCEPT

The 1970s and '80s saw a resurrection of Soviet ecological thought, which had in many ways led the world in the development of ecological science in the 1920s and '30s, only to degenerate in the decades that followed due to political and social factors.[7] However, with its renewal in the 1970s and '80s, Soviet ecology took on a new, distinctive character, seeing the ecological problem as related to the general question of civilization.[8] This was especially evident in an important collection on *Philosophy and the Ecological Problems of Civilisation*, edited by A. D. Ursul and published in 1983.[9] This volume included contributions by some of the USSR's leading scientists and philosophers. This led directly to the concept of ecological civilization, with a number of other works on the topic appearing in 1983 and 1984, and with the same notion entering almost immediately into Chinese Marxism, where it was to become a central category of analysis.[10]

Ecological civilization in the Marxian sense points to the struggle to transcend the logic of all previous, class-based civilizations, particularly capitalism, with its two-fold domination/alienation of nature/humanity. Writing in *Philosophy and the Ecological Problems of Civilisation*, P. N. Fedoseev, vice president of the USSR Academy of Sciences, addressed the issue of "rejection of the gains of civilization" implicit in many so-called green attempts to confront the ecological problem, often generating historically disembodied utopias, either backward-looking or technocratic.[11] Leading environmental philosopher Ivan Frolov, following Marx, emphasized that the human metabolism with nature was mediated by science and the labor and production process, and thus depended on the mode of production.[12] Philosopher V. A. Los' explored how "culture is becoming an antagonist . . . to nature" and referred to the need to construct a new "ecological culture" or civilization, reconstituting on more sustainable grounds the role of science and technology in relation to the environment. As he explained: "It is in the course of shaping an ecological culture that we can expect not only a theoretical solution of the acute contradictions existing in the relations between man and his habitat under contemporary civilization, but also their practical tackling."[13]

From an ecological Marxist standpoint, the emerging planetary crisis thus demanded an epochal transformation to create a new ecological civilization, in line with the long history of environmental analysis within Marxism, and a socialist path of development. Marx and Engels dealt extensively with the ecological contradictions of capitalism, going beyond simply their well-known discussions on the degradation of the soil and the division between town and country, to encompass such issues as industrial pollution, the depletion of coal and fossil fuels more generally (in terms of what Engels called the "squandering" of "past solar heat"), the clearing of forests, the adulteration of food, the spread of viruses due to human causes, and so on.[14] Marx's celebrated theory of metabolic rift, with which he addressed the ecological crises of his day, has been extended today to address capitalism's destruction of ecosystems and the disruption of nearly every aspect of the planetary environment.[15]

In twenty-first-century China, ecological Marxism has contributed to the development not only of a powerful critique of contemporary environmental devastation, but also to the promotion of ecological civilization as an answer. Aware that ecology ultimately constitutes a deeper materialist grounding for society than mere economics, Xi has emphasized, in his conceptions of ecological civilization and of a "beautiful China," that ecology is "the most inclusive form of public well-being."[16] He has stated: "Man and nature form a community of life; we, as human beings, must respect nature, follow its ways and protect it. Only by observing the laws of nature can humanity avoid costly blunders in its exploitation. Any harm we inflict on nature will eventually return to haunt us. This is a reality we have to face."[17] These words are closely connected to the classical ecological analysis of Marx and Engels, who forcefully argued that human beings are part of nature and need to follow nature's laws in carrying out production, while referring to the "revenge" of nature on those who disregard its laws.[18]

The concept of ecological civilization being implemented in China today is seen as representing a new, revolutionary, and transformative model of civilization. Prior civilizations are viewed, in accordance with Marxist analysis, as tied to class society, but historically giving rise to new stages of development. In this view, ecological civilization is a stage in the development of "a great modern socialist society" that, unlike capitalism, does not sacrifice people and the planet to profits.[19] In contrast to the dominant capitalist notion of sustainable development, ecological civilization is understood as incorporating the domains of politics and culture, leading to a "five-in-one approach" that goes beyond the standard triad of environmental, economic, and social factors that has come to characterize liberal sustainable development. Ecological civilization conceived in this way is aimed at *sustainable human development*, giving more emphasis to the non-economic definition of well-being, and putting politics in charge.[20]

As Chen Xueming noted in *The Ecological Crisis and the Logic of Capital*, the basic principles underlying the socialist ecological modernization associated with ecological civilization are "prevention, innovation, efficiency, non-equivalence, dematerialization, greenification, ecologization, democratic participation, pollution fees and win-win scenarios between economy and environment."[21] The eight priorities for the establishment of ecological civilization are categorized as (1) spatial planning and development; (2) technological innovation and structural adjustment; (3) sustainable use of land, water, and other natural resources; (4) ecological and environmental protection; (5) regulatory systems for ecological civilization; (6) monitoring and supervision; (7) public participation; and (8) organization and implementation of environmental policy/planning.[22]

In the Chinese case, such revolutionary-scale ecological reforms are being attempted even in a context of rapid economic growth aimed at bringing China up to a level with the West. Integrated planning to protect the environment is being incorporated in all economic development plans. The seriousness with which ecological civilization is being pursued is reflected in the clear acknowledgment that, in the implementation of these ecological plans, economic growth will need to be slowed somewhat in relation to earlier decades.[23] This environmental focus can be seen in the radical transformations that China has been introducing in such areas as pollution reduction, reforestation and afforestation, development of alternative energy sources, imposing restrictions in sensitive river areas, rural revitalization, food self-sufficiency through collective means, and many other areas.[24] China has made dramatic progress in reducing the degree of its reliance on coal, but it has partly regressed in this respect over the last few years due to the pandemic and world crises.[25] Nonetheless, it has set definite dates for the implementation of ecological civilization, including having the main components of its ecological civilization in place by 2035, establishing a beautiful China by 2050, and reaching net-zero carbon emissions by 2060.[26]

The struggle to create an ecological civilization in China would, of course, mean very little if it were simply a top-down program, which would almost certainly lose its impetus and succumb to economic and bureaucratic forces. The radical nature of the transformation is safeguarded by the fact that, in China's post-revolutionary society, the ecological metamorphoses are emanating from both above and below, drawing on struggles for rural reconstruction in response to the rural-urban divide. For example, Yin Yuzhen, a peasant woman living in the desert in Uxin Banner in Inner Mongolia, decided to reclaim the desert, entering into a thirty-seven-year struggle in which she and her family have planted 500,000 trees. She has become a respected expert on the greening of deserts. Peasants in the region joined in the afforestation effort and nearly 6,700 square kilometers of barren sand were turned green. Yun Jianli, a former high school teacher, successfully organized against water pollution. In 2002, she founded Green Han River, an environmental protection organization to protect the Han River from pollution, producing countless environmental reports and opposing factory owners and managers. The organization has more than 30,000 volunteers. By 2018, they had organized over a thousand field trips to investigate pollution sources along the Han River, traveling over 100,000 kilometers altogether. The object is to mobilize the whole society for environmental protection. Wang Pinsong of Shangri-La by the Gold Sand River in southwest China—an area that is the home of fifteen ethnic groups—took the lead in mobilizing her village in opposition to a dam-building project in Tiger Leap Grove, which would have displaced 100,000 villagers and engulfed 33,000 acres of fertile land by the riverbanks. Environmental organizing at the grassroots level, based on the self-mobilization of the population, is a powerful force in today's China, pointing to the development of a new ecological communism.[27]

A major indication of China's approach to environmental issues and threats is its successful response to COVID-19, which has resulted in a mortality rate of four deaths per million people, as compared to the United States' COVID mortality rate of 3,507 per million (as of September 2023). China's achievement in protecting its population and, in a win-win situation, also protecting its economy, has been widely misconceived in the West as simply the result of an authoritarian set of lockdowns imposed from the top of society. Nevertheless, the secret to China's achievement, especially in the early stages, was adopting the model of people's revolutionary war: enlisting the self-mobilization of the entire population in the fight against COVID and the resurrection of the mass line, connecting the population to the state and party.[28]

CHINA AND ECOLOGICAL REVOLUTION

China faces enormous ecological contradictions internal to its society, as does world production as a whole. In terms of annual carbon emissions, China is the world's largest polluter. However, much of this is devoted to producing manufactured products to be consumed in the West, while China's historic carbon emissions are still far exceeded by the United States and Europe, with the United States responsible for seven times as much per capita of the carbon dioxide concentrated in the atmosphere. In terms of annual per capita carbon dioxide emissions, China today produces less than half the U.S. level.[29] In *Will China Save the Planet?*, Barbara Finamore, senior strategic director for Asia of the Natural Resources Defense Council in the United States, contends that while "China is still the largest GHG [greenhouse gas] emitter, it is arguably doing more than any other country to try to reduce global carbon emissions—though it continues to face enormous challenges."[30] There is no doubt that China's struggles to create an ecological civilization are revolutionary when placed against the efforts of other countries. This is largely due to its role as a post-revolutionary social formation that retains a large element of economic planning capability, state direction, and collective values, invigorated by continual popular mobilization in both rural and urban areas.

This brings us back to the question that Lent implicitly asked in the passage quoted at the beginning of this chapter. Why is it so impossible that a U.S. or European head of state could have referred, as Xi did, to a present and future goal for society couched not in terms of mere economic growth, but stressing the importance of creating an ecological civilization? The answer to this is not simply, as Lent would have us believe, that China has regenerated its traditional ecological values, or that the West is wedded to a culture, going back thousands of years, geared to the "conquest of nature." Rather, the fundamental division is between a post-revolutionary society that has adopted Marxism with Chinese characteristics—embracing the ecological critique emanating from classical historical materialism and treating it as central to the entire long revolution of socialism—and an unalloyed capitalist order in which the sole mantra is "Accumulate, accumulate! That is Moses and the prophets!"[31]

There is no possibility that the ruling-class interests in a core capitalist country like the United States, which has long cultivated an "imperial mode of living" and production, mainly benefiting the very top of society, will somehow turn around and advocate a low-carbon, "simple, moderate, green" way of life or oppose excessive consumption and inequality as advanced in the Chinese notion of ecological civilization.[32] Rather, the main radical proposal in the West to deal with the global ecological threat is that of a state-sponsored Green New Deal, usually articulated in terms of market mechanisms, technological change, and climate jobs, which will allow production to continue essentially unchanged. Yet the prospect of a Green New Deal, given the extent of opposition to fossil capital that it would require, has gone virtually nowhere in the United States or Europe, since even this is conceived as a dire threat to the ruling interests.[33] The result is that saving the planet as a place for human habitation is, ironically, left in contemporary capitalism almost entirely up to the private sector, which is the historical source of global ecological destruction, while the environmental reform effort has been reduced to creating state-financed green markets for private corporations and new forms of the financialization of nature.[34] Hence, the capitalist juggernaut continues in its forward motion, destroying in its path the very conditions of the human future.

In terms of sheer capacity, the wealthy, developed, technologically advanced countries at the core of the world capitalist system could easily lead the way in addressing the ecological problem. Their political inability to do so is linked to the weakness of socialist, collective, and ecological principles in capitalist commodity society, the virtual absence of planning (outside the military), and the ruling class's fears of the self-mobilization of populations, which is necessary if revolutionary-scale transformations in our economic relation to the environment are to be effected. What is needed to carry out an ecological revolution directed at human survival is not simply environmental reform, but a much broader ecological and social revolution aimed at transcending the logic of capitalism itself.

REVOLUTIONARY ECOSOCIALISM AND THE FUTURE

So far in this chapter, I have emphasized the importance of revolutionary ecosocialism or ecological Marxism in the conception of ecological civilization. It is no accident that the notion of ecological civilization first appeared in the 1980s in the Soviet Union and that it is being implemented as a guiding principle and central project in China, while it is scarcely discussed elsewhere in the world. This cannot be attributed solely to China's traditional culture, though it has played a part. Nor does it make sense to connect this to the notion of postmodern culture, which has had no real material relevance in this regard.[35] Rather, the notion of ecological civilization is inconceivable in any meaningful sense outside of a society engaged in building socialism, and thus actively engaged in combating the primacy of capital accumulation as the supreme measure of human progress. It is exactly here that Marxian ecology has had a huge role to play.

Ecological Marxism has developed in China in terms of its own "vernacular revolutionary tradition," where new critical concepts are seen as directly problem-oriented and immediately put in operation.[36] This is distinct from its conceptualization in the West, where ecosocialist researchers are more removed from praxis and have generally been engaged in wider, and often more abstract, theoretical developments. A principal concern of Marxian ecology in the West (as in much of the rest of the world) has been the reconstruction of Marx's theory of metabolic rift, and how to enhance the continuing critique of capital in this respect. Bringing this renewed ecological critique emanating from classical historical materialism to bear on the problems of building ecological civilization in China therefore ought to be a priority—and in fact many scholars in China are currently engaged in this.

In terms of what we have learned in the recent renewal and elaboration of Marxian ecology, a number of concepts are crucial. Chief among these is Marx's triad of concepts of the "universal metabolism of nature," "social metabolism," and the "irreparable rift in the interdependent process of social metabolism"—or the *metabolic rift* brought on by capitalist development.[37] The concept of the universal metabolism of nature recognizes that human beings and human societies are an emergent part of nature. Social metabolism expresses how humanity interacts with and transforms nature through production. And the metabolic rift reflects the fact that an alienated social metabolism, aimed at the expropriation of nature as a means of the exploitation of humanity and the accumulation of capital, necessarily produces an ecological crisis, driving a wedge between this alienated social metabolism and the universal metabolism of nature of which we are a part.

Marx himself provided a penetrating definition of what we now call *sustainable human development*. No one—not even all of the people or all of the countries in the world—he argued, owns the earth; rather, we are obligated to hold it in usufruct as good managers of the household, sustaining it for the chain of human generations.[38] Genuine progress on this score, overcoming the alienation of nature and humanity associated with the processes of expropriation and exploitation, has to embrace the notion not simply of an *economic proletariat* (and economic peasantry) as the principal force for change, but, in a more inclusive materialism, of an *environmental proletariat* (and ecological peasantry). Indeed, the three categories that we started with—ecological civilization, ecological revolution, and ecological Marxism—hardly make sense without this *fourth term* of the environmental proletariat.

Our relation to the earth is our most fundamental material relation out of which our production, history, and social relations emerge. Those who are most alienated, exploited, and degraded by the system in their relations to nature and the earth constitute both the force and means for change in the twenty-first century.[39] In what Marx called the "hierarchy of [human] needs," our relation to the earth necessarily comes first, since it constitutes the basis of survival and of the development of life itself.[40]

7

Marxian Ecology, East and West: Joseph Needham and a Non-Eurocentric View of the Origins of China's Ecological Civilization

ECOLOGICAL MATERIALISM, OF WHICH ecological Marxism is the most developed version, is often seen as having its origins exclusively within Western thought. But if that is so, how do we explain the fact that ecological Marxism has been embraced as readily (or indeed, more readily) in the East as in the West, leaping over cultural, historical, and linguistic barriers and leading to the current concept of ecological civilization in China? The answer is that there is a much more complex dialectical relation between East and West in relation to materialist dialectics and critical ecology than has been generally supposed, one that stretches back over millennia.

Materialist and dialectical conceptions of nature and history do not start with Marx. The roots of an "organic naturalism" and "scientific humanism," according to Joseph Needham (李約瑟), in his magesterial work *Science and Civilization in China*, can be traced to the sixth to third centuries BCE both in ancient Greece, beginning with the pre-Socratics and extending to the Hellenistic

philosophers, and in ancient China, with the emergence of Daoist
and Confucian philosophers during the Warring States Period of
the Zhou Dynasty.[1] As Amin indicated in his *Eurocentrism*, the
"philosophy of nature [as opposed to metaphysics] is essentially
materialist" and constituted a "key breakthrough" in tributary
modes of production, both East and West, beginning in the fifth
century BCE.[2]

In *Within the Four Seas: The Dialogue of East and West* in 1969,
Needham noted the absolute alacrity with which "dialectical materialism" was taken up in China during the Chinese Revolution and
how this was treated as a great mystery in the West. Nevertheless,
the sense of mystery, he contended, did not extend in the same
way to the East itself. He wrote: "I can almost imagine Chinese
scholars" confronted with Marxian materialist dialectics, "saying
to themselves 'How astonishing: this is very like our own *philoso-
phia perennis* integrated with modern science at last come home to
us.'"[3] The Marxian materialist dialectic, with its deep-seated ecological critique rooted in ancient Epicurean materialism, was in
Needham's view so closely akin to Chinese Daoist and Confucian
philosophies as to create a strong acceptance of Marxian philosophical views in China, particularly since China's own perennial
philosophy was in this roundabout way integrated with modern
science. If Daoism was a naturalist philosophy, Confucianism was
associated, he wrote, with "a passion for social justice."[4]

The Needham convergence thesis—or simply the Needham
thesis, as I am calling it here—was thus that Marxist materialist
dialectics had a special affinity for Chinese organic naturalism
as represented especially by Daoism, which was similar to the
ancient Epicureanism that lay at the foundations of Marx's own
materialist conception of nature. Like other Marxist scientists and
cultural figures associated with what has been called the "second
foundation of Marxism," centered in Britain in the mid-twentieth century, Needham saw Epicureanism as providing many of
the initial theoretical principles on which Marxism as a critical-
materialist philosophy was based.[5] It was the similar evolution

of organic materialism East and West, but which in the case of Marxism was integrated with modern science, that explained dialectical materialism's profound impact in China.[6]

The Needham thesis, as presented here, can also throw light on the spurious proposition, recently put forward by Lent in *The Patterning Instinct*, that the Chinese conception of *ecological civilization* is derived entirely from China's own traditional philosophy, rather than Marxism.[7] Lent's argument fails to acknowledge that ecological civilization as a critical category was first introduced by Marxist environmental thinkers in the Soviet Union in its closing decades, and immediately adopted by Chinese thinkers, who were to develop it more fully.[8] For environmental philosophers and scientists in postrevolutionary societies who were familiar with dialectical materialism, it was natural to see the answer to ecological problems as demanding a new ecological civilization, constituting a necessary evolutionary development of socialism. This was further propelled by the fact that China, according to Needham, had avoided the disassociation of thought characteristic of the West through the identical opposites of abstract idealism/ theology and mechanistic materialism. Hence, from the critical standpoint introduced by Needham, the concept of ecological civilization can be seen as an organic outgrowth of philosophies of dialectical naturalism in both East and West to which Marxism added a crucial scientific component.

Of course, the Needham thesis may seem obscure at first from the standpoint of the Western left, since it relies on an Epicurean-Marxist interpretation of the origins of historical materialism, and at the same time sees this in relation to a radical conception of Chinese science and civilization over the millennia that is unfamiliar to Western eyes. This double disconnect from prevailing views has to do with the well-known alienation of the Western Marxist tradition from both science and materialism, coupled with a deep Eurocentrism that has affected Marxism in the West, associated with the systematic downplaying of colonialism and imperialism.[9]

All of this suggests that the Needham thesis, which sees

dialectical materialism as having roots in materialist and ecological ideas that arose separately and with quite different histories in East and West, but leading to a special affinity for Marxism in China, is well worth discussing in our time of planetary crisis, given the need for the reunification of humanity on more ecorevolutionary terms.[10] However, addressing the ancient philosophies underlying ecological materialism in both East and West, and the relation of this to the development of ecological Marxism today, requires that we strive to overcome the Eurocentric and other culturalist barriers that stand in the way of the emergence of the ecology of praxis on a planetary scale.

EUROCENTRISM AND MARXISM

The concept of Eurocentrism as constituting a definite ideological form first arose within the Marxist tradition. It was introduced by Needham in *Within the Four Seas* and was later employed by Amin in the preface to the first edition of his *Eurocentrism*. For both Needham and Amin, Eurocentrism is defined as the notion that European culture is the *universal culture* to which all other cultures must conform, given that non-Western cultures are reduced simply to being *particular cultures*.[11] As Needham argued, "The basic fallacy of Eurocentrism is therefore the tacit assumption that because modern science and technology, which grew up indeed in post-Renaissance Europe, are universal, everything else European is universal also."[12] Likewise, Amin writes: "Eurocentrism . . . claims that imitation of the Western model by all peoples is the only solution to the challenges of our time." Eurocentrism both projects itself as the universal culture and rejects the true universalism of peoples.[13]

Viewed in this way, classical Marxist thought and socialism in general have always been radically opposed to Eurocentrism as the ideology of Western colonialism. This is as true of Marx and Engels, particularly in their later years, as it was of Lenin and Luxemburg. In the twentieth century, moreover, the impetus for revolution shifted to the Global South and its struggle against imperialism, generating in the process new Marxist analyses in the works of figures as distinct as Mao Zedong, Amílcar Cabral, and Che Guevara, all of whom insisted on the need for a world revolution.

To be sure, one can point to traces of European ethnocentrism in some of Marx's early work, which was affected by the sources that he had available at the time, most of which came from European colonial reports. Nevertheless, it has been recognized by Marxist theorists of underdevelopment for decades—initially in the work of Horace B. Davis in the United States, Kenzo Mohri in Japan, and Suniti Kumar Ghosh in India—that by the late 1850s, Marx had become increasingly focused on the critique of colonialism, actively supporting anticolonial rebellions, and was progressively more concerned with analyzing the material and cultural conditions of non-Western societies.[14] Marx's growing attention to noncapitalist societies was a product of his identification with various revolts against colonialism, and was further facilitated by the "revolution in ethnological time" with the discovery of prehistory and the rise of anthropological studies, occurring in tandem with Darwin's theory of evolution.[15] He made a massive effort to research the history and cultures of societies on the periphery of Europe, leading to his studies of the Russian language, his exploration of the Russian peasant commune, and his research into social formations in Algeria, India, China, Indonesia, and the Indigenous nations of the Americas. He was, at least initially, a strong supporter of the Taiping Revolution in China.[16]

In this respect, Saito's important work *Marx in the Anthropocene* constitutes a noted deviation from the growing scholarship demonstrating that Marx was never Eurocentric (in the terms discussed above) and had moved decisively away from any residual European ethnocentrism by the late 1850s and early '60s. In support of his contrary view, Saito points to what he refers to as the statement in the preface to the first edition of *Capital* where Marx "notoriously" informs his German readers that "the tale is told of you," meaning German bourgeois development would follow the basic path already laid out by English bourgeoisie. For Saito, this establishes Marx's *Capital* was Eurocentric in assuming all countries everywhere had to follow the same linear European path. Yet, the question of the non-European world was altogether absent from the argument in the preface to *Capital*, which was directed solely at conditions in Western Europe, and specifically at the significance of the British developments for what was to come in Germany. Marx later clarified this in his 1878 letter in response to Nikolai Mikhailovsky and in his 1881 letter to Vera Zasulich (as well in the various drafts to that letter) by indicating that the argument on linear development in *Capital* was specific to Western Europe, and that fundamentally different lines of development were possible in Russia and in other noncapitalist societies.[17]

Saito seeks to back up his charge of Eurocentrism in the first volume of *Capital* by highlighting Marx's contention that non-capitalist village communities in Java and elsewhere in Asia were to be viewed as economically *unchanging*, or stagnant. Quoting Marx's use of the phrase the "unchangeability of Asian societies," Saito says this constitutes evidence not only of Eurocentrism but Orientalism. Yet, when viewed in context, it is clear that Marx was concretely addressing the economic tendency of *village communities* in Java, where a developed exchange economy did not yet exist, to reproduce themselves on the basis of simple, rather than expanded, reproduction. Thus, Marx quotes his source, T. Stamford Raffle's *History of Java* (1817), as saying that the "internal economy" of the village communities "remains unchanged" despite all the political shifts going on within their larger societies, which in this respect were hardly static. Hence, with respect to the economically unchanging character and stagnation of village communities in Java and elsewhere, which Marx places against the backdrop of the continual upheavals within these same societies, he was clearly referring to concrete, material productive forms/ relations within peasant communities at the base of the society. Naturally, the simple reproduction of such village communities stood out when contrasted to the constantly expanding economies and incessant technological revolutions of the accumulative societies of the West at the time of the Industrial Revolution. For Marx, such differences were to be understood in historical and materialist, not culturalist, terms.[18]

The issue of the "Great Divergence" between East and West at the time of the Industrial Revolution was a major issue in the late eighteenth and nineteenth centuries, one for which explanations were sought not only by Marx but by all of the classical political economists. Moreover, this same debate remains fundamental to today's historiography.[19] There is no doubt that the East, for a time, stagnated economically relative to the West. For example, China in 1800 accounted for a third of the world's industrial potential. By 1900 this had fallen to 6.3 percent (and in 1953 to a mere 2.3 percent).[20] Marx explained this historical divergence between East and West, already evident in his time, in terms of specific productive forms/modes, and as a product to a considerable degree of European colonialism. In the first volume *Capital* he described the terrible effects of Dutch colonial slavery in Java and how it served to undermine the village communities. None of this was developed in cultural nationalist or racist terms, as was the case in the dominant colonial-Eurocentric tradition within the West.[21]

But if Marxism, as classically represented initially by Marx and Engels, and later by figures such as Lenin and Luxemburg, was strongly opposed to any kind of Eurocentrism and Western colonialism/imperialism, explaining developments in materialist rather than culturalist terms, later *Western Marxism* as a distinct philosophical tradition has often been ambivalent with respect to imperialism and deeply ethnocentric in its approach to Marxism, viewing Marxism in the West, as Needham critically observed, as having a kind of "*a priori* superiority," despite the fact that revolution has long since shifted to the periphery of the capitalist world system.[22] This has gone hand in hand with Western Marxism's denial of the dialectics of nature, and thus science, nature, and any kind of ontological materialism. In many post-Marxist analyses, notions of class and socialism were also abandoned.[23]

The primary challenge confronting ecosocialism in the West is reconnecting Marxism to its materialist roots. A materialist conception of history cannot exist in a meaningful way apart from a materialist conception of nature (and vice versa). Marx's theory of metabolic rift depended on this much broader conception. Nor could Marxism exist in purely ideational form separate from the critique of class and imperialism or divorced from the new revolutionary vernaculars emerging throughout the Global South. In this sense, the parallels with respect to the materialist conception of nature and organic materialism that Needham pointed to with respect to pre-Socratic and Hellenistic Greece and the Warring States Period in China are crucial to understanding both the history and the future of ecological Marxism. Most importantly, the Chinese concept of ecological civilization needs to be put in this context of the rediscovery of the roots of an organic-ecological materialism.

Epicureanism and Daoism

To better understand the Needham thesis on the affinity of Marxism with traditional Chinese philosophy it is necessary to recognize that Needham, like many of the other scientists and cultural theorists associated with the second foundation of Marxism, saw Epicurean materialism as the key to the Marxian materialist conception of nature, and as underlying dialectical materialism. The essence of the materialist view, common to both Epicureanism and Daoism, and the basis of all scientific humanism, was that nature could be understood on its own terms, as spontaneously originating. For Daoism, "The Tao [the Way of nature] came into existence of itself"; meanwhile, for Epicureanism, "Nature loosed from every haughty lord /And forthwith free, is seen to have done all things / Herself and through herself of her own accord / Rid of all gods."[24] Chinese culture, Needham argued in *Science and Civilization in China*, had retained "an organic philosophy of Nature . . . closely resembling what modern science has been forced to adopt [most fully within dialectical materialism] after three centuries of mechanical materialism."[25] "Naturalism in the *Dao De Jing*," P. J. Laska indicates in the introduction to his English translation of this work,

> is similar to the naturalism that evolved in ancient Greek philosophy, beginning with the Presocratics, and continuing through the atomic systems of Democritus and Epicurus. What is distinctive about the naturalism of ancient China, [however], is the addition of the concept of *Dao*, meaning "the Way" the cosmic *process* that encompasses both Being and Non-Being. Ancient Greek materialism lacks this proto-ecological concept.... What the naturalism of East and West have in common is the debunking of anthropogenic projections that turn natural occurrences into supernatural agents. . . .
>
> In the *Dao De Jing* natural order is seen as developing spontaneously from the interaction of the various "beings" that comprise "the One."

The result was a "holistic naturalism," one built, like Epicurean materialism and Marxian dialectical naturalism, on the basis of conceptions of the unity of opposites and unending process.[26]

Marx noted that for Epicurus, in whose work an "immanent dialectic" in accord with nature was found, the "world is my *friend*."[27] Likewise, for Daoism, Needham insisted, "The natural world was not something hostile or evil, which had to be perpetually subdued by will-power and brute force, but something more like the greatest of living organisms, governing principles of which had to be understood so that life could be lived in harmony with it."[28] Thus, "the Order of Nature was a principle of ceaseless motion, change, and return. . . . This was a concept not of non-action [*wu wei*], but of no action contrary to Nature." In Chinese thought, "matter disperses and reassembles in forms ever new."[29] In the West, Epicureanism provided a similar materialist view, leading to notions of emergence and integrative levels and providing a critical realism that was to be developed most fully with Marxian-influenced materialist dialectics. Like Daoism, Epicureanism saw *sufficiency* (the principle of enough) as a key value. "Today," Needham stated, "we are all Taoists and Epicureans."[30]

If Epicurean materialism was an organic materialism, akin to Daoism, its more radical and environmental elements, for Needham, had been lost in the prevailing culture in the West, where it had been overtaken by a mechanistic materialism and a one-sided conception of the "domination of nature"—what he called, following Theodore Roszak, a "mechanistic imperative" and a "scientization of nature" that had become destructive. In response to this mechanistic view (and to abstract idealism), Marxian dialectical materialism, Alfred North Whitehead's process philosophy, and the new philosophies of emergence were the main counterforces, representing the highest levels of development of scientific thought.[31]

In contrast to the dominant mechanistic and idealist dualism of the West, China had in many ways retained its organic naturalism and was able to incorporate this with modern science by making use of Marxian dialectical materialism, with its more complex understanding of the relation of humanity to evolutionary ecology, mediating between Western science and traditional Chinese philosophy. Traditional Chinese natural philosophy reached its highest level, according to Needham, in the twelfth century with Neo-Confucianism, which was "in fact, an organic conception of Nature, a theory of integrative levels, an organic naturalism...closely allied to the conceptions of dialectical materialism." One of "the most profound of Neo-Confucian ideas," he wrote, is that embodied "in the famous phrase *wu chi erh thai chi*, 'that which has no Pole and yet itself is the supreme Pole,' namely the conception of the whole universe as an organic unity, in fact, as a single organism."[32]

Bertrand Russell, Needham suggested, was simply paraphrasing the second part of the *Dao De Jing* in his book *The Problem of China* in portraying the Chinese philosophy as "production without possession, action without self-assertion, development without domination."[33] As an expression of the human social relation to nature, this was deeply ecological. With its very different relation to the natural world, China, Needham pointed out, had avoided some of the worst aspects of the metabolic rift in soil fertility, critically analyzed by figures such as Liebig and Marx, through the continued "use of human excreta as fertilizer," preventing "the losses of phosphorous, nitrogen, and other soil nutrients which happened in the West."[34]

ECOLOGICAL CIVILIZATION AS MARXIAN ECOLOGY WITH CHINESE CHARACTERISTICS

According to what I have called the Needham thesis, Marxist dialectical naturalism, which developed as an organic-materialist ontology with deep roots in ancient Greek materialist philosophy, had a special affinity with traditional Chinese philosophy, since this form of scientific humanism had not been supplanted in China, as in the West, by a hegemonic dualism of mechanistic materialism and abstract idealism/theology. The fact that the Chinese Revolution was a peasant-based revolution also meant that it was rooted in very different material conditions than those that had governed bourgeois civilization in the West. These ideational and material conditions made China, as Needham argued in the 1970s, more open both to Marxism in its dialectical-materialist form, and to revolutionary ecological conceptions arising from both this and from traditional Chinese philosophy. Socialism with Chinese characteristics, from Mao to the present, thus included a dialectical-ecological component, which has become more, rather than less, evident, and is today exemplified by the notion of ecological civilization.

The concept of ecological civilization, as we have seen, arose in the final decade of the Soviet Union, as a natural extension of socialism. According to the Soviet environmental philosopher Frolov writing in 1983, Marx's approach to the unity/alienation of humanity and nature began with recognizing that human beings as social beings regulate the metabolism between themselves and nature as a whole, through their production, and their development of a "second nature" within society. The alienated nature of production under capitalism created various contradictions between human beings and nature, now referred to as the metabolic rift.[35] The answer, Frolov argued, was the "humanization of science" and the development of a "scientific humanism" in accord with socialized production, pointing to the need for a new ecological culture. As Los' put it:

> It is in the course of shaping an ecological culture [ecological civilization] that we can expect not only a theoretical solution of the acute contradictions existing in the relations between man and his habitat under contemporary civilisation, but also their practical tackling. Society, which has created an ecological culture, is, as Marx explained, "the complete unity of man with nature—the true resurrection of nature—the accomplished naturalism of man—and the accomplished humanism of nature."[36]

The idea of ecological civilization was quickly adopted by the Chinese thinker Ye Qianji in 1987 and became central to the definition of socialism with Chinese characteristics under Hu Jintao in the first decade of this century.[37] Ecological civilization is often seen as little more than a socialist counterpart of capitalist ecological modernization. However, in fact, it is radically removed from the general conception of industrial civilization in the West. Rather, it is conceived as a form of genuinely sustainable human development, exemplifying the goals of socialism with Chinese characteristics. It is an outgrowth of Marx and Engels's classical ecological critique plus the culture and historical conditions of China itself.[38] As Chen Xueming wrote in *The Ecological Crisis and the Logic of Capital*, "Unlike capitalist society, socialist society does not lead [the] human being to become an 'economic animal' who only knows how to fulfill himself with respect to material life. The aim of socialism is not to develop the way of life under capitalist conditions, but to create a new way of life. . . . The essential characteristics and core values of socialism consist of creating a way of being, which, unlike the capitalist way of life, aims at realizing the whole-sided development of the human being."[39]

But if Marxian dialectical and historical materialism, based particularly on the classical ecological critique introduced by Marx, has played a central part in the development of the Chinese concept of ecological civilization, the natural synergy of this (as expressed in the Needham thesis) with traditional Chinese thought is not to be ignored; to do so would in fact be Eurocentric. The complex, dialectical relation of the concept of ecological civilization to socialism with Chinese characteristics can be seen in Xi's thought in this area . As Huang Chengliang has explained, the "Theoretical Origins of Xi Jinping's Thought on Ecological Civilization" can be traced to five sources: (1) Marxist philosophy, integrating "the three fundamental theories of 'dialectics of history, dialectical materialism and dialectics of nature'"; (2) traditional Chinese ecological wisdom on "man-nature unity and the law of nature"; (3) the actual historical context of ecological governance in China in response to the ecological crisis; (4) struggles to develop a progressive and ecological model of sustainable development; and (5) the articulation of ecological civilization as the governing principle of the new era of socialism with Chinese characteristics.[40]

Hence, characteristic of Chinese understanding of ecological civilization today, as exemplified in Xi's thought, is a Marxian ecological dialectics and political economy interwoven with compatible elements taken from Daoism, Confucianism, and Neo-Confucianism, creating a powerful organic, ecological-materialist philosophy. Rather than simply an ideational product, the concept and implementation of ecological civilization is determined by the ecological crisis, struggles for ecologically sustainable development, and the new era of socialism with Chinese characteristics in which the development of a mature socialism characterized by a new ecological way of life becomes the primary goal.

This is apparent today in some of Xi's most famous pronouncements on ecological civilization. Thus, one can see Marxian and traditional Chinese ecological values wedded when he declares:

Man and nature form a community of life; we, as human beings, must respect nature, follow its ways, and protect it. Only by observing the laws of nature can humanity avoid costly blunders in its exploitation. Any harm we inflict on nature will eventually return to haunt us. This is the reality we have to face. The modernization we pursue is one characterized by harmonious coexistence between man and nature. . . . We should have a strong commitment to socialist eco-civilization and work to develop a new model of modernization with humans developing in harmony with nature.[41]

This was coupled with declarations that China would "encourage simple, moderate, green, and low-carbon ways of life, and oppose extravagance and excessive consumption."[42] In his April 2020 speech, "Build an Eco-Civilization for Sustainable Development," Xi starts out by quoting Engels: "Let us not however flatter ourselves overmuch on account of our human victories over nature. For each such victory nature takes its revenge on us." Xi concludes: "We must understand fully how humanity and nature form a community of life and step up efforts on all fronts to build an eco-civilization."[43]

In Xi's analysis, the traditional Chinese emphasis on the harmony of humanity and nature, or the view that "the human and heaven are united in one," is wedded to Marxian ecological views with a seamlessness that can only be explained in terms of Needham's thesis of the correlative development of organic materialism both East and West, with Marxism as the connecting link.[44] From this perspective, the Chinese notion of ecological civilization, due to its overall theoretical coherence and coupled with China's rise in general, is likely to play an increasingly prominent role in the development of ecological Marxism worldwide. Needham wrote: "China has in her time learnt much from the rest of the world; now perhaps it is time for the nations and the continents to learn again from her."[45]

-------- 8 --------

Extractivism in the Anthropocene

OVER THE LAST DECADE and a half, the concept of *extractivism* has emerged as a key element in our understanding of the planetary ecological crisis. Although the development of extractive industries on a global scale has been integral to the capitalist mode of production since its onset, commencing with the colonial expansion of the long sixteenth century, this took on a much larger worldwide significance with the advent of the Industrial Revolution of the late eighteenth and nineteenth centuries, marking the beginning of the age of fossil capital. Nevertheless, it was only with the Great Acceleration, beginning in the mid-twentieth century and extending to the present, that the *quantitative* expansion of global production and of resource extraction in particular led to a *qualitative* transformation in the human relation to the Earth System as a whole. This has given rise to the Anthropocene Epoch in geological history, in which anthropogenic (as opposed to non-anthropogenic) factors for the first time in Earth history constitute the major force in Earth System change.[1] In the Anthropocene, extractivism has become a core symptom of the planetary disease of late capitalism/imperialism, threatening humanity and the inhabitants of the earth in general.

The Great Acceleration is dramatically depicted by the Anthropocene Working Group of the International Commission on Stratigraphy in the form of a series of twenty-four charts, each showing a hockey stick–shaped curve of economic expansion, resource depletion, and overloading planetary sinks, representing a sudden speeding-up and scaling-up of the human impact on the earth, similar to the famous hockey stick chart on increases in global average temperature associated with climate change.[2] Viewed in this way, the Great Acceleration is seen as having brought the Holocene Epoch of the last 11,700 years of geological history to a sudden end, ushering in the Anthropocene Epoch and the current planetary crisis.

Recent research has shown two separate periods where global resource use—including all biomass, minerals, fossil fuel energy, and cement production—has increased much more rapidly than global carbon emissions: the first resource-use acceleration occuring in 1950–70 and the second acceleration in 2000–15.[3] The first resource acceleration is associated with the rapid economic expansion of North America, Western Europe, and Japan after the Second World War; the second resource acceleration coincided with the rapid growth of China, India, and other emerging economies beginning around 2000. In the case of the wealthy capitalist countries or "developed economies," resource use per capita has tended to level off in recent years, while remaining at levels far beyond overall sustainability from a limits-to-growth perspective. Yet, much of this apparent leveling off in per capita natural resource use in the Global North has been due to the outsourcing of world industrial production to the Global South, while world consumption of goods and services remains highly concentrated in the Global North, associated with an "imperial mode of living."[4] In 2016, the *Global Material Flows and Resource Productivity Report* of the UN Environmental Programme indicated that "since 1990 there has been little improvement in global material efficiency [that is, efficiency in the extraction of primary materials]. In fact, efficiency started to decline around 2000."[5] Global extraction of materials

tripled in the four decades prior to the 2016 report.[6] These conditions have resulted in an acceleration of extractivist pressures in
key regions throughout the earth, particularly in the Global South.

In many countries in the Global South, particularly in Latin
America and Africa, primary commodities, including both agriculture and fossil fuels/minerals, dominate the export economy,
reminiscent of an earlier age. In 2019, percentages of primary
commodities in merchandise trade exports were as high as 67
percent in Brazil and 82 percent in both Chile and Uruguay. In
Algeria, export dependence on fossil fuels is almost complete, now
accounting for 94 percent of the value of its merchandise trade
exports.[7] In Latin America, in particular, the import-substitution
industrialization era of the early post–Second World War years,
which promoted manufacturing, has been succeeded in the new
era of accelerated resource extraction and by a new dependence on
primary commodities, including both agricultural goods and fuels/
minerals. In 2017, natural resource rents (including mineral, oil,
natural gas, and forestry rents) accounted for 43 percent of GDP in
the Republic of Congo.[8] In Africa, the drive for resources and new
agricultural lands has fueled vast land grabs throughout the con-
tinent, made possible by the failure of the decolonization process
in securing the rights to the land for Indigenous populations.[9] In
island nations around the globe, fishing and resource rights over
vast ocean territories have been ceded to multinational corporations as the ocean commons are being intensively exploited.[10] New
technologies have led to a race for new rare minerals, as in the case
of lithium mining.[11] A vast financialization of the earth, in which
international finance based in the Global North is taking over the
commodification and management of ecosystem services, primarily in the Global South, is now underway.[12]

Nor is this acceleration of resource extraction and extractive
infrastructure confined simply to the periphery of the capitalist
world economy. The United States is now the world's largest oil producer as well as the world's largest oil consumer. There are 730,000
miles of oil and gas pipelines worldwide, equal to thirty times the

circumference of the earth. The United States and Canada alone account for about 260,000 miles of fossil fuel pipelines, or over a third of the world's total.[13] In Canada, primary commodities in 2019 accounted for 43 percent of export value in merchandise trade, while in Australia it was 81 percent.[14]

The ecological consequences of all these trends are catastrophic, extending all the way from the devastation of the land and communities up to climate change and the destruction of a human-habitable planet. Fifty years after *The Limits to Growth* report was published by the Club of Rome, resource depletion is following what it referred to as its threatening "standard scenario," with the result that the very existence of planet Earth as a home for humanity and innumerable other species is endangered.[15]

In Latin America in particular these conditions and their effects on the ground have led to the development of extractivism as a critical concept, which in recent theoretical discussions has often taken on an expansive meaning, encompassing wide aspects of capitalism and forms of exploitation. Numerous academic analyses have sought to stretch the notion to account for the entire set of economic, political, cultural, and ecological problems of modern times, largely displacing capitalism itself, encompassing questions as varied as modernity, violence, production, exploitation, environmental destruction, digitalization, and the new "ontological assemblages" of the so-called "new materialists."[16] For such thinkers, extractivism is viewed as the insatiable source of capitalist modernity's destructive and nonreproductive drive to commodify and consume all life and all existence, what some theorists refer to as "total extractivism" or the "world eater." Such views end up displacing the critical concept of capital accumulation itself, as well as removing attention from the very concrete popular struggles occurring at the ground level against extractive capitals.[17]

For this reason, the Uruguayan thinker Eduardo Gudynas, a leading Latin American analyst of extractivism, has insisted that the concept be approached in relation to modes of production/appropriation, giving extractivism a very definite meaning directed at the development of a broad political-economic-eco-logical critique. Gudynas specifically objects to what he sees as the loose academic approach that now proposes vague and ambiguous "labels for extractivism such as 'financial,' 'cultural,' 'musical,' and 'epistemological,'" creating endless sources of confusion, and removing the concept from its basis in political economy and ecological critique. "Extractivism," he writes, "cannot be used as a synonym for development or even for an exporting primary economy. There is no such thing as extractivist development. . . . Extractivisms . . . do not account for the structure and function of an entire national economy, which includes many other sectors, activities and institutions."[18]

Gudynas's own theory of extractivisms, which will be a central focus of what follows, can be seen as having arisen out of the broad historical materialist tradition. Thus, to understand the significance of his work, it is necessary to situate it within a larger historical materialist tradition, going back to the classical analysis of Marx and Engels, related to issues of the appropriation/expropriation of nature, extractive industries, and the metabolic rift. In this way, it is possible to provide the foundations for a critique of extractivism in the Anthropocene.

MARX AND THE EXPROPRIATION OF NATURE

The notion of "extractive industry" dates back to Marx in the mid-nineteenth century. He divided production into four spheres: extractive industry, agriculture, manufacturing, and transport. Extractive industry was seen by him as constituting the sector of production in which "the material for labour is provided directly by Nature, such as mining, hunting, fishing (and agriculture, but only insofar as it starts by breaking up virgin soil)."[19] In general, Marx drew a line between extractive industry and agriculture, insofar as the latter was not dependent on raw materials from outside agriculture, but was capable of building up from within, given agriculture's reproductive, as opposed to nonreproductive, characteristics. This, however, did not prevent him, in his theory of metabolic rift, from seeing capitalist industrial agriculture as expropriative, and in ways that we now call extractivist.

Some of Marx's most critical comments with regard to the capitalist mode of production are directed at mining as the quintessential extractive industry. In his discussion of coal mining in *Capital*, volume 3, he treats the absolute neglect of the conditions of the coal miners, resulting in an average loss of life of fifteen people a day in England. This led him to comment that capital "squanders human beings, living labour, more readily than does any other mode of production, squandering not only flesh and blood but nerves and brains as well."[20] But the destructive effects of extractive industry and of capital in general, for Marx, were not restricted to the squandering of flesh and blood, but also extended to the squandering of raw materials.[21] Moreover, Engels, in writing to Marx, famously discussed the "squandering" of fossil fuels resources, and coal in particular.[22]

In interviews that he gave responding to radical and Indigenous movements against extractivism, Ecuadorian president Rafael Correa rhetorically asked: "Let's see, *Señores marxistas*, was Marx opposed to the exploitation of natural resources?" The implication was that Marx would not have opposed contemporary extractivism. In response, ecological economist Joan Martinez-Alier pointed to Marx's famous analysis indicating that "capitalism leads to a 'metabolic rift.' Capitalism is not capable of renewing its own conditions of production; it does not replace the nutrients, it erodes the soils, it exhausts or destroys renewable resources (such as fisheries and forests) and non-renewable ones (such as fossil fuels and minerals)." On this basis, Martinez-Alier contends that Marx, though he did not live to see global climate change, "would have sided with Climate Justice."[23] Indeed, the extraordinary growth of the Marxian ecological critique, building on Marx's analysis in *Capital* of the "negative, i.e., destructive side" of capitalist production in his theory of metabolic rift, has provided the world with penetrating insights into every aspect of the contemporary planetary crisis.[24]

Not only was the expropriation of land and bodies recognized in Marx's analysis, but the earth itself could be expropriated in the sense that the conditions of its reproduction were not maintained, and natural resources were "robbed" or "squandered."

Key to a historical materialist analysis of extractivism is Marx's analysis of what he called "original expropriation," a term that he preferred to what the classical-liberal political economists called "previous, or original accumulation" (often misleadingly translated as "primitive accumulation").[25] For Marx, "so-called primitive [original] accumulation," as he repeatedly emphasized, was not *accumulation* at all, but rather *expropriation* or appropriation without equivalent.[26] Taking a cue from Karl Polanyi—and in line with Marx's argument—we can also refer to expropriation as appropriation without reciprocity.[27] Expropriation was evident in the violent seizure of the common lands in Britain. But "the chief moments of [so-called] primitive accumulation" in the mercantilist era, providing the conditions for "the genesis of the industrial capitalist," lay in the expropriation of lands and bodies through the colonial "conquest and plunder" of the entire external area/periphery of the emerging capitalist world economy. This was associated, Marx wrote, with "the extirpation, enslavement, and entombment in mines of the Indigenous population" in the Americas, the whole transatlantic slave trade, the brutal colonization of India, and a massive drain of resources/surplus from the colonized areas that fed European development.[28]

Crucial to this analysis was Marx's very careful distinction between *appropriation*, understood in its most general sense, as the basis of all property forms and all modes of production, and those particular forms of appropriation, such as *expropriation* and *exploitation* that characterized the regime of capital. Marx conceived *appropriation* in general as rooted in the free appropriation from nature, and thus as a material prerequisite of human existence, leading to the formation thereby of *various* forms of property, with private property constituting only one such form, which became dominant only under capitalism. This general historical theoretical

approach gave rise to Marx's concept of the "mode of appropriation" underlying the mode of production.[29] These distinctions were
to play an important role in Marx's later ethnological writings, and
his identification with the active resistance to the expropriation of
the lands of Indigenous communities in Algeria and elsewhere.[30]

Not only was the expropriation of land and bodies recognized
in Marx's analysis, but the earth itself could be expropriated in
the sense that the conditions of its reproduction were not maintained, and natural resources were "robbed" or "squandered."[31]
This was particularly the case with capitalism, in which the appropriation of nature generally took a clear, expropriative form. In
Marx's analysis, the free appropriation of nature by human communities, constituting the basis of all production, was seen as
having metamorphosed under capitalism into the more destructive form of "a free gift of Nature *to capital*," no longer geared
primarily to the reproduction of life, the earth, and community
as one largely indivisible whole, but rather dedicated solely to the
valorization of capital.[32] The "robbery" of the earth and the metabolic rift—or the "irreparable rift in the interdependent process
of social metabolism" between humanity and nature—were thus
closely interwoven.[33] Although some contemporary theorists have
attempted to define extractivism as meaning the nonreproduction
of nature, it is much more theoretically meaningful to view this in
line with Marxian ecology in terms of what Marx called the robbery or expropriation of nature, of which extractivism is simply a
particularly extreme and crucial form.

GUDYNAS AND THE EXTRACTIVIST SURPLUS

These conceptual foundations arising out of Marx's classical
ecological critique allow us to appreciate more fully the pathbreaking insights into extractivism provided by Gudynas in his
Extractivisms. A crucial point of departure in his analysis is the
concept of *modes of appropriation*. In his pioneering 1985 work
Underdeveloping the Amazon, environmental sociologist Stephen

G. Bunker introduced the concept of "modes of extraction" to address the issue of extractive industry and its nonreproductive character, contrasting this to Marx's larger concept of "modes of production."[34] Gudynas claims that Bunker was generally on the right track. However, in contrast to Bunker, Gudynas does not adopt the notion of modes of extraction. Nor does he retain Marx's notion of modes of production, arguing unaccountably that Marx's concept has been "abandoned," citing anthropologist and anarchist activist David Graeber. Rather, Gudynas turns to the concept of "modes of appropriation," while seemingly unaware of the theoretical connection between *appropriation* and *production* and between *modes of appropriation* and *modes of production* that Marx had constructed in the *Grundrisse*, and how this is related to current Marxian research into these categories.[35] Still, Gudynas's modes-of-appropriation approach allows him to distinguish between human appropriation from the natural environment in general and what he refers to as "extractivist modes of appropriation," which violate conditions of natural and social reproduction.

Gudynas defines extractivism itself in terms of processes that are excessive as measured by three characteristics: (1) *physical indicators* (volume and weight), (2) *environmental intensity*, and (3) *destination*, with extractivism seen as inherently related to colonialism and imperialism, requiring that the product be exported in the form of primary commodities.[36] Not all appropriation of nature carried out by extractive industries is extractivist. This is perhaps clearest in his short piece, "Would Marx Be an Extractivist?" As in Martinez-Alier's response to Correa, Gudynas states:

> Marx did not reject mining. Most of the social movements do not reject it, and if their claims are heard carefully, it will be found that they are focused on a particular kind of enterprise: large scale, with huge volumes removed, intensive and open-pit. In other words, don't confuse mining with extractivism. . . . Marx, in Latin America today, would not be an extractivist, because that would mean abandoning the goal of transforming the modes of

production, becoming a bourgeois economist. On the contrary, he would be promoting alternatives to [the dominant mode of] production, and that means, in our present context, moving toward post-extractivism.[37]

Today's global extractivism, what Martin Arboleda has called *The Planetary Mine*, is identified with "generalized-monopoly capital" and conditions of "late imperialism."[38] A central concern of Gudynas's work is a critique of the renewed imperial dependency in the Global South resulting from neo-extractivism, raising the question of "delinking from globalization" as perhaps the only radical alternative.[39] A similar view was powerfully developed by James Petras and Henry Veltmeyer in their *Extractive Imperialism*, which described the new extractivism as a new imperialist model, forcing countries into a new dependency, the ground for which had been prepared by the neoliberal restructuring that virtually annihilated many of the earlier forces of production in agriculture and industry.[40]

Gudynas's signal contribution, however, lies in his attempt to connect extractivism to the concept of surplus in order to explain the economic and ecological losses associated with the reliance on extractivist modes of appropriation. Here, he relies on the concept of economic surplus developed by Paul A. Baran in *The Political Economy of Growth* in the 1950s, which was designed to operationalize Marx's surplus value calculus in line with a critique that had rational economic planning as its yardstick.[41] Gudynas notes that in Baran's surplus concept, in conformity with Marx's surplus value, "ground rent and interest on money capital" are components of total surplus rather than production costs. In introducing the concept of economic surplus, Baran sought to reveal forms of surplus value that were, in capitalist accounting, as Gudynas puts it, disguised forms of "what is essentially an appropriation of the surplus."[42]

Employing this idea, Gudynas seeks to add to the economic or *social* dimension of surplus, based on the exploitation of labor, two environmental dimensions of the surplus in the context of extractivist modes of appropriation. The first of these, the *environmental renewable surplus*, is seen as related to the classic Ricardian-Marxian theory of agricultural ground rent focused primarily on renewable industry. It is meant to capture surplus not only associated with monopoly rents and thus integrated directly into the economic calculus, but also, according to Gudynas, to grapple with how ecosystem services such as pollination are extractively appropriated/ expropriated. Gudynas indicates that a larger "monetized surplus" is created for corporations by neglecting such crucial environmental aspects as soil and water conservation, thus generating an artificially large surplus based on the extractivist appropriation of renewable resources. This is related to what Marx called the "robbing" or expropriation of the earth, part of his theory of metabolic rift.[43]

According to Gudynas, the third dimension of the surplus (the second environmental dimension) is the *environmental nonrenewable surplus* related to nonrenewable resources, such as minerals and fossil fuels. "The key distinction here," he writes, "is that the resource will be exhausted sooner or later, and therefore the surplus captured by the capitalist will always be proportional to the loss of natural heritage that cannot be recovered. Similarly, the space occupied by a mining enclave will be impossible to use for another purpose, such as agriculture." Whatever extractivist surplus is obtained has to be set against the loss of natural wealth associated with resource depletion, something that is disguised by the common employment of the concept of "natural capital," conceived today not, as in classical political economy, in terms of use-value, but rather, in accord with neoclassical economics, in terms of exchange value and substitutability.[44]

The current planetary ecological crisis has to be seen in terms of the generation of a destructive expropriation of nature, which needs to be transcended in the process of going beyond capitalism.

In Marx and Engels's classical historical materialism, a very similar analytical approach was adopted with respect to the expropriation of nonrenewable resources to that presented by Gudynas in his analysis of the environmental nonrenewable surplus. For Marx and Engels, the destructive expropriation of nonrenewable resources could not be treated as a straightforward case of *robbing*, as in the case of the soil, forests, fishing, and so on. Hence, they approached extractivism with respect to nonrenewable resources under the rubric of the *squandering* of such resources, a concept that was especially used in relation to the avaricious expropriation of minerals and fossil fuels, particularly coal, but also applied to the extreme "human sacrifices" in extractivist industries, related to what is nowadays sometimes called the "corporeal rift."[45] Capitalism's relation to both renewable and nonrenewable resources was thus seen in the classical historical materialist perspective as pointing to the destructive expropriation of the earth, either as the "robbing" or the "squandering" of nature—an approach that closely corresponds to Gudynas's two forms of extractivist surplus appropriation/expropriation.

Gudynas's approach to what he calls the "extractivist surplus" associated with his two environmental dimensions of surplus is meant to encompass externalities, highlighting the fact that the "actual surplus" appropriated—to use Baran's terms—is, in some cases, artificially high, in relation to a more rational "planned surplus," as it does not account for depletion of fossil fuels and other natural resources.[46] This basic approach is employed in the remainder of Gudynas's analysis to engage with struggles on the ground over this bleeding of the extractivist economies and its relation to late imperialism, which carries out such bleeding on ever-larger scales to the long-term detriment of the relatively dependent peripheral or semiperipheral (that is, emerging) economies. As he argues in *Extractivisms*, this ultimately becomes a question of "extractivism and justice."[47]

EXTRACTIVISM AND THE CRISIS OF THE ANTHROPOCENE

Given that the Anthropocene, though still not official, has been defined as that epoch in which anthropogenic rather than non-anthropogenic factors, for the first time in geological history, are the primary drivers determining Earth System change, it is clear that the Anthropocene will continue as long as global industrial civilization survives. The current Anthropocene crisis, defined as an "anthropogenic rift" in the biogeochemical cycles of the Earth System, is closely associated with the system of capital accumulation and is pointing society toward an Anthropocene-extinction event.[48] To avoid this, humanity will need to transcend the dominant "accumulative society" imposed by capitalism.[49] But there will be no progressive escaping from the Anthropocene itself in the conceivable future, since humanity, even in an ecologically sustainable socialist mode of production will remain on a razor's edge, given the current planetary-scale stage of economic and technological development, and the fact that the limits of growth will need to be accounted for in the determination of all future paths of sustainable human development.

It was the recognition of these conditions that led Soriano, writing in *Geologica Acta*, to propose the *Capitalian* as the name of the first geological age of the Anthropocene Epoch.[50] According to this outlook, the current planetary ecological crisis has to be seen in terms of the generation of a destructive expropriation of nature, which needs to be transcended in the process of going beyond capitalism and the Capitalian Age. Others independently proposed the name *Capitalinian Age* for this new geological age, while also pointing to the notion of a *Communian Age*—standing for *communal, community, commons*—as the future geological age of the Anthropocene; one that needs to be created in coevolution with nature, necessitating a "great climacteric" by the mid-twenty-first century.[51]

In the present century, combating the capitalist expropriation of nature and in particular the extractivism that is more and more dominating our time—along with surmounting the present accumulative system itself—has to take priority at all levels and in all forms of social struggle. In the classical historical materialist perspective, production as a whole—not simply extractive industry, but also agriculture, manufacturing, and transportation—needs to be confronted in order to transcend the contradictions of class-based capital accumulation. In this regard, the insights of the broad historical materialist tradition are crucial. As Marx observed:

> Since *actual* labour is the appropriation of nature for the satisfaction of human needs, the activity through which the metabolism between man and nature is mediated, to denude labour capacity of the means of labour, the objective conditions for the appropriation of nature through labour, is to denude it, also, of the *means of life*. Labour capacity denuded of the means of labour and the *means of life* is therefore absolute poverty as such.[52]

With the growth of accumulation, denuding labor of its role as the direct mediator of the metabolism between humanity and nature, and substituting capital in this role through its control of the objective conditions of the appropriation of nature, has meant that the *means of life* on the planet are altogether being destroyed. The only answer is the creation of a higher form of society in which the associated producers directly and rationally regulate the metabolism between humanity and nature, in accord with the requirements of their own human development in coevolution with the earth as a whole.

9

Socialism and Ecological Survival

> The issue of survival can be put into the form of a fairly rigorous
> question: Are present ecological stresses so strong that—if not
> relieved—they will sufficiently degrade the ecosystem to make
> the earth uninhabitable by man? If the answer is yes, then
> human survival is indeed at stake in the environmental crisis.
> Obviously no serious discussion of the environmental crisis can
> get very far without confronting this question.
> —BARRY COMMONER, *THE CLOSING CIRCLE* (1971)

CAPITALISM HAS BROUGHT THE world to the edge of the abyss.[1] We
are rapidly approaching a planetary tipping point in the form of a
climate Armageddon, threatening to make the earth unlivable for
the human species, as well as innumerable other species. Such an
absolute catastrophe for civilization and the human species as a
whole is still avoidable with a revolutionary-scale reconstitution of
the current system of production, consumption, and energy usage,
though the time in which to act is rapidly running out.[2]

Nevertheless, while it is still possible to avoid irreversible climate change through a massive transformation in the mode of
production, it is no longer feasible to circumvent accelerating

environmental disasters in the present century on a scale never seen before in human history, endangering the lives and living conditions of billions of people. Humanity, therefore, is facing issues of ecological survival on two levels: (1) a still reversible but rapidly worsening Earth System crisis, threatening to undermine civilization as a whole and make the planet uninhabitable for the human species; and (2) accelerating extreme weather and other ecological disasters associated with climate change that are now unavoidable in the coming decades, affecting localities and regions throughout the globe. Social mobilization and radical social change are required if devastating near-term costs to people and communities, falling especially on the most vulnerable, are to be prevented.

Six decades after the threat of accelerated global warming was first raised by scientists, the situation has only gotten worse. In August 2021, UN Secretary General António Guterres declared that it is "Code Red for Humanity."[3] His warning coincided with the UN Intergovernmental Panel on Climate Change's (IPCC) release of the *Physical Science Basis* report of Working Group I of its *Sixth Assessment Report* (AR6). In this report, five primary scenarios were provided with respect to climate mitigation. Among the most significant findings was that *even in the best-case scenario* (SSP1-1.9), requiring at this point nothing less than a rapidly escalating transformation of the entire global system of production and consumption, the world will surpass a 1.5°C increase in global average temperature after 2040, and will not get below that temperature again until the very end of this century.[4]

The second scenario (SSP1-2.6) points to an increase in global average temperature at the end of the century of 1.8°C (still well below the guardrail of 2°C). The threat of irreversible planetary catastrophe is represented by the next three IPCC scenarios. The fifth scenario (SSP5-8.5) points to an increase in the global average temperature of 4.4°C (best estimate)—spelling the collapse of civilization and absolute disaster for the human species. To avoid such a prospect, given the direction in which the world is now headed, it is necessary to reverse "business as usual," transcending the prevailing logic of an "unsustainable" capitalist system.[5]

At the same time, the IPCC report makes it clear that it is no longer conceivable to prevent accelerating climate disasters this century, even in the best-case scenario, in which an irrevocable planetary tipping point would be avoided. The decades immediately ahead will therefore see the proliferation of extreme weather events that will compound one another: heavy precipitation, megastorms, floods, heatwaves, droughts, wildfires, and failing monsoons. Sea-level will continue to rise throughout this century and beyond, regardless of the actions taken by humanity—though the *rate* of sea-level rise can still be affected by the world's actions. Massive global crop failures are to be expected.[6] Climate refugees will be in the hundreds of millions.[7] All of this is further complicated by the fact that climate change is not the only planetary boundary that capitalism is currently crossing or threatening to transgress. Others include the loss of biological diversity (marking the sixth extinction), ocean acidification, disruption of the nitrogen and phosphorus cycles, loss of ground cover (including forests), loss of freshwater resources, chemical pollution, and radioactive contamination.[8]

Up to now, the ecological, including ecosocialist, strategy with respect to climate change has focused almost entirely on mitigation, aimed at stopping greenhouse gas emissions, particularly carbon emissions, before it is too late. Yet, this general approach has all too often been rooted in a type of reformist environmentalism that does not seriously challenge the parameters of the present system, allowing the ecological crisis to deepen and expand. Mitigation—but today necessarily of a far more revolutionary character—still has to play the leading role in any global climate strategy, since it is essential for the continuation of civilization and survival of the human species (and most of the known species on Earth). However, it is now also necessary, given the inevitable degradation of the earth this century, to *mobilize immediately for survival at the level of communities, regions, nations, and whole peoples.* The harsh reality is that during the next few decades, which according to even the IPCC's most optimistic scenario will involve breaching the 1.5°C threshold—at least for a time—humanity will inevitably see the proliferation of environmental catastrophes at all levels and throughout the planet. This requires that populations organize, plan, and create spaces of ecological sustainability and substantive equality designed to protect what Marx called "the chain of human generations."[9]

Self-mobilization of populations in order to protect lives, communities, and local and national environments, while carrying out revolutionary changes at all levels of existence as part of completely reorganizing production, consumption, and energy usage, now constitutes the pathway to ecological survival. Yet, this new strategic moment—in which mitigation has to be accompanied by environmental disaster management aimed at protecting populations in the community in the present as well as future—has not yet been fully mapped. A broad revolutionary ecological and socialist strategy has to be articulated that transcends the dominant liberal refrains of individual "adaptation" and "resilience," which largely deny the realities of class, race, gender, and imperialism—along with the metabolic rift between capitalism and the environment.[10]

The only meaningful, radical approach to these unprecedented challenges and multiple levels of catastrophe is that of *socialism as a pathway to ecological survival*. It is now widely understood within natural science that the Holocene Epoch in the geological history of the earth of the last twelve millennia has ended and that the planet entered into the Anthropocene Epoch around 1950.[11] The latter is defined as the geological epoch in which *anthropogenic*, rather than *non-anthropogenic* factors (as in the entire prior history of the earth), now largely determine the rate of Earth System change. In what might be called the *Capitalinian Age*, the first geological age of the Anthropocene, the world is characterized by an Anthropocene crisis associated with "anthropogenic rifts" in the biogeochemical cycles of the planet, brought on by the Great Acceleration of the industrial impact on the planet under mature monopoly capitalism.[12] What is needed in these circumstances is the creation of a novel mode of production ushering in a new geological age of the Anthropocene (since the Anthropocene itself is now a permanent feature of geological history, as long as industrial civilization continues).

In this book and elsewhere, we have dubbed this potential future geological age of the Anthropocene the *Communian Age*, standing for *community*, *communal*, and the *commons*. The advent of the Communian Age would mark the historical development of a new, higher, more sustainable human relation to the earth, one that could only come about through ecological, collective, and socialist action. This transition to the second age of the Anthropocene, transcending the present Capitalinian, must begin as soon as possible to protect lives, coordinate environmental disaster management strategies, and undercut the momentum associated with the accelerating trends of ecological disaster.[13] Such revolutionary, socialist transformations constitute the necessary foundation for survival, moving forward in this century.

THE GREAT ACCELERATION AND THE GREAT ECOLOGICAL REVOLT

The advent of the Anthropocene Epoch is associated in natural science with the Great Acceleration of economic impacts, energy use, and pollution, marking the changed physical relation to the environment arising from anthropogenic factors. However, the Great Acceleration and the advent of the Anthropocene also correspond to the emergence of the modern environmental movement in the late twentieth and early twenty-first century, which might be seen as signifying the beginnings of a Great Ecological Revolt, still emerging on a planetary level in the present century.[14]

Modern environmentalism, or the ecological revolt of the post–Second World War years, is usually said to have begun in 1962 with the publication of Rachel Carson's *Silent Spring*. It is more accurate, however, to see its point of origin in the response to the disastrous U.S. thermonuclear test carried out under the code name

"Castle Bravo" at Bikini Atoll in the Marshall Islands on March 1, 1954. The Castle Bravo hydrogen bomb test was intended to have a yield of no more than six megatons, but, due to an error by the scientists involved, it had an explosive power of fifteen megatons, about two and a half times what was expected and a thousand times that of the atomic bomb that the United States dropped on Hiroshima. The detonation resulted in ten million metric tons of coral being radiated and absorbed into the fiery mushroom cloud that climbed over 100,000 feet into the air and spanned over seventy-five miles.[15]

The Castle Bravo test released an enormous, unexpected level of radiation, with the fallout extending over 11,000 square kilometers. Traces of radioactive materials, which had entered the atmosphere and stratosphere, were detected all over the globe. Marshall Islanders on the inhabited atolls were covered with a fine, white-powdered substance (calcium precipitated from the vaporized coral) containing radioactive fallout. Decades after the Castle Bravo test, most of the children and many adults on Rongelap Island had developed thyroid nodules, some of which proved malignant. The crew of a Japanese fishing boat, the *Lucky Dragon*, which at the time of the test was some eighty-two nautical miles from Bikini, well outside the official danger zone, were coated in radioactive fallout. By the time the boat reached Japan, members of the crew were already exhibiting radiation sickness, setting off a world alarm.[16]

The Dwight Eisenhower administration refused to release information on the effects of radioactive fallout and exposure in the face of the Castle Bravo disaster, downplaying the issue for almost a year. However, the veil that hid the fallout problem was pulled aside. Alarmed scientists immediately began to research the effects of radioactive fallout and how it was distributed by air, water, and living organisms throughout the global ecosystem. This work revealed how the operations of the Earth System resulted in fallout being concentrated in the Arctic, despite this region being far removed from where nuclear testing was taking place. It documented how iodine-131 adversely affected the thyroid gland. It was discovered how plants and lichen absorbed strontium-90, which then moved throughout the food web, where this radio-active isotope was incorporated into bones and teeth, increasing cancer risks. These studies raised fears of a planetary ecological crisis, whereby the world's population would share a common environmental fate from the spread of radiation, threatening survival everywhere, as dramatized in fictional form in Nevil Shute's 1957 dystopian nuclear holocaust novel *On the Beach*.

All of this was to contribute to the inception of the Great Ecological Revolt or the worldwide emergence of environmental movements. Disturbed by the spread of radionuclides in the biosphere, scientists began protesting against above-ground nuclear tests, led by such left/socialist figures as Bernal, Virginia Brodine, Commoner, Du Bois, Albert Einstein, H. J. Muller, Linus Pauling, and Russell.[17] Reflecting on these issues, Leo Huberman, the editor of *Monthly Review*, remarked in 1957 that "time is running out. . . . The tests [of these bombs] are dangerous to the health of the world. We must make the movement to ban the bomb encompass not just the Left who are already aware of the dangers, but *all* of our countrymen."[18]

Commoner, as a biologist and a pioneer in ecological thought, helped organize in 1958 the St. Louis Citizens' Committee for Nuclear Information (later the Committee for Environmental Information) that brought scientists and citizens together to share accurate information regarding nuclear issues and concerns, including the dangers of exposure to radioactive fallout. This group famously initiated the baby tooth study in 1958, which involved coordinating with community organizations to recruit participants to collect teeth from young residents in the region to examine the absorption and prevalence of strontium-90. By 1970, approximately 300,000 teeth had been analyzed, revealing that the presence of strontium-90 in teeth rose in direct correspondence to an increase in atmospheric bomb tests, only to decline following the end of such above-ground tests. Given the rich findings, similar studies were done in other parts of the United States, Canada, and Germany, further documenting how radioactive isotopes were readily incorporated into specific parts of the body, contributing to an increase in childhood cancer.

Carson herself entered into this ecological movement initially through her concern over bioaccumulation (concentration of contaminants like radionuclides and other toxins within organisms) and *biomagnification* (the magnified concentration of contaminants at higher levels within the food chain). She offered an extensive analysis of the dangers that accompanied the widespread use of synthetic pesticides, explaining that the "chemical war," poisoning, and ecological degradation were driven by "the gods of profit and production."[19]

In the context of the Great Ecological Revolt, both before and after the publication of Carson's *Silent Spring*, socialist environmentalists were generally distinguished by their more thoroughgoing critiques and far-reaching analyses of the fundamental threat that the capital accumulation system posed to the global environment, and by their insistence on the need for the formation of a revolutionary ecological movement for human survival.[20] Three classic works in this respect are Commoner's *Science and Survival* (1963); Charles H. Anderson's *The Sociology of Survival: Social Problems of Growth* (1976); and Rudolf Bahro's *Socialism and Survival: Articles, Essays, and Talks 1979–1982* (1982).[21] Commoner's and Anderson's books both addressed the multiple critical ecological thresholds, such as climate change, that were being crossed as a result of the profit-driven production system.[22] The red-green theorist Bahro, building on the analysis of British Marxist historian E. P. Thompson, insisted in "Who Can Stop the Apocalypse?" that capitalism was leading to "exterminism," or the systematic death of multitudes. He called for the mobilization of a massive, global ecological "conversion movement" aimed at transcending the system of capital accumulation.[23]

As Commoner, Anderson, and Bahro all emphasized, there were two existential crisis tendencies facing humanity—a reality that remains true today. One is associated with the nuclear arms race and the threat of a global thermonuclear exchange, ushering in nuclear winter.[24] The other is the crossing of planetary boundaries, constituting a direct threat to ecological existence, due to the inherent drive of the system of capital accumulation in the Anthropocene.[25] The only answer is to build, locally and globally, a strong socialist and ecological, or ecosocialist, movement that ensures the survival of populations and communities in the present while safeguarding the future of humanity and the earth.

ECOSURVIVAL AND ECOSOCIALISM

Born in 1917, Commoner was a child of the Great Depression and of the socialist and communist movements of the time. He was strongly influenced by the mass movements supporting the Republican cause in the Spanish Civil War and by protests against lynchings in the U.S. South. Drawn early on to socialist, dialectical-materialist approaches to science, he was a close reader of Engels's *Anti-Dühring* and the *Dialectics of Nature*. He was to be a lifelong ecosocialist. He once declared, ironically, that "the Atomic Energy Commission made me an environmentalist."[26] In "To Survive on the Earth," the closing chapter of *Science and Survival*, Commoner warned:

> As a biologist, I have reached this conclusion: we have come to a turning point in the human habitation of the earth. The environment is a complex, subtly balanced system, and it is this integrated whole which receives the impact of all the separate insults inflicted by pollutants. Never before in the history of this planet has its thin life-supporting surface been subjected to such diverse, novel, and potent agents. I believe that the cumulative effect of these pollutants, their interactions and amplification, can be fatal to the complex fabric of the biosphere. And, because man is, after all, a dependent part of this system, I believe that continued pollution of the earth, if unchecked, will eventually destroy the fitness of this planet as a place for human life. . . .

I believe that world-wide radioactive contamination, epidemics, ecological disasters, and possibly climatic changes would so gravely affect the stability of the biosphere as to threaten human survival everywhere on the earth.[27]

Commoner was deeply concerned with "the assault on the biosphere." Already in *Science and Survival*, he presented the basic nuclear winter hypothesis in which a general thermonuclear exchange would result, due to the lofting of smoke and soot into the stratosphere, in a drastic reduction in global average temperatures imperiling all of humanity.[28] In the same work, he pointed to climate change, warning of the effects of accelerated carbon dioxide accumulation in the atmosphere, the consequences of this on the biosphere, and the "catastrophic floods" arising from sea-level rise. In the mid-1960s, he observed: "Control of this danger [global warming] would require the modification, throughout the world, of domestic furnaces and industrial combustion plants. . . . Solar power, and other techniques for the production of electrical power which do not require either combustion or nuclear reactors, may be the best solution. But here . . . massive technological changes will be needed in all industrial nations." Nevertheless, technology itself was not the answer. As Commoner went on to state, "Technology has not only built the magnificent material base of modern society, but also confronts us with threats to survival which cannot be corrected unless we solve very great economic, social, and political problems. . . . Science can reveal the depth of this [ecological] crisis, but only social action can resolve it."[29]

In 1971, in the chapter on "The Question of Survival" in *The Closing Circle*, Commoner made a similar declaration, writing:

My own judgment, based on the evidence now at hand, is that the present course of environmental degradation, at least in industrialized countries, represents a challenge to essential ecological systems that is so serious that, if continued, it will destroy the capability of the environment to support a reasonably civilized human society. . . . One can try to guess at the point of no return—the time at which major ecological degradation might become irreparable. . . . It is now widely recognized, I believe, that we are already suffering too much from the effects of the environmental crisis, that with each passing year it becomes more difficult to reverse, and that the issue is not how far we can go to the brink of catastrophe, but how to act—now.[30]

The ultimate problem was the mode of production itself. As Commoner stated in the introduction to the 1992 edition of *Making Peace with the Planet*, "If the environment is polluted and the economy is sick, the virus that causes both will be found in the system of production."[31]

Anderson, who was deeply influenced by Commoner's work, was a Marxian sociologist and political economist, author of *The Political Economy of Social Class* (1974). In the mid-1970s, he developed a powerful ecosocialist degrowth analysis, focusing on the planetary environmental crisis and issues of human ecological survival. His major work, *The Sociology of Survival*, argued that the alienated capitalist growth economy was destroying the environmental conditions of human existence. "The stakes involved in this crisis of survival," he wrote, "are in the extreme sense nothing less than the physical continuation of human beings on the planet."[32]

Operating in the tradition of Paul A. Baran and Paul M. Sweezy's *Monopoly Capital*, Anderson saw capitalism in its mature state as prone to economic stagnation, manifested in a tendency toward slower growth and higher levels of unemployment/underemployment and excess capacity. But stagnation (what Herman Daly was to call a "failed growth system") in many ways only served to intensify the system's thrust against the environment, since a "stagnating capitalism is a doomed system and everything must be directed toward restoring growth, including industrial and technological innovation and change, regardless of need or impact." Hence, a capitalism prone to stagnation becomes more intensively destructive of "earthly life" relative to the level of output.[33] This has been partially confirmed by research on the effects of economic slowdowns on carbon emissions. Thus, an empirical study by environmental sociologist Richard York showed that, as the capitalist economy declines in terms of overall output in recessions, carbon emissions do not decrease proportionately, but rather increase in intensity.[34]

Focusing on the core ecological problem posed by the exponential accumulation of capital, Anderson argued: "With ever increasing speed and force, humanity presses forward upon the unknown limits of its own life-support systems. The breaking point, or a point of irreversible 'no return,' approaches in such major life-giving systems as the atmosphere, hydrology, nitrogen cycles, and photosynthesis. It is the nature of living systems to have threshold levels, meaning that things may appear to be going quite all right until virtually all of a sudden the system is in a state of irreversible decline."[35]

An important part of Anderson's argument was the danger to human survival represented by climate change, in which he argued that "a mere two degrees centigrade increase" in average global temperature due to the concentration of carbon dioxide in the atmosphere "could destabilize or melt the polar ice caps, raising the ocean 50 meters and flooding coastal populations and agricultural areas."[36] He insisted that in the rapacious capitalist growth economy "nothing grows faster in the growth of society than energy consumption"—a view that continues to be borne out in the twenty-first century, with the U.S. Energy Information Administration projecting in 2021 that world energy consumption will rise by 50 percent from 2020 to 2050, despite the urgent need to reach net zero carbon emissions by 2050.[37]

A crucial aspect of Anderson's argument was his emphasis on "environmental debt."[38] Inherently unable to adopt a sustainable approach to nature, requiring relations of ecological reciprocity incompatible with its economic expropriation of the planet, capitalism was in effect drawing down the resources of the earth needed for human survival. As he cogently explained, referring to what is now known as Marx's theory of metabolic rift: "Modern agriculture, charged Marx, is as guilty of soil exploitation as it is of labor exploitation; the capitalist extracts a fictitious surplus from the soil by taking more wealth out than he restores. Thus, just as workers produce more value than they are paid in return, and thus perform unpaid labor, so has nature been forced to yield up its capital stock at a rate far in excess of actual or restorative costs. The unpaid costs to the environment underlie the ecological challenge to survival."[39]

For Anderson, the extraction and depletion of resources was even more evident in the underdeveloped nations of the Third World or Global South, given imperialistic relations. Resources in the periphery of the capitalist world system were expropriated without any concern for restoration or reciprocity while the economic surplus generated in those countries was siphoned off by the rich countries in the capitalist core. In the case of poor, underdeveloped countries, therefore, growth remained necessary, but it was also crucial to implement a more "balanced growth" in the periphery and internationally, organized on a socialist, equitable, and sustainable basis aimed at addressing real needs. Here, growth is related to advancing human social development, establishing social relations with nature that mend ecological rifts, and preventing further "environmental debts."[40] Such a transformation necessitated strongly confronting capital.

Monopoly capitalism, for Anderson, was a system of economic and ecological waste in both production and consumption. It included a massive sales effort, which penetrated into the production process, high levels of military spending, and financial speculation—all of which reinforced its unsustainable tendencies and intensified its wasteful operations. Science and technology themselves took alienated forms. This generated "an openly exploitative and destructive science and technology geared toward the maximization of surplus wealth and the minimization of immediate financial cost."[41] The result was an anti-ecological system, which became more unecological the further accumulation proceeded.

Growth beyond a certain "point, particularly artificially forced growth," he wrote, "may be seen to reverse previous progress, destroying the foundation upon which a socialist society and culture could be constructed." Nevertheless, there was no possibility of a shift away from growth/accumulation by capitalism, since to "give up growth" would be to "give up everything that really matters to the capitalist class *qua* class."[42]

The critique of unlimited capitalist economic growth, for Anderson, did not mean that "social growth" or human development could not continue. "Growth becomes what it must become: social growth.... True socialism provides the conditions for growth in knowledge, art and literature, music, science and technology, ties with nature, sociality, individuality, bodily activity and spiritual appreciation—available for all and pursued with everyone's well-being and personal dignity in mind."[43]

"Socialism and survival," in Anderson's view, were, "in effect, synonymous." But survival was not simply about preserving human existence; it was also about the quality of that existence, and for this too socialism was required. Such a view stressed not only the "danger inherent in existing economic, technological, environmental, resource, population, and agricultural conditions . . . but also . . . the kind of social reconstruction" crucial to overcoming capitalism's existential ecological crisis. Ecological survival means a thoroughgoing transformation of the mode of production. "The manner in which people organize their materially productive activities"—in other words, their metabolic relations with nature—he explained, constitutes "the crucial linkage between the social quality of life people experience and the reproductive viability of the physical life-support system." Above all, this requires the "liberation of time," both work time and leisure time, so they promote human development and sustainability, and neither are aimed at profits. The breakdown of the "work-leisure dichotomy" is essential since it is "the heart of the growth system."[44]

Bahro, a socialist dissident from East Germany who became a leader of the red-green movement within West Germany, articulated in his *Socialism and Survival* a sense of real urgency associated with the need to stop the planetary devastation and deepening social contradictions brought on by the "so far unstoppable process of capital accumulation."[45] Capitalism, he contended, raises the question of survival, which only an ecological, socialist, peace movement, involving a new material and spiritual relation to the earth, can solve.[46]

For Bahro, following Thompson's earlier analysis, exterminism meant the destruction of industrial civilization along with human multitudes. "To express the extermination thesis in Marxian terms," he wrote,

one could say that the relationship between productive and destructive forces is turned upside down. Like others who looked at civilisation as a whole, Marx had seen the trail of blood running through it, and that "civilisation leaves deserts behind it." In ancient Mesopotamia it took 1500 years for the land to grow salty, and this was only noticed at a very late stage, because the process was slow. Ever since we began carrying on a productive material exchange with nature, there has been this destructive side. And today we are forced to think apocalyptically, not because of culture-pessimism, but because this destructive side is gaining the upper hand.[47]

Capitalism, precisely because its motor and purpose are found in the process of endless, exponentially increasing capital accumulation, can only proceed down the exterminist path. Hence, there is "*no* Archimedean point [a place to stand and move the world] within existing institutions which could be used to bring about even the smallest change of course." Turning to G. W. F. Hegel, Bahro explained that the prevailing "economic principle of surplus-value production" means that social advance is defined in terms of the narrowest of quantitative criteria associated with the gains of capital. Significantly, "Hegel used to speak in such cases of a 'bad infinity,' by which he meant a process which involved no more than adding 1 to 1, and did not lead in its own context to a decisive qualitative leap. This kind of progress must cease, for the share of the earth's crust that can be ground up in the industrial metabolism is limited, despite all possible and senseless expansion, if the planet is to remain habitable."[48]

"The enormous ecological destabilisation" in the Global South, Bahro argued, "is primarily a symptom of Western structural penetration into 'indigenous' social and natural conditions."[49] The result of this global capitalist exterminist expansion is "a crisis of human civilisation in general. There has never been anything comparable in the whole past history of our species on the earth." In fact, "exterminism is expressed in the destruction of the natural basis of our existence as a species."[50] The control exerted by the system over the working class is a product of capitalism's ability constantly to create an internal dependence of workers on the system, which the combined ecological and economic crisis is now weakening. But the movement of resistance that is needed has to be organized primarily through the merger of the ecological and peace movements and their relation to the working class, rather than on traditional productivist grounds. Ecology, given the scope and depth of the planetary crisis and the undermining of the conditions of life, becomes the common material ground "affecting more people in their existential interests than in any other contradiction."[51]

To advance on a path of sustainability and survival therefore would mean a revolutionary break with the logic and institutions of capitalism, out of which the ecosocialist transition was to emerge. Capitalism, in Bahro's view, was not all-inclusive, in the sense that it is often depicted in contemporary ideology as constituting the entirety of the present-day world. It continued to have an external area, which, as in the conception of Arnold Toynbee, gave rise to an "external proletariat" occupying the periphery and precarious parts of the capitalist world. This existed alongside the "internal proletariat" of the advanced capitalist world, which, by definition, was never fully incorporated within the system.[52]

"The oldest stratum of civilisation involved in the present crisis," Bahro argued, following Engels, "is that of patriarchy, with ten millennia behind it."[53] Many of the distinctive tendencies of contemporary civilization, including forms of oppression, thus run deeper than present-day capitalism. There were cultural and spiritual resources that were resistant to capitalist exterminism. All of this created the potential that "the capitalist industrial system" could be "driven back and destroyed by an unstoppable manifold movement of humanity," defined in ecological and socially reproductive more than "purely economic terms."[54]

A central reality of capitalism, in this view, was the inability of the capitalist state itself to change course or to reverse the ecological devastation generated through its own operations. The capitalist state governed by industrial and financial interests, Bahro wrote, "is obviously so very much wedded to exterminism that it doesn't permit itself to be used as an emergency brake. . . . No government which could be constituted on the present 'place' of the state [within the existing socioeconomic order] could be anything but a bad emergency government."[55]

The essence of the problem was the juggernaut of capital itself, which the capitalist state only sought to accelerate, never to apply the brakes, heading therefore toward a collision with the earth. This, Bahro said, would especially impact "the marginalised and excluded, those with their backs to the wall, [who] now [however] have an unbeatable ally in this very wall that they have their backs against. This wall is formed by the limits of the earth itself, against which we really shall be crushed to death if we do not manage to brake and bring to a halt the Great Machine that we have created before this finally bumps against it." The answer clearly could not be seen as lying in a capitalist "emergency state," which would only make things worse for the vast majority, and for the earth itself, but in a revolutionary "salvation government" in which the material struggle for survival coupled with the struggle for human liberation—the end of alienation and the focus on essential human needs—would generate a new emergent reality.[56]

However, this revolutionary ecological critique offered by social-ist ecologists, premised on the rejection of capitalism's relentless destruction of humanity and the earth, and therefore on the link-ing of the struggle for survival to the struggle for human freedom, did not come to dominate the environmental movement—even though it played a critical role in the ecological struggles of the time. The environmental movement, and even much of ecosocial-ist thought, in the tamer periods that followed the initial revolt, gravitated toward a radical reformism, in which the full urgency of the struggle for survival was forgotten, despite the rapidly accelerating planetary ecological crisis. A stage of environmen-tal denialism—not of the whole environmental problem but of its worst threats and their inherent relation to capitalism—set in on the left. Hence, the understanding of the existential crisis stem-ming from the ecological deficits of capitalism that thinkers such as Commoner, Anderson, and Bahro raised—not apocalyptically, but in terms of an ecosocialism of survival demanding revolution-ary social change—is now needed more than ever.

Existential Crisis Now!

The IPCC reports, representing the world scientific consensus with respect to climate change, serve to illuminate how the imper-atives of capitalism are pushing the world into the inferno looming before us. The more optimistic IPCC scenarios—those resulting in a growth of global average temperature this century of well below 2°C—point to the actions necessary to reach net zero carbon emis-sions (as well as reducing other greenhouse gas emissions), thus avoiding irrevocable climate change. The remaining scenarios, representing the continuation of "business as usual," depict how the ongoing accumulation of greenhouse gases in the atmosphere will drive an increase in the average global temperature, resulting in abrupt changes in the Earth System that undermine the con-ditions of life for humanity and other species. Unfortunately, the capitalist "business-as-usual" trends persist, pointing to hellish consequences. Thus, with each new IPCC report, the situation is ever more dire, and the possibility of pulling away from disaster requires ever more revolutionary change, given both the increasing physical scale of the problem and the diminishing time scale. This represents the existential crisis that now lies before the entire world.

In the best-case scenario (SSP1-1.9) provided by the *Physical Science Basis* assessment in part 1 of AR6, written by Working Group I, global average temperature, as we have seen, is expected to surpass a 1.5°C increase above pre-industrial levels after 2040, rising to 1.6°C and not declining below the 1.5°C threshold again (returning to 1.4°C) until the end of the century. But in order for this scenario to hold, global carbon emissions must peak within a few years, with net zero emissions achieved by 2050. Still, even in this scenario—the most optimistic one now provided by the IPCC—the world will continue to experience the propagation of extreme weather events, heavy precipitation, flooding, drought, heatwaves, wildfires, glacial melting, and sea-level rise, which will affect every region of the earth while threatening billions of people.[57]

The IPCC's *Impacts, Adaptation and Vulnerability* assessment, written by Working Group II of AR6, released in February 2022, documents the observed consequences of climate change so far, detailing the vulnerabilities and projected risks in the coming decades. The "Summary for Policymakers" of Working Group II highlights the range of changes in the Earth System, which have already increased the risks that much of humanity experiences and which are decreasing the quality of existence in general. Among the "observed impacts," it emphasizes that:

Human-induced climate change, including more frequent and intense extreme events, has caused widespread adverse impacts and related losses and damages to nature and people, beyond natural climate variability. . . . Across sectors and regions, the most vulnerable people and systems are observed to be disproportionately affected. The rise in weather and climate extremes has led to some irreversible impacts as natural and human systems are pushed beyond their ability to adapt.[58]

Heat- and drought-related conditions have increased tree mortality and wildfires. The warming of the ocean has resulted in "coral bleaching and mortality" and the "loss of kelp forests." Half of the species considered are already migrating toward the poles or moving to higher elevations. Climate change is also increasing irreversible conditions such as species extinctions. In comparison to previous estimates in prior assessments, "the extent and magnitude of climate change impacts are [now] larger."[59]

Climate change is negatively affecting both the physical and mental health of people. For example, "extreme heat events have resulted in human mortality and morbidity"; "the occurrence of climate-related food-borne and water-borne diseases has increased"; "the incidence of vector-borne diseases has increased from range expansion and/or increased reproduction of disease vectors"; and "animal and human diseases, including zoonoses, are emerging in new areas." Populations around the world are experiencing greater trauma from extreme weather events. They are also contending with "climate-sensitive cardiovascular and respiratory distress" due to "increased exposure to wildfire smoke, atmospheric dust, and aeroallergens." Heatwaves are amplifying air pollution events. Climate change and extreme weather events are reducing "food and water security." It is estimated that up to 3.6 billion people currently reside in places that "are highly vulnerable to climate change," which is contributing to the overall humanitarian crisis.[60]

The "Summary for Policymakers" report of Working Group II of AR6 is clear that the current socioeconomic system that organizes production and consumption is unsustainable, "increasing exposure of ecosystems and people to climate hazards." In fact, "unsustainable land-use and land cover change, unsustainable use of natural resources, deforestation, loss of biodiversity, pollution, and their interactions, adversely affect the capacities of ecosystems, societies, communities and individuals to adapt to climate change." Short-term interests, focused on increasing profits, drive poor management of resources, habitat fragmentation, pollution of ecosystems, and overall ecological degradation.[61]

Between now and 2040, it is absolutely necessary to keep warming below the rapidly approaching 1.5°C threshold (or at the very worst *well below* 2°C), otherwise the climate-related "losses and damages" to both ecosystems and society will dramatically multiply. Surpassing this threshold will result in extreme high risks associated with biodiversity loss, a dramatic decline in snowmelt water availability for irrigating crops, a severe reduction in above-ground and groundwater availability, declining health of soils, widespread food insecurity, flooding of "low-lying cities and settlements," accelerated proliferation of disease risks, even more intense and frequent weather events, and extensive heatwave conditions. "Many natural systems are near the hard limits of their natural adaptation capacity," whereby additional warming will result in irreversible changes that undermine essential ecosystem services that support life. The overall damages, threats, and problems "will continue to escalate with every increment of global warming." It will only become more and more difficult to intervene and manage the compounding risks that will cascade throughout the world, depending on the magnitude of the overshoot.[62]

Hence, the "Summary for Policymakers" of Working Group II in AR6 focusing on *Impacts, Adaptation and Vulnerability* concludes that "there is a rapidly narrowing window of opportunity" to forge a radically different future. It warns:

It is unequivocal that climate change has already disrupted human and natural systems. Past and current development trends . . . have not advanced global climate resilient development. . . . Societal choices and actions implemented in the next decade [will] determine the extent to which medium- and long-term pathways will deliver higher or lower climate resilient development. . . . Importantly, climate-resilient development prospects are increasingly limited if current greenhouse gas emissions do not rapidly decline, especially if 1.5°C global warming is exceeded in the near term.[63]

The leaked scientific-consensus draft of the "Summary for Policymakers" by Working Group II of AR6, received by Agence-France Presse in June 2021, included the following statement: "We need transformational change operating on processes and behaviours at all levels: individual, communities, business, institutions and governments. We must redefine our way of life and consumption." This transformation requires coordinated action, massive public mobilization, political leadership and commitment, and urgent decision-making to change the global economy and support an effective and accelerated mitigation-adaptation strategy.[64] Unfortunately, such action has been consistently thwarted by capital and global political leaders, who managed to remove the statement from the final published Working Group II report, where it is nowhere to be found.

In May 2022, the carbon dioxide concentration in the atmosphere measured 421.37 parts per million, marking a new high. Peter Tans, a climate scientist at the National Oceanographic and Atmospheric Administration, explained that in "this last decade, the rate of increase has never been higher, and we are still on the same path. So we are going in the wrong direction at maximum speed."[65] As climate breakdown accelerates, the conditions of life are rapidly deteriorating, creating numerous health problems, some of which manifest as corporeal rifts, undermining bodily existence.[66]

Corporeal challenges, which could be viewed as indications of a corporeal rift in which climate change disrupts human bodily functions, have received additional attention given the brutal heatwaves and record-breaking temperatures in India and Pakistan in spring 2022. On May 1, the temperature in Nawabshah, Pakistan, was 49.5°C (120.2°F). What made this heatwave, along the coasts and the Indus River Valley in these countries, particularly unbearable was that it was accompanied by high levels of humidity.[67] Together, these can create dangerous levels of heat stress, which can result in death. This issue is particularly important to consider in regard to global warming, as climate change increases heat and the amount of water vapor in the atmosphere. Furthermore, warmer air holds more moisture, making humidity worse. Heat and humidity are additive, generating conditions in the form of wet-bulb temperatures (combining normal dry-bulb temperature and humidity) that exceed the capacity of people to survive.Under such conditions, one of the important issues is that nighttime temperatures are also high, making it difficult or impossible for the body to recover partially overnight—worsening the situation. This is part of the reason that, as heatwaves progress, it becomes increasingly difficult for people to function physically.

In the article "The Emergence of Heat and Humidity Too Severe for Human Tolerance," published in *Science Advances*, Colin Raymond, Tom Matthews, and Radley M. Horton explain that what are called dry-bulb temperatures, measurements obtained from an ordinary thermometer, are not adequate in ascertaining the dangers to human health associated with heat stress.[68] Instead, it is necessary to measure the wet-bulb temperature—heat and humidity. This is obtained by placing a wet cloth on the thermometer and blowing air on it. Human beings cool themselves or shed their metabolic heat at high temperatures via sweat-based latent cooling. But once the wet-bulb temperature reaches 35°C (or 95°F), this cooling mechanism ceases to be effective. Under such conditions, human beings are not able to cool themselves by sweating, even if they are in the shade, wearing little clothing, and drinking plenty of water. When outside and exposed to such wet-bulb temperatures for six hours, even young, healthy individuals will perish from this heat stress. In humid regions, and for populations whose physical conditions are less than optimal, it is possible for lives to be threatened even with lower wet-bulb temperatures, between 26°C and 32°C, as was the case in the heatwaves that hit Europe in 2003 and Russia in 2010, killing thousands of people, especially the elderly and other vulnerable populations.[69]

Raymond and his colleagues stress that "extreme heat remains one of the most dangerous natural hazards" and "a wet-bulb temperature . . . of 35°C marks our upper physiological limit." Thus, it is not possible simply to adapt to progressively warmer temperatures, when heat and humidity surpass the point of what is survivable. These worrying wet-bulb temperature conditions are occurring a few hours at a time in coastal and major river regions of South Asia, the Middle East, Mexico, and Central America. Such conditions are likely to become more regular and to last longer in these regions over the next few decades, or even years, with even more deadly consequences, while spreading across larger terrestrial stretches, rendering parts of the world uninhabitable. In the second half of the century, if "business-as-usual" trends continue, the likely consequences are too horrific to imagine.[70]

In the opening scene of *The Ministry for the Future*, science-fiction novelist and socialist Kim Stanley Robinson tries to imagine what could happen to human beings under the unbearable heat and humidity associated with wet-bulb temperatures. The population of a town in India is suffering from an intense heatwave. People are panicking, immersing themselves in the lake, trying to cool down, but to no avail, as the water provides no relief. It is noted that the people are being poached in the water. Before too long, the lake is filled with corpses—"all the children were dead, all the old people were dead."[71] It is a hellish scene, but it captures the gravity of exterminism that is unfolding and the urgency of the fight for survival. This is the sobering reality of the current ecological moment, as the leaked draft of the "Summary for Policymakers" of Working Group II stated (though this was removed, probably by governments, from the published report): "Life on Earth can recover from a drastic climate shift by evolving into new species and creating new ecosystems. Humans cannot."[72]

THE STRUCTURAL CRISIS OF CAPITAL AND THE FAILURE OF ENVIRONMENTAL REFORM

The failure of capital to face up to the rapidly increasing ecological crisis, even as the earth as a home for humanity is fast approaching an irreversible tipping point, is often attributed to the growth of neoliberalism, as if this were simply a contingent fact of history determined by political swings and policy changes.[73] The advance of neoliberalism, however, was itself a response of the capitalist system to the insurmountable structural crisis of capital that first emerged in the mid-1970s, leading to the restructuring of this system. This included not only the reduction of the relative autonomy of the state, but also the restructuring of the capital-labor relation through the globalization of production and the financialization of the capitalist system.[74] In these changed circumstances, the centrality of what was dubbed the "environmental state," introduced as the capitalist system's response to the deepening environmental crisis, experienced an early death. It was to be replaced under neoliberalism by a more diffuse system of "environmental governance," involving both the private and public sectors, ensuring that the accumulation of capital always took complete precedence over the sustainability of the natural environment.[75]

The initial Great Ecological Revolt of the early post–Second World War years was largely radical in inspiration, strongly critical of capitalism, drawing its strength from the grassroots, and raising the essential question of human survival. However, these radical environmental challenges to the system were soon contained and co-opted through the rise to prominence of the capitalist environmental state, allowing the Great Acceleration of economic impacts on the environment to expand largely unhindered. The notion of the environmental state stood for a patchwork system of environmental regulations and statutory laws introduced by the state within the limits allowed by the powers that be, thereby precluding any major challenges to the process of capital accumulation. The dominant state-directed environmental reformism that emerged in these years, combating isolated cases of extreme pollution and environmental degradation at the local level, was commonly presented in received ideology as a logical outgrowth of capitalist modernization, viewed as an extension of the logic of the welfare state. Capitalism, it was claimed, followed a path whereby environmental spending increased at higher levels of economic development, ameliorating the negative effects of growth.[76]

All of this has proved to be a dangerous illusion. The environmental state as a central actor within the system was at best a very short-term affair, soon overshadowed by the structural crisis of capitalism that emerged only a few years later by the mid-1970s. The economic restructuring of the late 1970s and early '80s was a response to the deepening stagnation of capital accumulation, evident in a slowdown in economic growth and rising unemployment/underemployment and idle capacity.[77] Although there was no solution to the economic malaise of the mature capitalist economies, the ruling class was able to extend its power, in a context of "disaster capitalism," through the promotion of a more predacious system that brought the state more firmly within the rules of the market.[78] These developments were accompanied by the globalization of production and the financialization of the economy, ushering in a new phase of globalized monopoly-finance capital, made possible in part by new systems of communication and surveillance.

By the 1990s, even those proponents of capitalist ecological modernization, who were the most enthusiastic cheerleaders of the environmental state, were forced to point to the counterpressures being imposed on it by capital, while more recently, they have acknowledged its virtual demise.[79] In the context of this rapid decline of the state-directed system of environmental regulation (the environmental state), the notion of environmental governance was introduced as the new reform-oriented concept to take its place. Environmental governance was meant to refer to the much greater role assumed by private interests, including corporations, corporate foundations, non-governmental organizations, international financial institutions, and intergovernmental organizations, in determining the realm of environmental regulation, which in many areas, such as various certification processes, carbon markets, and the financialization of nature/conservation, generated new markets for capital accumulation, legitimated in terms of so-called green capitalism.[80] The environmental *nation-state*, a notion that in the international context represented a further distancing from the concept of the domestic environmental state, was seen as subject to *intergovernmental* agreements such as the 2015 Paris Accord on climate change.[81]

Nonetheless, the phases of limited environmental reform, presided over initially by the capitalist environmental state and more recently by so-called environmental governance under direct corporate and ruling-class dominance, have seen the acceleration of the destruction of the earth as a home for humanity. According to the world scientific consensus, ecological catastrophes, on scales never before seen by humanity, are now fast approaching. Marginal attempts by the present political-economic system to address the planetary ecological emergency have proven entirely ineffectual because the capitalist juggernaut always takes priority. The world is now on a runaway train to disaster, rapidly approaching the edge of the cliff. As Engels once remarked, capitalism is ruled by "a class under whose leadership society is racing to ruin like a locomotive whose jammed safety-valve the driver is too weak to open."[82] The ruin, when it comes, will be ecological as well as political-economic and will fall most heavily on the vulnerable and future generations.

This deadly trajectory is evident everywhere, underscoring the failure of capitalist ecological reform. According to the *UN Emissions Gap Report 2021*, the present voluntary national climate pledges of countries in accordance with the Paris Agreement would generate a 2.7°C increase (66 percent probability) in global average temperature this century, as opposed to the well-below 2°C increase, which is the goal of the Accords, and far above the scientific-consensus goal of 1.5°C, which is the most important threshold for planetary climate security.[83] Presently, there are more than four hundred ongoing fossil fuel extraction projects in process in the world (40 percent of which have not yet commenced extraction), currently advanced by corporations and supported by governments, known as "carbon bombs." Each of these represents at least one gigaton of carbon emissions, which, if they are all carried out, "will exceed the global 1.5°C carbon budget by a factor of two."[84] There is no sign anywhere that the necessary limits will be imposed by capitalism to protect the planetary environment. Rather, the signs all point to the opposite as a frenzy for fossil fuels is developing. The G7 leading capitalist countries, meeting in May 2022, agreed eventually to "phase out" "unabated coal" but put forward no date for doing so, with the discussion dominated instead by the need for vast new fossil fuel sources in the context of the Ukraine war, setting aside all climate objectives.[85]

Perhaps the greatest single example of the collective duplicity of governments within the dominant capitalist world system in the face of the planetary ecological emergency is the rewriting of the scientific-consensus "Summary for Policymakers" of Working Group III in the IPCC's AR6 report on *Mitigation*, published in April 2022. A comparison of the scientific-consensus version of the "Summary for Policymakers," leaked in August 2021, with the later published version—which was censored and completely rewritten by governments in consultation with corporate lobbyists, carried out in line with the IPCC process—demonstrates a complete betrayal of science and humanity. Removed from the report were the collective pronouncements of the scientists on the need to: (1) eliminate all unabated coal-fired plants worldwide by 2030 in order to avoid greatly surpassing the 1.5°C target; (2) carry out immediate, rapid transformational change in the political-economic regime affecting production, consumption, and energy use; (3) shift to low-energy solutions; (4) implement plans for "accelerated mitigation"; and (5) support mass social movements against climate change rooted in the most vulnerable sectors of society, advancing a radical just transition. All criticisms of the "vested interests," including the term itself, were erased from the report. Flatly contradicting the scientific-consensus "Summary for Policymakers," the redacted governmental-consensus report went so far as to claim that the number of coal-fired plants could be increased due to the promise of carbon capture and sequestration—a view that the scientists had rejected.

Governmental leaders also eliminated statements in the scientific-consensus "Summary for Policymakers" regarding how: (1) the wealthiest 10 percent of the global population are responsible for around ten times the greenhouse gas emissions of the poorest 10 percent (despite the fact that this was a very conservative estimate of the emissions gap); (2) the top 1 percent of air travelers account for 50 percent of aviation-based emissions; and (3) some 40 percent of the emissions from developing countries are linked to export production for core nations.[86]

Indeed, the entire critique of the fossil capital regime presented in the scientific-consensus "Summary for Policymakers" was excluded by governments in the interest of keeping the accumulation process, the motor of the capitalist system, going. In nearly every line of the final published "Summary for Policymakers" by Working Group III of AR6, the *Mitigation* report, the betrayal of the global population by the world's governments is present, as the latter, operating together, eviscerated the IPCC's scientific consensus, undermining any meaningful actions and policies. When the *Mitigation* report was published in April 2022, Guterres remarked that the current moment is one of "climate emergency," marked by "a litany of broken climate promises," constant "lies," and "empty pledges [by the vested interests] that put us firmly on track toward an unlivable world."[87] The consequence of this is to further promote what Engels called "social murder," but now on a planetary scale, threatening the entire chain of human generations.[88]

The U.S. federal government's prioritization of capital accumulation, including that of the fossil fuel industry, over not only human lives in the present, but the future of humanity as a whole, is evident in the nonstop battles of the Barack Obama, Donald

Trump, and Joe Biden administrations against the federal law-suit of *Juliana vs. the United States*, in which twenty-one young plaintiffs have challenged the U.S. government for wrongfully promoting the fossil fuel industry in violation of what is known as the public trust doctrine within the common law, affirmed in a famous 1892 decision involving the Illinois Central Railroad company, as applicable to the U.S. Constitution. Applying the public trust doctrine to the federal government, the lawsuit declares that the executive and legislative branches in Washington knowingly violated the public trust with respect to climate change by allowing the undermining of the "survival resources" on which the lives of people in the present and future depend, putting human survival in question. As Oregon District Court judge Ann Aiken ruled in 2016, "I have no doubt that the right to a climate system capable of sustaining human life is fundamental to a free and ordered society." *Juliana vs. the United States* is based on the presumption that statutory law with respect to the climate is too narrow and is not enforced, requiring that the federal government be mandated on constitutional grounds to cease its support of the fossil fuel industry.[89]

In response, successive Democratic and Republican administrations have done everything they could to stop this lawsuit, which has been subject to more "exceptional legal tactics" than any other federal lawsuit in history—including "six rulings on the notorious shadow docket," where legal opinions are not published and the justices' votes are not made public. The Biden administration's Department of Justice has made it evident that it will use every procedural tool available to arrest the progress of the lawsuit, killing it at the earliest opportunity.[90] The goal is to allow the fossil fuel industry to continue to accumulate and expand by preventing any obligation of the U.S. federal government to protect the present and future of humanity.

Not only has the U.S. federal government put capital accumulation and the fossil fuel industry before human life as a whole, promoting social murder on a global scale, or exterminism, it has

also neglected to take proactive and comprehensive action to protect the population, particularly the most vulnerable, in the face
of accelerating ecological catastrophes. The U.S. government's
program of disaster relief is based in the Federal Emergency
Management Agency (FEMA). But FEMA at present is under-
funded and geared primarily to protecting high-end private
property, thus leaving the mass of the population with little or
no protection—and without any coordinated programs aimed at
reducing risk associated with environmental disasters. Under the
Obama administration, proposals were made, as articulated by
FEMA director Craig Fugate, to put FEMA on a fully capitalist
basis along the lines of the private insurance industry, complete
with deductibles. FEMA assistance was thus to be determined
largely by whether the private insurance industry had decided to
ensure a given structure, an approach that would inevitably have a
detrimental effect on the poor.[91]

With record-breaking hurricanes, wildfires, and other extreme
weather disasters presenting themselves in 2020, coupled with
the COVID-19 pandemic, FEMA and the U.S. government in
general, as explained by *Scientific American*, proved itself utterly
incapable of addressing the growing natural and epidemiological
disasters. This brought "into stark relief problems of capacity and
inequity—[with] people of color and low-income communities"
getting "hit disproportionately hard." "All emergency agencies" in
the United States taken together do little in advance to prepare for
disasters, while FEMA programs have been shown to "entrench
and exacerbate inequities because they focus on restoring private
property. This approach favors higher income, typically majority
white areas with more valuable homes and infrastructure over
people of color and low-income communities, which are disproportionately affected by disaster and least able to recover from
it." A precondition of FEMA disaster relief is "cost matching,"
which systematically and structurally favors wealthier over poorer
communities. The comprehensive failure of the United States to
address the COVID-19 pandemic, resulting in more than a million

deaths, is a manifestation of the complete lack of an infrastructure, including public health facilities, equipped to cope with disasters in general, particularly where the most vulnerable populations are concerned. Instead, the capitalist system has enshrined the principle of the devil take the hindmost.[92]

ECOLOGICAL CIVILIZATION OR EXTERMINISM

In the 1860 edition of his *Trades' Unions and Strikes*, the English Chartist and trade unionist Thomas Joseph Dunning wrote:

> Capital is said . . . to fly turbulence and strife, and to be timid, which is very true; but this is very incompletely stating the question. Capital eschews no profit, or very small profit, just as Nature was formerly said to abhor vacuum. With adequate profit, capital is very bold. A certain 10 per cent will ensure its employment anywhere; 20 per cent certain will produce eagerness; 50 per cent positive audacity; 100 per cent will make it ready to trample on all human laws; 300 per cent, and there is not a crime at which it will scruple nor a risk it will not run, even to a chance of its owner being hanged. If turbulence and strife will bring a profit, it will freely encourage both. Smuggling and the slave-trade have amply proved all that is here stated.[93]

Trampling over all other social considerations, it is this innate drive of capital depicted by Dunning in the nineteenth century that helps explain why, even in the face of the certain ruination of contemporary civilization, humanity, and to a considerable extent life as a whole, it nonetheless proceeds down that same road of creative destruction. Capital is not deterred from burning all existing fossil fuel reserves, and thus the catastrophic heating up of the climate, as long as the short-term profits are ample. Its "solutions" to the environmental crisis increasingly take the form of the financialization of nature, aimed at buying up the "environmental services" of the entire planet, operating under the senseless presumption that if there is a global ecological crisis it is due to the failure to incorporate nature fully into the market.[94]

Consequently, a whole new revolutionary ecological civilization and mode of production, dedicated to sustainable human development, one in which the associated producers regulate the metabolism between humanity and nature, is now necessary for survival and for life. This requires revolutionary transformative actions to mitigate climate change, in order to protect the planet as a safe place for human habitation and life in general. But in seeking to protect the earth as a home for the future of the chain of human generations, it is also necessary to protect current generations. At issue today is not only the long-term issue of the survival of humanity as a species, but also the more immediate imperative of ensuring the lives and living conditions of twenty-first-century populations, including whole communities, nations, and peoples, and especially those whose lives and living conditions are most exploited, precarious, and vulnerable.

This *two-level movement*, to protect the earth both as a home for humanity (and innumerable other species) well into the future and for the defense of human communities in the present, is most fully addressed in the world today, though not without contradictions, in those societies with a more socialist bent.[95] It is socialist, post-revolutionary societies that are better able to resist the logic of capital, despite the continuing dominance of the capitalist world economy, by introducing ecological as well as economic planning, and facilitating alternative forms of social metabolic reproduction. We can see this in Cuba, which has developed an ecosocialist model of degrowth, in the sense, designated by U.S. Green Party analyst Don Fitz, of a social order identified with a revolutionary "reduction of unnecessary and destructive production by and for rich countries (and people)," while also providing for the "growth of production of necessities by and for poor countries (and people)."[96]

Cuba has not only repeatedly been designated by international indicators as the most ecological nation on the earth, but also as the one most prepared for disasters. Cuba in 2017 was "the only country in the world," Fitz explained, "with a government-led plan (Project Life, or *Tarea Vida*) to combat climate change" based on a century-long projection. In September 2017, Maria, a category 5 hurricane, hit Puerto Rico, a U.S. colony, resulting in almost three thousand deaths. In that same month, Irma, another category 5 hurricane, hit Cuba, causing only ten deaths. Cuba's low mortality was the result of comprehensive disaster protection measures introduced from the beginning of its revolution and built into the entire structure of the society. Cuba put in place a national plan to protect the population from COVID-19 prior to the first death there from the pandemic. It has also developed highly effective COVID-19 vaccines, which have been used to vaccinate its entire population and to help other countries at low cost.[97]

In terms of the wider issues of climate change, Cuba, rather than following the dominant capitalist strategy of promoting maximum energy usage and simply converting to "alternative" energies (which are also extremely damaging to the environment at higher levels of energy generation), has chosen energy conservation, seeking to minimize both energy usage and the resultant negative effects. As Cuban energy advisor Orlando Rey Santos has observed: "One problem today is that you cannot convert the world's energy matrix, with current consumption levels, from fossil fuels to renewable energies. There are not enough resources for the panels and wind turbines, nor the space for them. There are insufficient resources for all this. If you automatically made all transportation electric tomorrow, you would continue to have the same problems of congestion, parking, highways, heavy consumption of steel and cement."[98]

In "Cuba Prepares for Disaster," Cuban analyst Fitz explains that

a poor country with a planned economy can design policies to reduce energy use. Whatever is saved from [energy efficiency] can lead to less or low-energy production, resulting in a spiraling down of energy use. In contrast, in accordance with the well-known Jevons Paradox, competition drives capitalist economies toward investing funds saved from EE [energy efficiency] toward economic expansion resulting in perpetual growth.

Such endless accumulation leads to mounting ecological contradictions.

As Fitz goes on to observe: "What is amazing is that Cuba has developed so many techniques of medical care and disaster management for hurricanes and climate change, despite its double impoverishment from colonial days and neocolonial attacks from the U.S.," including the permanent embargo imposed by Washington as a form of economic siege warfare.[99] Following the demise of the Soviet Union and its fossil fuel subsidies to Cuba, Cuba's Special Period forced Havana, which was also faced with a tightening U.S. embargo, to develop agroecology and urban farming at very high levels, resulting in Cuba's eco-revolutionary transformation into a model of sustainable human development.[100]

Cuba's successes in promoting sustainable human development fed the anti-communist ire of Washington. Relying on new means of financial warfare, the Trump administration introduced 243 additional financial sanctions directed at Cuba, while the Biden administration extended those further. This generated increased shortages in food and other basic items, made worse by the COVID-19 pandemic. In July 2021, popular protests emerged in Cuba for the first time in a generation. The increases in global food prices and wheat shortages in early 2022, associated with the pandemic, profiteering, and the Russia-Ukraine war, have only exacerbated these conditions.[101] This crisis has resulted in critical debates in Cuban society that, while intense, are mostly taking place *within the revolution* rather than outside of it, suggesting that Cuba will continue to carry out a process of socialist construction and reconstruction that will defy all those who are seeking its demise.[102]

Venezuela's Bolivarian Revolution, although in a different way than Cuba, has also moved toward an ecological society, promoting communes that put resources and production back in the hands of associated producers, ensuring that basic needs are met. Government resources are being transferred to communes and organized communities in both rural and urban areas with the objective of enhancing food security and sovereignty partly through such agencies as the Pueblo a Pueblo (People to People) Plan, promoting an "assembly culture, planned consumption and participatory democracy." All of this points in the direction of ecosocialism.[103]

Although still one of the world's largest polluters, the Chinese economy has made rapid ecological advances in line with its goal—outside the capitalist framework—of promoting an *ecological civilization*, a concept that originated with socialist environmentalists in the final decades of the Soviet Union and that has now taken on Chinese characteristics.[104] Although still a developing country in the sense of having a low per capita income relative to the developed capitalist states, China has set 2060 as its target to reach net zero carbon emissions. Meanwhile, it has become the world leader in solar power—both production and consumption—and in reforestation/afforestation. China was able to protect its population from the COVID-19 pandemic, with 5,272 total deaths as of September 20, 2023, versus 1,175,395 total deaths in the United States, a country with less than a quarter of the population. With only 10 percent of the world's arable land and 20 percent of the global population, China currently produces 25 percent of the world's grain. In the decade from 2003 to 2013, China increased its total grain output by about 50 percent. Most farms are largely organized on a semi-communal, cooperative basis, with the land held in common and distributed among producers by the community. From 2013 to 2019, the number of towns with supply-marketing cooperatives in rural China increased from 50 percent to 95 percent, as part of the revitalization of the countryside, contributing to the elimination of extreme poverty in the country.[105]

The global struggle for sustainable human development can also be seen in places within the advanced capitalist core, including the United States, where considerable opposition is exhibited in some locations to the dominant logic of the political-economic system. Cooperation Jackson, based in Jackson, Mississippi, is engaged in a revolutionary, transformative project as part of building ecosocialism, in order to protect and advance the survival of existing communities and to create an "ecologically regenerative," sustainable future. Kali Akuno, the co-founder and co-director of Cooperation Jackson, explains that the continuing realities of racial capitalism have led to extreme forms of inequality, control of knowledge by private capital, and uneven development, whereby Jackson, Mississippi, has largely been organized around resource extraction to serve capital accumulation for distant vested interests. This exploitative system "is rapidly destroying all of the vital, life giving and sustaining systems on our planet."[106] Thus, it is urgent to forge an alternative productive system.

Through collectively organizing, mobilizing, and working with "structurally under- and unemployed sectors of the working class, particularly from Black and Latino communities," Cooperation Jackson seeks to "replace the current socio-economic system of exploitation, exclusion and the destruction of the environment with a proven democratic alternative." It promotes a radical form of social organization built on equality, cooperation, worker democracy, and environmental sustainability, aimed at providing meaningful work through living-wage jobs, while reducing racial and other inequities, and building the public wealth of the community. This is all seen as part of a "transition to ecosocialism."[107]

Cooperation Jackson has as its goal collectively owning and controlling the means of production. Akuno explains that this involves "control over processes of material exchange and energy transfer," including the "processes of distribution, consumption, and recycling and/or reuse" to ensure that the social metabolism operates within natural limits and advances "sustainability and environmental justice."[108] Through self-organization, self-determination, and self-management, human beings will gain social control over their productive lives, allowing them democratically and collectively to make decisions focused on how to meet human needs, rather than those of capital. This approach

serves as the basis on which to "upend" the dictates of the exploi-
tive class-hierarchical system. It seeks to eliminate the artificial
scarcity, rooted in waste, destruction, and inequality imposed by
capital, generating the potential for abundance, while remaining
"within ecological limits." Human interactions with nature need
to be focused on conservation and "preservation of the environment and ecology," fixing and "repairing the damage done," while
creating new efforts to "regenerate the bounty of life on our planet,
in all its diversity."[109]

Despite the extreme capitalism promoted by U.S. corporations,
the wealthy, and the servile state, which constitutes its environment, Cooperation Jackson has begun and plans to implement a
series of concrete, integrative projects that serve as the means to
accomplish their larger goals. This includes forming a nonprofit,
community land trust focused on removing as much land as possible from the "capitalist market" in order to "decommodify" it.
Under these conditions, the community serves as the steward.
It also establishes a basis with which to help block gentrification
processes that have been premised on expanding capital accumulation at the expense of the local community. This revolutionary
transformation involves creating an alternative currency, a system
of mutual credit, and "community-controlled financial institutions
ranging from lending circles to credit unions" in order to expand
the overall capacity and support of citizens.

Building on these foundations, Cooperation Jackson has gone
on to establish urban farm co-ops, a restaurant/grocery store, and
a lawn-care team. Compost from the store and lawns is used as
fertilizer on the farms, returning important nutrients to the soil
as part of metabolic restoration. There are plans to create a series
of cooperatives focused on housing, recycling, construction, child
care, retrofitting homes, and solar energy. All of these efforts are
organized as "non-reformist reforms" to improve the quality of
people's lives, expand the power of the citizens, and confront capital by subverting its very logic and operations. The goal is to foster
"the development of a non-capitalist alternative" that will "socialize

every step of the productive process required to create, distribute, and recycle a product"; forging "collective ownership and democratic management"; and increasing "the effective scale and scope of the solidarity economy."[110] Rather than promoting fashionable ideas of "resilience," which fail to challenge the dominant system, Cooperation Jackson can be regarded as a microcosm of ecological and social revolt, as part of the struggle for survival while advancing sustainable human development and ecosocialism.

The most radical and comprehensive strategy with respect to the planetary ecological emergency emanating from North America is the Red Nation's *The Red Deal: Indigenous Action to Save Our Earth*. In the words of the U.S. Indigenous movement organization Red Nation:

> Rather than taking an explicitly conservationist approach, the Red Deal instead proposes a comprehensive, full-scale assault on capitalism, using Indigenous knowledge and tried-and-true methods of mass mobilization as its ammunition. . . . We must be straightforward about what is necessary. If we want to survive, there are no incremental or "non-disruptive" ways to reduce emissions. Reconciliation with the ruling classes is out of the question. Market-based solutions must be abandoned. We have until 2050 to reach net-zero carbon emissions. That's it. Thirty years. The struggle for a carbon-free future can either lead to revolutionary transformation or much worse than what Marx and Engels imagined in 1848, when they forewarned that "the common ruin of the contending classes" was a likely scenario if the capitalist class was not overthrown. The common ruin of entire peoples, species, landscapes, grasslands, waterways, oceans, and forests—which has been well underway for centuries—has intensified more in the last three decades than in all of human existence.[111]

Survival in these terms requires the growth of what could be called an *environmental proletariat*, bringing together the global revolt against the capitalist expropriation of nature and exploitation of labor, thereby uniting the struggles over the economy and the earth. This means learning from Indigenous, colonized, and historically enslaved peoples while embracing issues of social reproduction. A revolt by the world's environmental proletariat conceived in these terms, in which hundreds of millions, even billions, of people will inevitably take part, is destined to come about in the coming decades as a result of the struggle for ecological survival. It will lead to new microcosms of existence and an assault on the macrocosm of capital and its state. But this struggle by an emerging new power can only succeed in the end if it takes the form of a revolutionary transformation directed at the creation of a socialist ecological civilization, drawing on the rich reservoirs of human knowledge and community. In the words of the great Irish revolutionary James Connolly: "We only want THE EARTH."[112]

Planned Degrowth: Ecosocialism and Sustainable Human Development

> All important concepts are dialectically vague at the margins.
> —HERMAN E. DALY

THE WORD *DEGROWTH* STANDS for a family of political-economic approaches that, in the face of today's accelerating planetary ecological crisis, reject unlimited, exponential economic growth as the definition of human progress.[1] To abandon economic growth in wealthy societies means to shift to zero net capital formation. With continual technological development and the enhancement of human capabilities, mere replacement investment is able to promote steady qualitative advancements in production in mature industrial societies, while eliminating exploitative labor conditions and reducing working hours. Coupled with global redistribution of the social surplus product and reduction of waste, this would allow for vast improvements in the lives of most people. Planned degrowth, which specifically targets the most opulent sectors of the world population, is thus directed at the enhancement of the living conditions of the vast majority while maintaining the environmental conditions of existence and promoting sustainable human development.[2]

Science has established without a doubt that, in today's "full-world economy," it is necessary to operate within an overall Earth System budget with respect to allowable physical throughput.[3] However, rather than constituting an insurmountable obstacle to human development, this can be seen as initiating a whole new stage of ecological civilization based on the creation of a society of substantive equality and ecological sustainability, or ecosocialism. Degrowth, in this sense, is not aimed at austerity, but at finding a "prosperous way down" from our current extractivist, wasteful, ecologically unsustainable, maldeveloped, exploitative, and unequal, class-hierarchical world.[4] Continued growth would occur in some areas of the economy, made possible by reductions elsewhere. Spending on fossil fuels, armaments, private jets, sport utility vehicles, second homes, and advertising would need to be cut in order to provide room for growth in such areas as regenerative agriculture, food production, decent housing, clean energy, accessible health care, universal education, community welfare, public transportation, digital connectivity, and other areas related to green production and social needs.[5]

When the first systems of national income accounting were devised at the time of the Second World War, all increases in national income, regardless of source, were characterized as constituting economic growth. GDP became the primary measure of human progress.[6] Nevertheless, much of this was questionable from a wider social and ecological standpoint. According to the prevailing system of national economic accounting, anything that provides "value added," in accordance with the capitalist valorization process, represents "growth." This includes such things as war spending; the production of wasteful and toxic products; luxury consumption by the very rich; marketing (encompassing motivation research, targeting, advertising, and sales promotion); replacements of social consumption by private consumption, as in the substitution of the private automobile for public transportation; expropriation of the commons; business expenditures to enhance the exploitation of workers; legal costs related to the administration, control, and enhancement of private property; anti-union activities by corporate management; the so-called criminal justice system; rising pharmaceutical and insurance costs; financial sector employment; military spending; and even criminal activities.[7] Maximum extraction of natural resources is seen as crucial for rapid economic growth, since it draws on nature's "free gift . . . to capital."[8]

In contrast, non-market and subsistence production carried out throughout the world, domestic labor mainly performed by women, numerous expenditures for human growth and development (seen as relatively nonproductive), conservation of the environment, and reductions in the toxicity of production are all seen as "counting for nothing" or assigned a diminished worth, since they do not enhance productivity or directly promote economic value.[9]

Today, the elemental tragedy of this is all around us. It is now widely perceived that economic growth, based on nonstop capital accumulation, is the main cause of the destruction of the earth as a safe place for humanity. The Earth System crisis is evident in the crossing of planetary boundaries related to climate change, ocean acidification, destruction of the ozone layer, species extinction, disruption of the nitrogen and phosphorus cycles, loss of groundcover (including forests), depletion of fresh water, aerosol loading, and novel entities (such as synthetic chemicals, nuclear radiation, and genetically modified organisms).[10] The drive to capital accumulation is thus generating a "habitability crisis" for humanity in this century.[11]

The world scientific consensus, as represented by the UN Intergovernmental Panel on Climate Change (IPCC), has established that the global average temperature needs to be kept below a 1.5°C increase over pre-industrial levels this century—or else, with a disproportionately higher level of risk, "well below" a 2°C increase—if climate destabilization is not to threaten absolute catas-trophe as positive feedback mechanisms come into effect. In the IPCC's *Sixth Assessment Report* (AR6, released in its various parts over 2021–23), the most optimistic scenario is one of an end-of-the-century increase in global average temperature over pre-industrial levels of below 1.5°C. This requires that the 1.5°C boundary not be crossed until 2040, rising by a tenth of a degree to 1.6°C, and then falling near the end of the century back down to a 1.4°C increase. All of this is predicated on reaching net zero (in fact, real zero) carbon emissions by 2050, which gives a fifty-fifty chance that the climate-temperature boundary will not be exceeded.[12]

Yet, according to leading climate scientist Kevin Anderson of the Tyndall Center for Climate Change Research, this scenario is already out of date. It is now necessary, based on the IPCC's own figures, to reach the zero–carbon dioxide emissions point by 2040 in order to have the same 50 percent chance of avoiding a 1.5°C increase. "Starting now," Anderson wrote in March 2023,

> to not exceed 1.5°C of warming requires 11% year-on-year cuts in emissions, falling to nearer 5% for 2°C. However, these global average rates ignore the core concept of equity, central to all UN climate negotiations, which gives "developing country parties" a little longer to decarbonise. Include equity and most "developed" nations' need to reach zero CO2 emissions between 2030 and 2035, with developing nations following suit up to a decade later. Any delay will shrink these timelines still further.[13]

The World Meteorological Organization indicated in May 2023 that there is a 66 percent chance that the annual average near-surface global temperature will temporarily exceed a 1.5°C increase over pre-industrial levels during "at least" one year by 2027.[14]

Existing IPCC scenarios are part of a conservative process, designed to conform to the prerequisites of the capitalist economy, which builds continued economic growth in the wealthy countries into all scenarios while excluding any substantial changes in social relations. The sole device relied upon in such climate modeling is to assume price-induced shifts in technology. Existing scenarios thus necessarily rely heavily on negative emissions technologies, such as bioenergy and carbon capture and sequestration and direct carbon air capture, that do not presently exist at scale and cannot be instituted within the prescribed timeline, while also presenting enormous ecological hazards in themselves. This emphasis on essentially nonexistent technologies that are environmentally destructive (given their enormous land, water, and energy requirements) has been challenged by scientists within the IPCC. Thus, in the original "Summary for Policymakers" for the mitigation report, part 3 of AR6, the scientists agreed that such technologies are not viable in a reasonable time frame and suggested that low-energy solutions based on popular mobilization might offer the best hope of carrying out the massive ecological transformations now required. All of this, however, was excluded from the final published "Summary for Policymakers" as determined by governments during the normal IPCC process, which allows for the censorship of scientists.[15]

Price-induced technological solutions, which would allow continued economic growth and the perpetuation of current social relations, do not exist on anything like the required scale and tempo. Hence, major socioeconomic changes in the mode of production and consumption are needed, running counter to the reigning political-economic hegemony. "Three decades of complacency," Anderson writes, "has meant that technology on its own cannot now cut emissions fast enough." There is thus a drastic need for low-energy solutions based on changes in relations of production and consumption that also address deep inequalities. The necessary reductions in emissions are "only possible by re-allocating society's productive capacity away from enabling the private luxury of a few and austerity for everyone else, and toward wider public prosperity and private sufficiency. For most people, tackling climate change will bring multiple benefits, from affordable housing to secure employment. But for those few of us who have disproportionately benefited from the status quo," Anderson tells us, "it means a profound reduction in how much energy we use and stuff we accumulate."[16]

A degrowth/deaccumulation approach that challenges accumulative society and the primacy of economic growth is crucial here. Social provisioning for human needs and sharp reductions in inequality are essential parts of a shift to a low-energy transformation in the economy and the elimination of ecologically destructive forms and scales of output. In this way, the lives of most people can be improved both economically and ecologically. Accomplishing this, however, requires going against the logic of capitalism and the mythology of a self-regulating market system. Such a radical transformation can only be achieved by introducing significant levels of economic and social planning, through which, if carried to its fullest, the associated producers would work together in a rational way to regulate the labor and production process governing the social metabolism of humanity and nature as a whole.

Classical nineteenth-century socialism in the work of Marx and Engels saw the need for the institution of collective *planning* in response to the ecological and social contradictions of capitalism, as well as its economic ones. Engels's analysis insisted on the need for socialist planning to overcome the ecological rift between town and country, while Marx's theory of metabolic rift, operating on a more general level, insisted on the need for sustainable human development.

Planning has been crucial to all economies, both capitalist and socialist, in times of war. Giant monopolistic corporations have themselves instituted of their own accord what economist John Kenneth Galbraith called a "planning system," though operating largely *within*, rather than *between*, multinational conglomerates.[17] Nevertheless, the whole idea of economic planning is seen, in the received ideology, as antagonistic to the capitalist market and has been effectively banned from public discussion—declared unworkable and a form of despotism—following the triumph of capitalism in the Cold War and the demise of the Soviet Union.

This is now rapidly changing. As French economist Jacques Sapir recently noted, "plan and planning are back in fashion," due to the internal and external contradictions of the capitalist market system.[18] It is now clear that, without the return of planning and environmental-state regulation of the economy in a context of the degrowth/deaccumulation of capital, there is zero possibility of successfully addressing the present planetary emergency and ensuring the continuation of industrialized society and the survival of the human population.

MARX, ENGELS, AND ECOLOGICAL PLANNING

Marx and Engels were always reluctant to provide what Marx called "recipes . . . for the cook-shops of the future," demarcating what forms socialist and communist societies should take. As Engels put it, "To speculate on how a future society might organize the distribution of food and dwellings leads directly to *utopia*."[19] Nevertheless, they were clear throughout their writings that the reorganization of production under a society of associated producers would involve cooperative labor organized in accordance with a common plan.

In *Principles of Communism*, Engels wrote that in the future society "all . . . branches of production" would be "operated by society as a whole, that is, for the common account, according to a common plan, with the participation of all members of society." The same approach was adopted by Marx and Engels in the *Communist Manifesto*, where they singled out the need for the "extension of factories and instruments of production owned by the State; the bringing into cultivation of waste lands, and the improvement of the soil generally in accordance with a common plan."[20] Here, the problem of ending the division between town and country through the dispersal of the population more evenly across the country, so that it was no longer concentrated in the large industrial cities separating the urban and rural populations, was central to their idea of a common plan.

Much of Marx's analysis in the *Grundrisse* focused on the need

for the "economy of time, [which] in accord with the planned distribution of labour time among the various branches" of industry,
constituted "the first economic law on the basis of communal production."[21] As he wrote to Engels on January 8, 1868: "No form
of society can prevent the working time at the disposal of society
from regulating production one way or another. So long, however,
as this regulation is accomplished not by the direct and conscious
control of society over its working time—which is possible only
with common ownership—but by the movement of commodity
prices, things remain as you have already quite aptly described
them in *Deutsch-Französische Jahrbücher*"—referring to Engels's
"Outlines of a Critique of Political Economy" of 1843.[22] This
early work of Engels was greatly admired by Marx. In his 1843
"Summary of Engels's 'Outlines,'" Marx emphasized "the split
between the land and the human being," and thus the alienation of
nature, as the external basis of capitalist production.

In *Capital*, Marx argued with respect to planning that the part
of the social product destined for the reproduction of the means
of production is properly collective while the other part, devoted
to consumption, is divided among consumers individually. How a
given society carries out this all-important division is the key to the
entire mode of production and reflects the historical development
of society itself. Under socialism, labor time would necessarily be
apportioned "in accordance with a definite social plan" that "main-
tains the correct proportion between the different functions of
labour and the various needs of the associations" of labor. This was
only possible when "the practical relations of everyday life between
man and man, and man and nature generally present themselves . . .
in a rational form" as a result of historical development, making possible "production by freely associated [individuals] . . . under their
conscious and planned control."[23] As Marx explained, in response
to the Paris Commune, "cooperative societies" in the future society
would "regulate national production upon a common plan."[24] The
fact that such planning was both an *economic* problem and an *ecological* one was clear throughout his work.

"Freedom in this sphere," a higher society, Marx wrote in the third volume of *Capital*, "can consist only in this, that socialized man, the associated producers, govern the human metabolism with nature in a rational way, bringing it under their collective control . . . accomplishing it with the least expenditure of energy and in conditions most worthy and appropriate for their human nature."[25] The historical record of human-caused ecological destruction in forms such as deforestation and desertification, embodied, for Marx, unconscious "socialist tendencies" since demonstrating the *necessity of social control.*[26]

However, it was Engels in *Anti-Dühring* who most explicitly grounded the need for planning in relation to environmental conditions. For Engels, it was the negative externalities of capitalist production, associated with the division between town and country, a permanent housing problem, and the destruction of both the natural as well as social conditions of working-class existence, that most clearly called for large-scale planning. Modern industry, he argued, needed "relatively pure water," as opposed to what existed in "the factory town" that "transforms all water into stinking manure."[27] Extending themes present in both *The Condition of the Working Class in England* and the *Communist Manifesto*, he declared:

Abolition of the antithesis between town and country is not merely possible. It has become a direct necessity of industrial production itself, just as it has become a necessity for agricultural production and, besides, of public health. The present poisoning of the air, water and land can be put an end to only by the fusion of town and country; and only such fusion will change the situation of the masses languishing in the towns, and enable their excrement to be used for the production of plants instead of for the production of disease. . . . The abolition of the separation of town and country is therefore not utopian . . . in so far as it is conditioned on the most equal distribution possible of modern industry over the whole country.[28]

Organizing production collectively according to a "social plan," Engels argued, would "end the . . . subjection of men to their own means of production" characteristic of capitalist commodity production.[29] Under socialism, it would of course "still be necessary for society to know how much labour each article of consumption requires for its production." It would then "have to arrange its plan of production in accordance with its means of production, which include, in particular, its labour-powers. The useful effects of the various articles of consumption compared with one another and with the quantities of labour required for their production, will in the end determine the plan."[30] But beyond the rational and economical use of labor within industry, planning would be necessary to overcome the exhaustion of the soil in the country and the related pollution of the town. Engels wrote, "Only a society which makes it possible for its productive forces to dovetail harmoniously into each other on the basis of one single vast plan can allow industry to be distributed over the whole country in the way best adapted to its own development, and to the maintenance and development of the other elements of production."[31]

In the *Dialectics of Nature*, Engels was concerned in particular with the failure of classical political economy as "the social science of the bourgeoisie" to account for "human actions in the fields of production and exchange" that were unintended, external to the market, and remote. The anarchic and unplanned character of the capitalist economy thus amplified ecological disasters. "What cared the Spanish planters in Cuba," he wrote,

> who burned down forests on the slopes of the mountains and obtained from the ashes sufficient fertiliser for *one* generation of very highly profitable coffee trees—what cared they that the heavy tropical rainfall afterwards washed away the unprotected upper stratum of the soil, leaving behind only bare rock! In relation to nature, as to society, the present mode of production is predominantly concerned only about the immediate, the most

tangible result; and then surprise is expressed that the more remote effects of actions directed to this end turn out to be quite different, are mostly quite the opposite in character.[32]

In order to promote the interests of the human community as a whole, it was therefore necessary to carry out "planned action" and regulate production in line with science, taking into consideration the earthly environment, that is, in accord with nature's laws.[33]

Marx and Engels saw socialism as enhancing human productive capabilities in a quantitative as well as a qualitative sense, and Engels even referred in *Anti-Dühring* to how the advent of socialism would bring about "the constantly accelerated development of the productive forces and . . . a practically unlimited increase of production itself." However, the context in which they were writing was not today's "full-world economy," but rather a still early stage of industrialization. In the period of industrial development, extending from the beginning of the eighteenth century until the first Earth Day in 1970, world industrial productive potential increased in size around 1,730 times, which from a nineteenth-century perspective would have seemed "a practically unlimited increase." Today, however, it raises the issue of ecological "overshoot."[34]

Hence, the long-term ecological consequences of production emphasized by Engels have more and more come to the fore in our time. This is symbolized by the proposed Anthropocene Epoch in the Geologic Time Scale, beginning around 1950, representing the emergence of human-industrialized society as the primary factor in Earth System change. From this standpoint, what is perhaps most remarkable about Engels's statement on the development of the productive forces under socialism is that it was immediately followed—in the same paragraph and the one after—by the view that the goal of socialism was not the expansion of production itself, but rather the "free development" of human beings, which required a rational and planned relation to "the whole sphere of the conditions of life which environ man."[35]

Marx and Engels, therefore, viewed planning as crucial in the organization of socialist/communist society, freeing it from the domination of commodity exchange, and relying on a *common plan*. Nevertheless, they cannot be seen as envisioning the kind of central planning under a command economy that was to emerge in the late 1920s and '30s in the Soviet Union. Rather, they contended that planning by the direct producers would be democratic with respect to production itself.[36] The entire system of socialism, as Marx put it, "starts with the self-government of the communities" in a society where "cooperative labor" would be "developed to national dimensions and, consequently . . . fostered by national means."[37] The rational organization of human labor as communal or cooperative labor, moreover, could not occur without a planning system. "All directly social or communal labour on a larger scale requires, to a greater or lesser degree, a directing authority, in order to secure the harmonious co-operation of the activities of individuals, and to perform the general functions that have their origin in the total productive organism," as a system of *social metabolic reproduction*. Production therefore requires guidance, foresight, and management, in the sense of a "conductor" of an orchestra. Marx's vision of a planned economy, as Michael A. Lebowitz emphasized, was one run by "associated conductors" who would rationally govern the metabolism between humanity and nature.[38]

As Marx wrote in *Theories of Surplus Value* on the need for a noncapitalist, and thus a *non-exhaustive*, approach to labor and nature:

> *Anticipation* of the future—real anticipation—occurs in the production of wealth only in relation to the worker and to the land. The future can indeed be anticipated and ruined in both cases by premature over-exertion and exhaustion, and by the disturbance of the balance between expenditure and income. In capitalist production this happens to both the worker and the land. . . . What is expended here exists as δίναμις [the Greek word for

power, in Aristotle's sense of a causal force] and the life span of this δίναμις is shortened as a result of accelerated expenditure.[39]

Capitalism, according to the founders of historical materialism, promoted a negative, perverse dialectic of exploitation, expropriation, and exhaustion/extermination, the "common ruin of the contending classes." What was necessary, therefore, was the "revolutionary reconstitution of society as a whole."[40]

This negative dialectic of exploitation, expropriation, and exhaustion/extermination characterizing capitalism was vividly captured by Engels in terms of the notion of the "revenge" of nature, a metaphorical expression that Sartre in his *Critique of Dialectical Reason* was to convert into the concept of "counter-finality."[41] Human beings, through their class-based social formations, became anti-*physis* (anti-nature). This could be seen in the destruction of forests and the consequent floods (Sartre had in mind Chinese peasant production described in René Grousset's 1942 *Histoire de la Chine*), in which populations undermined their own existence and their own supposed victories over nature, leading to catastrophic results. "Nature," Sartre wrote, "becomes the negation of man precisely to the extent that man is made anti-*physis*" and thus "*antipraxis*."[42] The only answer to the problem of counter-finality for Sartre, as for Marx and Engels, was to alter the social relations of production that propel humanity forward to ultimate catastrophe. This required a *revolution of the earth* in the form of a new socialist praxis of sustainable human development in which life itself was no longer posited as the enemy of humanity: the reunification of nature and society.

The tradition of "degrowth communism" within Marxism goes back to William Morris, who argued that Britain could do with less than half the coal it used.[43] But it can also be seen as related to what Burkett called Marx's overall "vision of sustainable human development." Here, the accumulation of capital was to be displaced by advances in qualitative human development and dedicated to the production of use value (rather than exchange value) and the fulfillment of the needs of all individuals, moving from the most basic needs all the way to the most developed human and social needs, in harmony with the environment as a whole.[44]

The Efficacy of Central Planning

Upon taking power in the October Revolution in 1917, the Bolsheviks, as Baran observed, "had no intention of immediately establishing socialism (and comprehensive economic planning) in their hungry and devastated country."[45] They originally envisioned a strict regulation and control of the capitalist market under a worker-directed government and the nationalization of key enterprises, encompassing a long and slow transition to a fully socialist economy. In fact, no concrete notion of central planning or of a command economy existed at the time.[46] "The word 'planning,'" Alec Nove wrote in *An Economic History of the U.S.S.R.*,

> had a very different meaning [in the Soviet Union] in 1923–26 to that which it later acquired. There was no fully worked-out production and allocation programme, no "command economy." The experts in Gosplan . . . worked with remarkable originality, struggling with inadequate statistics to create the first "balance of the national economy" in history, so as to provide some sort of basis for the planning of growth. . . . The point is that what emerged from these calculations were not plans in the sense of orders to act, but "control figures," which were partly a forecast and partly a guide for strategic investment decisions, a basis for discussing and determining priorities.[47]

War Communism, which began in the middle of 1918, eight months after the October Revolution, was a desperate effort to cope with the chaos and ravages resulting from the Russian Civil War, including the invasion of the country by all the major imperial powers in support of the "White" forces. War Communism was not about planning, but about wholesale nationalizations, war production, a ban on private trade, partial elimination of prices, free rations, and the forced requisition of supplies and surpluses.[48] The revolutionary Soviet state won the Civil War, defeating the White armies and forcing the imperial powers to vacate the country. But the economy was devastated and the small industrial proletariat, which had been the backbone of the Revolution, was decimated, with only half as many industrial workers in 1920 as in 1914.[49] In 1921, faced with economic deterioration, famine, and the revolt of the Kronstadt sailors, Lenin organized a strategic retreat, reintroducing market trading in the New Economic Policy (NEP). Beginning in 1920, Lenin also took personal initiative in introducing a plan for the electrification within ten to fifteen years of all of Russia, building power stations and related infrastructure in all the major industrial regions. This was to prove to be the greatest accomplishment with respect to economic development in the early 1920s.[50]

The NEP was seen as a transitional period in the movement toward socialism. Lenin designated it as "state capitalism." The Soviet state retained control of the commanding heights of the economy, including heavy industry, finance, and foreign trade. In Lenin's initial conception, the NEP was a limited alliance with big capital with the goal of transforming production in accordance with its most developed form of monopoly capitalism, but under socialist control, together with an accommodation with the peasantry. "The Soviet state," Tamás Krausz wrote in *Reconstructing Lenin*, "gave preferential treatment to organized large-scale capital and market-oriented state property rather than anarchic private property, the uncontrollably chaotic economy of the petit bourgeois." Lenin utilized the concept of state capitalism to refer not only to the state sector in a mixed economy, but also to a definite social formation in the movement toward socialism, constituting the essence of the NEP.[51]

It was during the NEP that a level of development planning was first introduced into the economy. The Supreme Council of the National Economy had been established as early as 1917. However, it was under the NEP that Gosplan was set up as the main state planning commission. Gosplan developed the first system of balances for a national economy, providing control figures to guide investment decisions with limited directives to a few strategic sectors under state control. A nascent method of input-output tables was introduced in 1923–24, inspired by François Quesnay's *Tableau économique* and Marx's reproduction schemes in *Capital*.[52]

By 1925, the NEP had succeeded in restoring the prewar economy, and industrial production outside of agriculture was beginning to level off. Lenin had hinted in 1922 that the NEP might need to remain in place for a long time, with twenty-five years as "a bit too pessimistic."[53] But with his death in 1924 and the success of the NEP in restoring the economy, a Great Debate arose over socialist transformation and planning. Classical Marxist theory had been based on revolutions occurring first in the developed countries of Western Europe. The Russian Revolution was originally envisioned as sparking a wider European proletarian revolution, which, however, never materialized. Russia found itself an underdeveloped, primarily peasant country, existing in a state of political and economic isolation and faced by the continual threat of further imperial invasions.

All the major participants in the Great Debate agreed on the need to move toward a socialist planned economy, but disagreements arose over the nature and tempo of the change, and the degree to which the peasants should have their land expropriated. Some leading Bolsheviks, such as Nikolai Bukharin, argued for what was then the dominant line, insisting on a slower, balanced-growth approach based on the continuation of the NEP as a transitional period. In contrast, those like the economist E. A. Preobrazhensky, who was identified with the "left opposition," favored a much more rapid shift to a centrally planned economy and the expropriation of the peasantry through a process of socialist primitive accumulation.[54] The major figures of both the *left opposition*, including Preobrazhensky and Trotsky, and what Stalin was to characterize as the *right opposition*, associated with

Bukharin (with whom Stalin had been aligned during the Great Debate), were all eventually eliminated one after the other, leaving Stalin entirely in command.[55]

With Stalin's rise to power by 1928, a rapid industrialization course was adopted in line with the proposals originally advanced by the left opposition, which Stalin himself had at first opposed. The goal became one of building "Socialism in One Country" given the USSR's isolated position. This, however, took the form of a brutal socialist primitive accumulation and a top-down, bureaucratic command economy, commencing with the first five-year plan in 1929. In 1925–26, under the NEP, the state sector constituted 46 percent of the economy; by 1932, it had risen to 91 percent.[56]

The tragedy of Soviet planning lay in the dire historical circumstances in which it arose, leading to what the noted historian of the USSR, Moshe Lewin, called "the disappearance of planning in the plan."[57] Industrial output in 1928–29 under the NEP had grown at a rate of 20 percent. Yet that was not considered enough. Bukharin spoke out against plans being constructed by "madmen" who sought an annual economic growth rate twice what the NEP had delivered. The planning process was thus conceived from the first on unrealistic bases. A system of central planning arose that took the specific form of a *command economy*, with all directives over the allocation of labor and resources, inputs into production, specified targets, and so on being determined bureaucratically from the top. This was coupled with a perpetuation of the basic character of the capitalist labor process with the incorporation of Taylorist scientific management techniques, eliminating the possibility of bottom-up forms of organization or workers' control, as originally envisioned in the workers' Soviets.

The directives laid out in the first five-year plan were beyond all possibility of fulfillment, with the result that the plan was effectively shelved almost from the beginning. The command system that emerged was centrally and bureaucratically administered, while rational planning was hardly in evidence. Meanwhile, the "supertempo" of industrialization meant the massive confiscation of peasant property and forced collectivization, affecting millions. As Lewin wrote, "Stalin's antipeasant drive was an attack against the popular masses. It required coercion on such a large scale that the whole state had to be transformed into a huge, oppressive machine." Under such circumstances, the harsh regimentation of the population was inevitable.[58]

Nevertheless, with all of its shortcomings and barbarities, the crude, clunky, bureaucratic command economy that arose in the Soviet Union was hugely successful in its developmental effects. It was able to prioritize investment in heavy industry in a way never quite seen before. The average annual growth rate in industrial output for the years 1930–40 was officially 16.5 percent, which, in Lewin's words, was "certainly an impressive figure (and not much less impressive even if smaller assessments by Western economists are preferred)."[59] The Soviet Union leaped into industrialization, also expanding transportation and electrical generation, albeit with agriculture lagging behind. Other vast improvements occurred in education and urbanization.[60] Some eight thousand massive modern enterprises were constructed between 1928 and 1941.[61]

In 1928, the Soviet Union was still an underdeveloped country. By the Second World War, it had emerged as a major industrial power. There is no questioning Stalin's hard realism when he stated, in 1931, "We are 50–100 years behind the advanced countries. We have to traverse this distance in ten years. We will either accomplish it or else we will be crushed."[62] His calculations were correct. By the time the German *Wehrmacht* invaded Russia exactly ten years later, in 1941, with more than three million Axis troops organized in armored divisions and deployed on an 1,800-mile front, the invading forces found themselves confronted by a major industrial and military power quite unlike the Russia they had faced in the First World War. The Soviet forces carried out an extraordinary resistance far exceeding anything that Adolf Hitler and his advisors had conceived. The history of the modern world was to turn on that very fact, leading to the defeat of Nazi Germany.[63]

Yet, the weaknesses of the Soviet economy, with its centrally administered and planned production, were to haunt the system after the Second World War. Although maintaining fairly impressive growth rates and, in the post-Stalinist, particularly early Leonid Brezhnev era, able to provide both guns and butter in the context of the Cold War—in which it was confronted by a much larger and more aggressive counterpart in the United States—the weaknesses of the Soviet system became more and more evident.[64] The bureaucratic planned economy had led to a concentration of power and the emergence of a new ruling class of bureaucratic bosses, or *nachal'niki*, arising out of the *nomenklatura* system (exercising control over top-level nominees to the Party), which weighed on the system, preventing necessary changes.[65] Despite its early developments in input-output analysis, the Soviet command economy never integrated the methods of cybernetics and the possibilities for more optimal planning that emerged with the new computing revolution in the decades after the Second World War, despite some movements in this direction.[66] An overemphasis on new investment projects led to a neglect of replacement investment, with the result that production was carried on with obsolete equipment resulting in numerous work stoppages.[67] The proletarianization of labor, coupled with full employment and other guarantees, reduced the possibilities of economic coercion within the system compared to capitalism, leading to problems of material incentives for the workers.[68]

The Soviet system of enterprise management, as Che Guevara acutely recognized, was based on pre-monopoly capitalism, not monopoly capitalism, and thus relied more heavily on interfirm rather than intrafirm transactions. This meant that enterprises were dependent on external prices, with the ironic result that market relations undercut planning at the enterprise level in ways that did not occur within what Galbraith had called the "planning system" of monopolistic corporations in the West. At the same time, factory production was organized along the old Ford Motors model in which each division or syndicate made all the components, as opposed to the more developed monopoly capitalist production system with multiple suppliers, which prevented bottlenecks.[69] Most important, the Soviet command economy was reliant from the first on extensive, rather than intensive, development through the forced drafting of labor and resources, as opposed to the cultivation of dynamic efficiencies.[70] Consequently, once labor and resources began to be scarce, rather than abundant, the economy went into stagnation, creating widespread shortages.[71]

Still, even then the economy continued to grow, although more slowly, until the chaos of the Gorbachev era—while also providing extensive social welfare amenities to the population, which were enviable from the standpoint of most of the world, if lacking in mass consumerism and luxury goods.[72] In the end, it was the direction taken by the upper end of the social hierarchy associated with the *nomenklatura* system, which aspired to the same opulent lifestyle as the upper echelons in the West, that was to seal the fate of the Soviet system.[73]

As Harry Magdoff and Fred Magdoff explained in "Approaching Socialism":

> The shortcomings of the Soviet economy, which became evident not long after recovery from the Second World War, were not a result of the failure of central planning, but of the way planning was conducted. Central planning in peacetime does not need control by the central authorities over every detail of production. Not only are commandism and the absence of democracy not necessary ingredients of central planning, they are counterproductive to good planning.

Ironically, it was the class character of the Soviet system and rampant corruption that led to its demise.[74]

China's command economy period, following the 1949 Revolution, was much shorter, lasting essentially from 1953 to 1978. It launched its first five-year plan based on the Soviet model in 1953, with its planning phase lasting until it instituted "market reforms" a quarter-century later. During its central planning period, when it had also to deal with the U.S. threat and thus was forced to divert major needed resources to national defense, the People's Republic of China nevertheless logged impressive achievements, establishing the industrial and social base for the even more impressive economic development that was to follow with the opening up of the Chinese economy and its controlled integration with the world economy.

There is no doubt that the record of the Chinese command economy in its initial planning period was patchy. Central planning, as instituted in China, had many of the same weaknesses as it had in the Soviet Union, leading to imbalances and the same phenomenon of "the disappearance of planning in the plan." Nevertheless, huge accomplishments were made. Agriculture was put on a new foundation with collectives and social property.[75] "Few people are aware," Fred Magdoff wrote in his preface to Dongping Han's *The Unknown Cultural Revolution: Life and Change in a Chinese Village*,

of the visit to China in the summer of 1974, during the Cultural Revolution, by a delegation of U.S. agronomists. They traveled widely and were amazed by what they observed, as described in an article in the *New York Times* (September 24, 1974). The delegation was composed of ten scientists who were "experienced crop observers with wide experience in Asia." As Nobel Prize–winner Norman Borlaug put it—"You had to look hard to find a bad field. Everything was green and nice everywhere we traveled. I felt the progress had been much more remarkable than what I expected." The head of the delegation, Sterling Wortman, a vice president of the Rockefeller Foundation described the rice crop as ". . . really first rate. There was just field after field that was as good as anything you can see." They were also impressed with the increased skill levels of the farmers on the communes.

Wortman said, "They're all being brought up to the level of skills of the best people. They all share the available inputs." A detailed description of their observations on agriculture in China was published in the prestigious journal *Science* in 1975 by Dr. Sprague. Much of the progress in China's agriculture after the Cultural Revolution was made possible by the advances during that period. Even the increase in fertilizer use that occurred in the late 1970s and early 1980s was made possible by factories that were contracted for by China in 1973.[76]

Growth of industrial potential in China under Mao Zedong was "relatively rapid" when compared to almost all other developing countries.[77] Literacy and average life expectancy were completely transformed, placing China on a par with middle-income countries in terms of human development factors by the late 1970s, despite its still extremely low per capita income. The "net impact of planning" was a vast increase in "the rate of technical progress." As Chris Bramall wrote in his major 1993 work, *In Praise of Maoist Economic Planning*, "If one believes that capabilities are a better indicator of economic development than opulence, both China and Sichuan [Province] had developed a great deal by the time of Mao's death. That the World Bank chooses to place more emphasis on opulence is an entirely normative decision."[78]

Post-1978 China moved rapidly from an entirely centrally planned economy to a mixed economy system resembling Lenin's NEP. It could be structurally seen, in Marxist terms, as Samir Amin noted, as a "state capitalism" under the leadership of the Chinese Communist Party (although the terms *market socialism* and even *state socialism* have also been used).[79] This meant that there was a sharp turn to the market, while the state sector remained enormous, dominating the commanding heights of the economy and guiding the whole system, under "socialism with Chinese characteristics." China's GDP grew by thirty times between 1978 and 2015, far exceeding all the other historic "economic miracles" with respect to industrialization.[80]

Land, particularly in rural areas, remained for the most part under state/collective ownership. China, at present, has about 150,000 state-owned enterprises, about 50,000 of which are owned by the central government, and the rest by local governments. State-owned enterprises account for about 30 percent of total GDP (around 40 percent of nonagricultural GDP) and some 44 percent of national assets.[81] These firms are tightly controlled by the government (with general managers of the state-owned enterprises appointed by the Party's Central Organization Department). They are integrated with the market but receive state support and subsidies and are expected to fulfill government objectives beyond profit-maximization while also providing economic surpluses to the state, amounting to 30 percent of their profits. During the COVID-19 pandemic, the Party gave state firms a significant role.[82]

China continues to introduce five-year plans in which its control over the state sector is its major point of leverage in guiding the entire economy.[83] In 2002, there were six Chinese state-owned enterprises in the Global Fortune 500. By 2012, this had risen to sixty-five. It is explicitly recognized by the Chinese Communist Party that the market is a force that is heartless and brainless, requiring that the state play a direct role in guiding the economy. This has taken the form of what is known as "state regulation (a.k.a. planned regulation)" and the principle of "co-production" of state and market.[84]

As Yi Wen, economist and vice president of the Federal Reserve Board of St. Louis has noted, "China compressed the roughly 150 to 200 (or even more) years of revolutionary economic changes experienced by England in 1700–1900 and the United States in 1760–1920 and Japan in 1850–1960 into one single generation."[85] The Chinese economy retains a guiding state sector and therefore a much greater capacity of the state to regulate the economy—and, in effect, to plan shifts in the allocation of labor and resources. An important aspect of this is a much greater immunity to economic crises, which are generally confined to local disturbances in production.[86] Nevertheless, central contradictions of "socialism with

262   THE DIALECTICS OF ECOLOGY

Chinese characteristics" are to be found in the level of inequality
that has now almost reached U.S. proportions, and in the extreme
exploitation of migrant labor from rural areas employed in export
production for foreign multinationals. These have become major
areas of concern.[87]

The demise of the Soviet Union and the opening up of China
to the world economy were universally greeted in the West—particularly within orthodox economics as the ideological core of the
system—as offering definitive proof that economic planning was
unworkable and doomed to fail from the start. Socialism was identified entirely with planning, which, it was said, led to inevitable
failure. Implicit in this was the "assumption that Soviet practice
reveals the essential nature of a centrally planned economy."[88]

However, such a blanket condemnation of central planning in
all forms and circumstances, divorced from concrete analysis, had
no real theoretical basis and was contradicted by reality. Capitalist
economies had frequently resorted to emergency wartime central
planning. During the Second World War, for example, the United
States instituted an extensive system of national planning run
by the War Production Board and other agencies, which shifted
resources and production while instituting rationing and price
controls. Civilian automobile production, constituting the core
industrial sector of the country, was rapidly converted into the
production of armaments, tanks, and aircraft. There was a desperate need to produce warships and merchant ships. Military goods
were needed not only for the United States but also for its allies.[89]
This also demanded a massive expansion of and major shifts in
the labor force, as millions of men were drawn into military service. Paid employment of women grew by 57 percent during the
war; in 1943, women made up 65 percent of the work force in the
aircraft industry.[90] All of this required central planning, including
planning agencies, directives by the state, and fiscal and monetary controls. Government research in science and technology
was boosted, most famously in the Manhattan Project. The economic surplus generated by the society was massively redirected

to facilitate war production, while industry had to be coordinated to maximize specific military goods at the right time and tempo.[91] Central planning, as Michał Kalecki defined it, "embraces the volume of production, the wage fund, larger investment projects, as well as control over prices and the distribution of basic materials." U.S. wartime planning fits this definition to a considerable extent, demonstrating that a mixed economy was not incompatible in all circumstances with centralized planning.[92]

Without social and economic planning, the objectives of socialism aimed at substantive equality and ecological sustainability are impossible to achieve. Logic and historical experience show that without a planning system of some sort operating at various levels, from workplace to local to national, there is no conceivable way of effectively addressing the planetary ecological emergency or ensuring "*buen vivir* for all people."[93] This simply cannot be achieved in a society of "Accumulate, accumulate! That is Moses and the prophets!"[94] Planning, however, needs to be democratic if it is to attain socially optimal results. "There is nothing in central planning," Fred and Harry Magdoff observed in "Approaching Socialism,"

> that requires commandism and confining all aspects of planning to the central authorities. That occurs because of the influence of special bureaucratic interests and the overarching power of the state. Planning for the people has to involve the people. Plans of regions, cities, and towns need the active involvement of local populations, factories, and stores in worker and community councils. The overall program—especially deciding the distribution of resources between consumption goods and investment—calls for people's participation. And for that, the people must have the facts, a clear way to inform their thinking, and contribute to the basic decisions.[95]

A unified, multifaceted planned economy, which would encompass multiple levels and involve "whole-process democracy," does not demand the elimination of consumer markets or of the freedom of workers to work where they please (and thus a labor market in this sense).[96] It does, however, require control over investment in capital goods and of finance, and thus social controls allowing for the mobilization of the economic surplus in ways that benefit the population in its entirety (including future generations), ensuring egalitarian conditions, the fundamental bases of human development for all individuals, and protection of the natural environment.

In his essay "In Defense of Socialist Planning" in 1986, Ernest Mandel argued that the main advantage of economic planning is that decisions on allocation of resources and labor are made *ex ante* and then corrected by trial and error, rather than *ex post* through the mediating force of the commodity market (and its "rationing by the wallet"). Planning thus allows for decisions to be made directly on the basis of what Marx called the "hierarchy of . . . needs." This does not require that all decisions be made by a centralized bureaucracy; it is consistent with a socialized democracy based on the "*institutionalization* of popular sovereignty." The fundamental parameters of production would be established by the associated producers in a society organized on the principle of cooperation. Such a society "would *grow in civilization* rather than in mere consumption."[97]

SOCIALIST STATES AND THE ENVIRONMENT

There is a widely propagated notion, which became almost universally accepted after the demise of the Soviet Union, that the Soviet record on the environment was much worse than that of the West, and that this was attributable to socialism and central planning.[98] It is true that the USSR's record on the environment was deplorable in many respects. One only has to think of Chernobyl and the Aral Sea. In the Stalin era, many of the pioneering Soviet ecologists were purged, with major consequences for Soviet development. Yet, the dominant view erases Soviet environmental successes, manifested in its green belts around cities, its famous *zapovedniki* (scientific ecological preserves), its massive reforestation/afforestation campaigns, its leading role in promoting environmental agreements internationally, and its powerful environmental organizations, which exerted pressure on the government. The All-Russian Society for the Preservation of Nature, largely led by scientists, had 37 million members by 1987, making it the largest conservation advocacy organization in the world.[99]

As the Soviet Union industrialized and modernized while facing the need for high levels of military spending given the Cold War threat from the West, it naturally converged with Western levels of environmental destruction. Like the West, it eventually responded, though not without contradictions, to its environmental movements. Environmental protection and conservation were incorporated, if inadequately, into its overall planning system. The Soviet Union had a very extensive system of environmental laws, which were, however, insufficiently enforced. It was Soviet scientists, soon followed by U.S. scientists, who first raised the alarm on accelerated global warming.[100] Major efforts were also made in the area of soil conservation.[101] In the 1980s, the concept of "ecological civilization" first arose in the Soviet Union and was soon adopted in China, where it has become a core aspect of overall planning, as reflected in China's five-year plans.[102] Leading Soviet economists, such as P. G. Oldak, argued for a radical transformation of Soviet national-income accounting to integrate direct measures of environmental destruction. "'More,'" he argued, "is by no means always 'better.'"[103]

The Soviet Union's environmental record with respect to pollution, while hardly satisfactory, was generally favorable when compared to the United States, with roughly equal populations. The Soviet Union's per capita sulfur dioxide, nitrous oxide, particulate matter, and carbon dioxide emissions were all far below those of the United States, while its per capita carbon dioxide emissions actually declined in its final years. The per capita ecological footprint of the Soviet Union, the most comprehensive

measure of environmental impact, was far lower than that of the
United States, with the gap increasing in the 1980s, as the U.S. per
capita ecological footprint continued to grow while that of the
USSR leveled off. Moreover, this was true even though the United
States was able "to offload environmental harms on many other
countries." The United States was far wealthier and more technologically advanced, but also did much more damage to the global
environment.[104]

Although Soviet planning and that of other post-revolutionary societies had been directed at economic growth, mimicking
capitalism to some extent in this respect, the inner, class-based
drive for capital accumulation is not an inherent structural feature of a socialist planned society. For this reason, Sweezy argued
in 1989 that the actually existing planned economies offered the
best chance for humanity in terms of the rapid transformations in
production and consumption needed to confront the global environmental crisis.[105]

Cuba, though a poor country faced with a perpetual economic
blockade from the United States, has long been recognized as the
most ecological nation on Earth, according to the World Wildlife
Federation's *Living Planet Report*. Cuba was able to demonstrate
that a country can be rated highly on human development while
having a low ecological footprint. This is due to placing human
development for the population as a whole, including environmental conditions, at the forefront of its planning.[106]

The People's Republic of China, meanwhile, has made huge
strides in the direction of "ecological civilization"—despite its
attempt to bring up the per capita income of its population above
the current level, which is currently less than one-fifth that of the
United States (in market exchange terms), requiring high rates
of economic growth.[107] This has been accompanied though by a
continuing, if diminished, reliance on coal-fired plants as its main
source of energy. Still, China has forged ahead in sustainable
technologies (where it is the world leader), in rapid reductions in
pollution, and in global levels of reforestation/afforestation.[108]

In the current ecological climate, China and Cuba— along with other mixed, state-directed, semi-planned economies such as Venezuela, with its attempts, through its Bolivarian Revolution, to build a communal state and its extraordinary achievements in food security and food sovereignty—offer hope of ecological breakthroughs in the present planetary emergency, currently lacking in the opulent capitalist world.[109]

PLANNING SUSTAINABLE HUMAN DEVELOPMENT

If organized civilization is to survive, planned degrowth or deaccumulation and a shift to sustainable human development is now unavoidable in the wealthiest countries, whose per capita ecological footprints are non-sustainable on a planetary basis. The scale and tempo of the necessary ecological-energy transformation, as emphasized in scientific reports on climate change and other planetary boundaries, indicate that in order to avert disaster a revolutionary transformation of the entire system of production and consumption must be implemented under the principle "Better Smaller But Better."[110] Hence, the core capitalist/imperialist countries, which constitute the main source of the problem, must seek a "prosperous way down," focusing on use-value rather than exchange-value.[111] This requires moving toward much lower levels of energy consumption and gravitating to equal global per capita shares while simultaneously zeroing out carbon emissions.

At the same time, the poorer countries with low ecological footprints must be allowed to develop in a general process that includes *contraction* in throughput of energy and materials in the rich countries and the *convergence* of per capita consumption in physical terms in the world as a whole.[112] The downsizing of the rich economies will require a massive shift to sustainable technologies, including solar and wind energy. But no existing technologies by themselves can come anywhere near to solving the climate problem in the required timeline, not to mention addressing the planetary emergency in its entirety, while allowing for the continued unlimited exponential accumulation and maldistribution required by capitalism.[113]

What is objectively necessary at this point in human history is therefore a revolutionary transformation in social relations governing production, consumption, and distribution. This means a dramatic shift away from the system of monopoly capital, exploitation, expropriation, waste, and the endless drive to accumulation.[114] In its place, a revolutionary humanity based in the working population—an emergent environmental proletariat—will need to demand a new social formation that provides for the basic needs of all of the population, followed by community needs, including the developmental needs of all individuals.[115] This will be made possible by qualitative improvements in work, an emphasis on useful labor and care work, along with the sharing of abundant social wealth, itself the product of human labor. A sustainable relation to the earth is an absolute requirement without which there can be no human future. All of this necessitates going against the logic of capitalist accumulation in the present. Economic planning will need to be repurposed, not for economic growth or war on other countries, but in order to create a new set of social priorities aimed at human flourishing and a sustainable social metabolism with the earth.

A "socialist vision of the United States," Harry Magdoff wrote in 1995, would require decreases in the use of energy, production of civilian cars, and government subsidies to environmentally destructive firms. "A much simpler lifestyle would be needed in the rich countries for the sake of preserving the earth as a place of human existence." In order to achieve this, "growth would need to be curtailed or controlled." It would be essential in such a system to focus on basic needs, such as adequate and dignified housing for all. War spending geared to imperialism would have to cease and immigration restrictions would need to be eliminated. All of this requires social and economic planning. None of it could be achieved by relying primarily on the system of market prices, which invariably promotes inequality, environmental destruction, war, and exclusion.[116] As British sociologist Anthony Giddens wrote in *The Politics of Climate Change*, "planning of some sort is inevitable" in the face of the current planetary crisis.[117]

In the United States and other rich countries, the means already exist at present for such a massive, qualitative transformation of society in line with social priorities and the needs of the oppressed working class, while shifting away from imperialism and the global oppression of "the wretched of the earth." This can be easily seen by pointing to the now trillion-dollar-plus military budget, which could be repurposed to carry out those changes in the energy infrastructure necessary for human survival. But it can also be seen in the rising levels of expropriation of surplus from the direct producers. A study by the RAND Corporation estimated that $47 trillion (in 2018 dollars) was expropriated from the bottom 90 percent of the U.S. population between 1980 and 2018, calculated based on what they would have received if income had grown equitably within the economy over the period. This exceeds the entire current value of the U.S. housing stock, which in January 2022 was $43 trillion.[118] At the base of this enormous social surplus is social labor, which needs to be allocated on an economic and ecological basis, and no longer on the basis of private accumulation.[119]

Even the most cursory examination of the wider waste and exploitation in the system raises what Morris called the problem of "Useful work *versus* Useless Toil."[120] The massive economic surplus arising from social labor—measured not simply by profits, interest, and rent, but also by the waste, maldistribution, and elementary irrationality of the system—is already many times that which is necessary to carry out the vast changes needed to create a society of sustainable human development. It is capitalism itself that imposes scarcity and austerity on the population in order to compel workers to sacrifice their lives still further for an exploitative system that is now threatening a planetary habitability crisis for all of humanity along with innumerable other life forms.

Most degrowth strategies, even those promulgated by ecosocialists, defer to the reigning ideology, preferring not to raise the issue of planning even in the face of the planetary emergency. Indeed, there is a tendency to back off from such obvious measures as nationalization of energy companies and mandatory emissions cuts on corporations. Degrowth theorists instead generally propose a menu of "policy alternatives," like universal basic income, ecological tax reform, a shortened work week, increased automation, and so on, none of which come into direct conflict with the system or get close to addressing the enormity of the problem, in what are thought of as non-reformist reforms.[121]

Proposals for drastically reduced employment, not just shorter working hours, backed up in many degrowth schemes by a guaranteed basic income, seek to adjust the parameters of capitalism rather than transcend them, in an approach that would generate the kind of dystopian conditions described in Kurt Vonnegut's novel *Player Piano*.[122] As Huberman and Sweezy wrote when the notion of a guaranteed basic income was first floated in the 1960s, "our conclusion can only be that the idea of unconditionally guaranteed incomes is not the great revolutionary principle which the authors of 'The Triple Revolution' evidently believe it to be. If applied under our present system, it would be, like religion, an opiate of the people tending to strengthen the status quo. And under a socialist system . . . it would be quite unnecessary and might do more harm than good."[123]

Confronted with climate change, some non-degrowth socialists have succumbed to technology fetishism, proposing dangerous geoengineering measures that would inevitably compound the planetary ecological crisis as a whole.[124] There is no doubt that many on the left see the entire solution today as consisting of a Green New Deal that would expand green jobs and green technology, leading to green growth in a seemingly virtuous circle. But since this is usually geared to a Keynesian growth economy and defended in those terms, the assumptions behind it are questionable.[125] A more radical proposal, more in line with degrowth, would be a People's Green New Deal oriented toward socialism and democratic ecological planning.[126]

Under the monopoly-finance capital of today, whole sectors of the caring profession, education, the arts, and so on are affected by what is known as the Baumol cost disease, named after William J. Baumol, who introduced the idea in his 1966 book *Performing Arts: The Economic Dilemma*.[127] This applies when wages rise and productivity does not. Thus, as *Forbes* magazine declares without a trace of irony: "The output of a [string] quartet playing Beethoven has not increased since the 19th century," although their income has. The Baumol cost disease is seen as applicable mainly to those work areas where notions of quantitative increases in productivity are generally meaningless. Yet, how does one measure the productivity of a nurse treating patients? Certainly not by the number of patients per nurse, regardless of the amount of care each receives and their outcomes. The result of profit-centered goals in the highly financialized economy of today is underinvestment and institutionalization of low wages in precisely those sectors characterized as subject to the so-called Baumol cost disease, simply because they are not directly conducive to capital accumulation.

In contrast, in an ecosocialist society, where accumulation of capital is not the primary objective, it would often be those labor-intensive areas in the caring professions, education, the arts, and organic relations to the earth that would be considered most important and built into social planning.[128] In an economy geared to sustainability, labor itself might in some cases be substituted for fossil-fuel energy, as in small, organic, sustainable farming, which is more efficient in ecological terms.[129]

Writing in *The Political Economy of Growth* in 1957, Baran argued that the *planned economic surplus* might be intentionally reduced in socialist planning, in comparison to what was then possible, in order to ensure the "conservation of human and natural resources." Here, the emphasis would not be simply on economic growth, but on meeting social needs, including decreasing environmental costs—for example, by choosing to cut "coal mining."[130] All of this meant, in effect, prioritizing sustainable human development over destructive forms of economic growth. Today, elimination of fossil fuels, even if this means a reduction in

the economic surplus generated by society, has become an absolute necessity for the world at large, which is faced by what Noam
Chomsky has called "the end of organized humanity."[131] In the
words of Engels and Marx, it is necessary to release the "jammed
safety-valve" on the capitalist locomotive "racing to ruin." The
choice is one of *socialism or exterminism*, "ruin or revolution."[132]

As Diogenes of Oenoanda, faithfully inscribing the thoughts of
Epicurus in stone in the second century A.D., wrote: "The con-
fines of the inhabited world offer all men one common country,
the world, one common home, the earth."[133]

Notes

Preface

1. Richard Levins and Richard Lewontin, *The Dialectical Biologist* (Cambridge, MA: Harvard University Press, 1985), 269–70.

2. John Bellamy Foster, *Marx's Ecology: Materialism and Nature* (New York: Monthly Review Press, 2000). This book was immediately preceded by John Bellamy Foster, "Marx's Theory of Metabolic Rift," *American Journal of Sociology* 105, no. 2 (September 1999): 366–405. This work was closely aligned with Paul Burkett, *Marx and Nature* (Chicago: Haymarket, 1999, repr. 2014).

3. "The free movement in matter," Marx wrote, "is nothing but a paraphrase for the *method* of dealing with matter: that is, the *dialectic method*." Karl Marx, *Letters to Kugelman* (New York: International Publishers, 1934), 112.

4. Foster, *Marx's Ecology*, 250.

5. John Bellamy Foster, *The Return of Nature: Socialism and Ecology* (New York: Monthly Review Press, 2020).

6. Karl Marx and Frederick Engels, *Collected Works* (New York: International Publishers, 1976), 62–63; Karl Marx, *Capital* (New York: Monthly Review Press, 1976), 949; Karl Marx, *Early Writings* (London: Penguin, 1974), 355.

7. The notion that a "degrowth communism" was articulated in Marx's final unpublished writings in the late nineteenth century, including the *Critique of the Gotha Programme* and his draft letters to Vera Zasulich, has recently been advanced by Kohei Saito. Yet, concrete evidence of this—beyond Marx and Engels's commitment to sustainable human

development controlled by the associated producer, and their explicit recognition of the possibility of quite different paths of development then taking place in Europe (for example, based on the Russian *Mir*, or the form of communal peasant production)—is completely lacking. Although Marx was opposed to the class-based system of capital accumulation, nowhere does he take a principled stance against the growth of production as such, which would scarcely have made sense in a nineteenth-century global political-economic and ecological context. See Kohei Saito, *Marx in the Anthropocene* (Cambridge: Cambridge University Press, 2023).

8.   See Yrjö Hail and Richard Levins, *Humanity and Nature* (London: Pluto Press, 1992), 252.

Introduction

1.   Epigraph: Denis Diderot, *Rameau's Nephew* and *D'Alembert's Dream* (London: Penguin, 1966), 181. Richard Lewontin and Richard Levins, *Biology Under the Influence* (New York: Monthly Review Press, 2007), 185–86, at 110.

2.   Dipesh Chakrabarty, *The Climate of History in a Planetary Age* (Chicago: University of Chicago Press, 2021), 173, 205.

3.   Karl Marx, *Capital*, vol. 1 (London: Penguin, 1976), 283; *Critique of the Gotha Programme* (New York: International Publishers, 1938), 2; Karl Marx. *Early Writings* (London: Penguin, 1974), 328.

4.   Karl Marx and Frederick Engels, *Collected Works*, vol. 5 (New York: International Publishers, 1975–2004), 28.

5.   Marx, *Capital*, vol. 1, 283.

6.   Marx, *Capital*, vol. 1, 637.

7.   Clive Hamilton and Jacques Grinevald, "Was the Anthropocene Anticipated?" *Anthropocene Review* 2, no. 1 (2015): 6–7; Marx and Engels, *Collected Works*, vol. 25, 461.

8.   Karl Marx, *Grundrisse* (London: Penguin, 1973), 162; Marx, *Early Writings*, 389–90.

9.   Marx and Engels, *Collected Works*, vol. 30, 54–66.

10.   Marx and Engels, *Collected Works*, vol. 5, 28.

11.   Corrina Lotz, "Review of John Bellamy Foster's *The Return of Nature*," *Marx and Philosophy*, December 16, 2020; Marx and Engels, *Collected Works*, vol. 25, 123; Evald Ilyenkov, *Intelligent Materialism* (Chicago: Haymarket, 2018), 27; Immanuel Kant, *Critique of Pure Reason* (Cambridge: Cambridge University Press, 1997), 304.

12.   Marx and Engels, *Collected Works*, vol. 1, 30, 102, 407–9; Benjamin Farrington, *The Faith of Epicurus* (London: Weidenfeld and Nicolson, 1967).

13. William Leiss, *The Domination of Nature* (Boston: Beacon, 1974).

14. Marx and Engels, *Collected Works*, vol. 25, 460-61.

15. Karl Marx and Frederick Engels, *The Communist Manifesto* (New York: Monthly Review Press, 1964), 2.

16. Joseph Dietzgen, "Excursions of a Socialist in the Domain of Philosophy," in Joseph Dietzgen, *Philosophical Essays* (1887; repr., Chicago: Charles H. Kerr, 1912), 293; Georgi Plekhanov, *Selected Philosophical Works*, vol. 1 (Moscow: Progress Publishers, 1974), 421.

17. V. I. Lenin, "On the Significance of Militant Materialism," in Yehoshua Yakhot, *The Suppression of Philosophy in the USSR* (Oak Park, MI: Mehring, 2012), 233-40.

18. Marx and Engels, *Collected Works*, vol. 25, 110-32, 492-502, 606-8.

19. Marx and Engels, *Collected Works*, vol. 25, 117; Marx, *Capital*, vol. 1, 443.

20. Marx and Engels, *Collected Works*, vol. 25, 313; István Mészáros, *Marx's Theory of Alienation* (London: Merlin, 1975), 12.

21. V. I. Lenin, *Collected Works*, vol. 38 (Moscow: Progress Publishers, 1961), 227-31.

22. Lenin, *Collected Works*, vol. 38, 226; Mikhail Shirokov, *A Textbook on Marxist Philosophy*, ed. John Lewis (London: Left Book Club, 1937), 364-68. On the narrow interpretation of Lenin's dialectics as limited in comparison to Engels, see Z. A. Jordan, *The Evolution of Dialectical Materialism* (London: Macmillan, 1967), 226-27.

23. Yakhot, *The Suppression of Philosophy in the USSR*, 21-41.

24. Bukharin's *Historical Materialism* was based on a mechanistic theory of equilibrium. He subsequently attempted to develop a dialectical approach to philosophy and science, in many ways transcending the debates of his time. His last effort of this kind, his *Philosophical Arabesques*, which engaged with ecological conceptions, was written in 1937 in prison prior to his execution in 1938, with the manuscript long remaining in Stalin's safe and only being released to Stephen Cohen under Mikhail Gorbachev. See Nikolai Bukharin, *Philosophical Arabesques* (New York: Monthly Review Press, 2005).

25. Alex Levant, "Evald Ilyenkov and Creative Soviet Marxism," in *Dialectics of the Ideal: Evald Ilyenkov and Creative Soviet Marxism*, ed. Alex Levant and Vesa Oittinen (Chicago: Haymarket, 2014), 12-13.

26. David Bakhurst, *Consciousness and Revolution in Soviet Philosophy: From the Bolsheviks to Evald Ilyenkov* (Cambridge: Cambridge University Press, 1991), 34-41; Yakhot, *The Suppression of Soviet Philosophy in the USSR*, 22-26.

27. Marx and Engels, *Collected Works*, vol. 25, 527.

28. Yakhot, *The Suppression of Philosophy in the USSR*, 29-30.

29. William Seager, "A Brief History of the Philosophical Problem of
     Consciousness," in *The Cambridge Handbook of Consciousness*,
     ed. Philip David Zelazo, Morris Moscovitch, and Evan Thompson
     (Cambridge: Cambridge University Press, 2007), 23, 27. See also
     Georgi Plekhanov, "Marx," in *Essays on the History of Materialism*:
     Marx,   https://www.marxists.org/archive/plekhanov/1893/essays/3-
     marx.htm.

30. Bakhurst, *Consciousness and Revolution in Soviet Philosophy*, 45.

31. Yakhot, *The Suppression of Soviet Philosophy in the USSR*, 43–76;
     Bakhurst, *Consciousness and Revolution in Soviet Philosophy*, 47–51;
     George Kline, introduction to *Spinoza in Soviet Philosophy*, ed. George
     Kline (London: Routledge, 1952), 15–18; Helena Sheehan, *Marxism
     and the Philosophy of Science* (Atlantic Highlands: Humanities Press,
     1985), 191–96; I. I. Rubin, *Essays in Marx's Theory of Value* (Delhi:
     Aakar, 2008). It is worth noting that Georg Lukács, who was in the
     Soviet Union in 1930 working under David Riazanov, was not very
     sympathetic to the Deborinists at the time, considering some of the
     criticisms of them to be correct. Georg Lukács, "Interview: Lukács and
     His Work," *New Left Review* 68 (July–August 1971): 57.

32. Joseph Stalin, "Dialectical and Historical Materialism," in *History of
     the Communist Party of the Soviet Union—Bolshevik: Short Course*, by
     the Communist Party of the USSR (Moscow: Foreign Languages Press,
     1951), 165–206.

33. Z. A. Jordan, *The Evolution of Dialectical Materialism* (London:
     Macmillan, 1967), 252.

34. Mario Livio, "Did Galileo Truly Say 'And Yet It Moves'?," *Scientific
     American* (blog), May 6, 2020, https://blogs.scientificamerican.com/
     observations/did-galileo-truly-say-and-yet-it-moves-a-modern-
     detective-story/.

35. Karl Jacoby, "Western Marxism," in *A Dictionary of Marxist Thought*,
     ed. Tom Bottomore (Oxford: Blackwell, 1983), 523–26; John Bellamy
     Foster, *The Return of Nature* (New York: Monthly Review Press, 2020),
     16–21.

36. Herbert Marcuse, *Soviet Marxism* (New York: Columbia University
     Press, 1958), 143–45; Theodor Adorno, *Negative Dialectics* (New York:
     Continuum, 1973), 355; Lucio Colletti, *Marxism and Hegel* (London:
     Verso, 1973).

37. Adorno, *Negative Dialectics*, xix; Robert Lanning, *In the Hotel Abyss:
     An Hegelian-Marxist Critique of Adorno* (Leiden: Brill, 2014), 174. The
     contradictions and limitations of an exclusively idealist conception of
     dialectics "does not cardinally change," Ilyenkov writes, "if the emphasis
     is made on the 'negative,' while 'successes and achievements' are

ignored as it is done today by the distant descendants of Hegel such as Adorno or Marcuse. Such change of emphasis does not make dialectics more materialist. Dialectics here begins to look more like the trickery of Mephistopheles, like the diabolical toolbox for the destruction of all human hopes." Ilyenkov, *Intelligent Materialism*, 50.

38.   Ironically, the passage in Marx most often cited in defense of this interpretation ended not with the domination of nature as if a foreign enemy, but rather with the rational regulation of the social metabolism between humanity and nature by the associated producers, in line with the conservation of their energies and the development of human capacities: a model of sustainable human development. Karl Marx, *Capital*, vol. 3 (London: Penguin 1981), 959.

39.   Adorno, *Negative Dialectics*, 244, 355; Max Horkheimer and Theodor W. Adorno, *Dialectic of Enlightenment* (New York: Continuum, 1944), 254; Alfred Schmidt, *The Concept of Nature in Marx* (London: New Left Books, 1971), 156; John Bellamy Foster and Brett Clark, *The Robbery of Nature* (New York: Monthly Review Press, 2020), 196.

40.   Alfred Schmidt, *The Concept of Nature in Marx* (London: Verso, 1971), 164–66, 175–76, 195. Schmidt's reversal was a direct response to the famous debate in France between Jean Hippolyte and Jean-Paul Sartre, as critics of the dialectics of nature, and Roger Garaudy and Jean-Pierre Vigier as its defenders. Schmidt clearly lined up with Hippolyte and Sartre, distancing himself from his earlier professed views.

41.   See Sebastiano Timpanaro, *On Materialism* (London: Verso, 1975).

42.   Perry Anderson, *Considerations on Western Marxism* (London: Verso, 1976), 59.

43.   Perry Anderson, *In the Tracks of Historical Materialism* (London: Verso, 1983), 83.

44.   Roy Bhaskar, *Reclaiming Reality* (London: Routledge, 2011), 122.

45.   Frederick Engels, *Ludwig Feuerbach and the Outcome of Classical German Philosophy* (New York: International Publishers, 1941), 59.

46.   N. I. Bukharin et al., *Science at the Crossroads* (London: Frank Cass and Co., 1971), 7; Foster, *The Return of Nature*, 358–73; Sheehan, *Marxism and the Philosophy of Science*, 206–9.

47.   B. Zavadovsky, "The 'Physical' and the 'Biological' in the Process of Organic Evolution," in *Science at the Crossroads*, 75–76. Translation follows Needham's version, which substitutes *different* for *varied*. Joseph Needham, *Time: The Refreshing River* (London: George Allen and Unwin, 1943), 243–44; Joseph Needham, *Order and Life* (Cambridge, MA: MIT Press, 1968), 45–46; Richard Levins and Richard Lewontin, *The Dialectical Biologist* (Cambridge, MA: Harvard University Press, 1985), 180.

48. Needham, *Order and Life*, 44–48.

49. Joseph Needham, foreword to Marcel Prenant, *Biology and Marxism* (New York: International Publishers, 1943), v.

50. Foster, *The Return of Nature*, 24–72.

51. Peter Ayres, *Shaping Ecology: The Life of Arthur Tansley* (Oxford: Wiley-Blackwell, 2012), 43.

52. Foster, *The Return of Nature*, 300–357.

53. Foster, *The Return of Nature*, 337–39, 350–51, 390, 475, 367–412.

54. Foster, *The Return of Nature*, 417–56, 526–29; J. D. Bernal, "Dialectical Materialism," in Farrington, *The Faith of Epicurus*; Jack Lindsay, *Marxism and Contemporary Science* (London: Dennis Dobson, 1949).

55. M. Shirokov, *A Textbook of Marxist Philosophy*, ed. John Lewis (London: Left Book Club, 1937).

56. Needham, *Time*, 242.

57. Shirokov, *A Textbook of Marxist Philosophy*, 341, emphasis added to the word *emergence*, all other emphases in original. The sharp difference between the 1931 Shirokov text and the official view propounded by Stalin's 1938 "Dialectical and Historical Materialism" is evident in the fact that Part IV of the former is devoted to "The Negation of the Negation," which is entirely excluded in the latter.

58. Shirokov, *A Textbook of Marxist Philosophy*, 137, 328. On Epicureanism and emergence, see A. A. Long, *From Epicurus to Epictetus* (Oxford: Oxford University Press, 2006), 155–77; A. A. Long, "Evolution vs. Intelligent Design in Classical Antiquity," Berkeley Townsend Center, November 2006; John Bellamy Foster, Brett Clark, and Richard York, *Critique of Intelligent Design* (New York: Monthly Review Press, 2008), 49–64.

59. Bakhurst, *Consciousness and Revolution in Soviet Philosophy*, 17–22, 236–43.

60. Bakhurst, *Consciousness and Revolution in Soviet Philosophy*, 111–116, 236–43.

61. Evald Ilyenkov, *Dialectics of the Ideal* (Chicago: Haymarket, 2014), 78.

62. Andrey Maidansky, interview by Vesa Oittinen, "Evald Ilyenkov and Soviet Philosophy," *Monthly Review* 71, no. 8 (January 2020): 16.

63. Foster, *Capitalism in the Anthropocene*, 316–23; V. N. Sukachev and N. Dylis, *Fundamentals of Forest Biogeocoenology* (London: Oliver and Boyd, 1964); V. N. Sukachev, "Relationship of Biogeocoenosis, Ecosystem, and Facies," *Soviet Soil Scientist* 6 (1960): 580–81; Levins and Lewontin, *The Dialectical Biologist*, 184.

64. Theodosius Dobzhansky, 1949 foreword to *Factors of Evolution: The Theory of Stabilizing Selection*, by I. I. Schmalhausen (Chicago: University of Chicago Press, 1949, 1986), xv–xvii.

65. David B. Wade, 1986 foreword to *Factors of Evolution*, v–xii; Lewontin

and Levins, *Biology Under the Influence*, 75–80. The term *triple helix* is taken from Richard Lewontin, *The Triple Helix: Gene, Organism and Environment* (Cambridge, MA: Harvard University Press, 2000).

66. Schmalhausen, *Factors of Evolution*, xix; Marx and Engels, *Collected Works*, vol. 25, 492.

67. Lewontin and Levins, *Biology Under the Influence*, 77; Derek Turner and Joyce C. Havstad, "The Philosophy of Macroevolution," *Stanford Encyclopedia of Philosophy* (2019), https://plato.stanford.edu/entries/macroevolution/; Levins and Lewontin, *The Dialectical Biologist*, 169.

68. Lewontin and Levins, *The Dialectical Biologist*, 187.

69. Georgy S. Levit, Uwe Hossfeld, and Lennart Olsson, "From the 'Modern Synthesis' to Cybernetics: Ivan Ivanovich Schmalhausen (1884–1963) and His Research Program for a Synthesis of Evolutionary and Developmental Biology," *Journal of Experimental Zoology* 306B (2005): 89–106; Foster, *Capitalism and the Anthropocene*, 323–24.

70. A. D. Ursul, ed., *Philosophy and the Ecological Problems of Civilisation* (Moscow: Progress Publishers, 1983); Foster, *Capitalism in the Anthropocene* (New York: Monthly Review Press, 2022), 331–32, 449–51.

71. Georg Lukács, *History and Class Consciousness* (London: Pluto), 24. It became customary in Western Marxist thought to refer to Lukács's footnote as a "critique." But even considering the common watering down of the notion of critique, it could hardly be said that a critique of Engels on the dialectics of nature could be carried out, even by Lukács, in what in English comes to a mere 110 words.

72. Lukács, *History and Class Consciousness*, 207; Marx and Engels, *Collected Works*, vol. 25, 492.

73. Georg Lukács, *A Defense of History and Class Consciousness: Tailism and the Dialectic* (London: Verso, 2000), 102–7; Foster, *The Return of Nature*, 16–20.

74. Lukács, *History and Class Consciousness*, xvii; Lukács, "Interview: Lukács and His Work," 56–57. Riazanov was purged from his position later in 1931 and executed in 1938.

75. Georg Lukács, *The Ontology of Social Being 2: Marx's Basic Ontological Principles* (London: Merlin, 1978), 95; George Lukács, *The Ontology of Social Labour 3: Labour* (London: Merlin, 1980).

76. Henri Lefebvre, *Dialectical Materialism* (London: Jonathan Cape, 1968), 13–19, 142.

77. Henri Lefebvre, *Marxist Thought and the City* (Minneapolis: University of Minnesota Press, 2016), 121–22, 140; Marx, *Capital*, vol. 1, 637–38; John Bellamy Foster, Brian M. Napoletano, Brett Clark, and Pedro S. Urquijo, "Henri Lefebvre's Marxian Ecological Critique," *Environmental Sociology* 6, no. 1 (2019): 31–41.

78. Jean-Pierre Vigier, "Dialectics and Natural Science," in *Existentialism vs. Marxism*, ed. George Novack (New York: Dell, 1966), 243–57. Vigier made a point in his text of criticizing Stalin's "Dialectical and Historical Materialism" as "dogmatic and mechanistic," 151.

79. Carles Soriano, "Epistemological Limitations of Earth System Science to Confront the Anthropocene Crisis," *Anthropocene Review* 9, no. 1 (2020): 112, 122.

80. Goethe and Hegel quoted in Johann Peter Eckermann, *Conversations with Goethe* (London: Penguin, 2022), 559–60.

81. Joseph Needham, *Within Four Seas: The Dialogue of East and West* (Toronto: University of Toronto Press, 1969), 27, 97.

82. Richard Levins, "Touch Red," in *Red Diapers: Growing Up in the Communist Left*, ed. Judy Kaplan and Linn Shapiro (Urbana: University of Illinois Press, 1998), 264; Lewontin and Levins, *Biology Under the Influence*, 366–67.

83. Richard Levins, "Science of Our Own: Marxism and Nature," *Monthly Review* 38, no. 3 (July–August 1986): 5.

84. Levins and Lewontin, *The Dialectical Biologist*, 279; Lewontin and Levins, *Biology Under the Influence*.

85. Stephen Jay Gould, *The Hedgehog, the Fox, and the Magister's Pox* (New York: Harmony, 2003) 201–3; Richard York and Brett Clark, *The Science and Humanism of Stephen Jay Gould* (New York: Monthly Review Press, 2011), 95–96.

86. Stephen Jay Gould interviewed in Wim Kayzer, *A Glorious Accident* (New York: W. H. Freeman, 1997), 83, 99–100, 104.

87. John Vandermeer and Ivette Perfecto, *Ecological Complexity and Agroecology* (London: Routledge, 2018); John Vandermeer, "Ecology on the Heels of the Darwinian Revolution: Historical Reflections on the Dialectics of Ecology," in *Science with Passion and a Moral Compass: A Symposium Honoring John Vandermeer*, Publication No. 1, Ecology and Evolutionary Biology, University of Michigan, Ann Arbor, 2020; John Vandermeer, "Objects of Intellectual Interest Have Real Impacts: The Ecology (and More) of Richard Levins," in *The Truth Is the Whole: Essays in Honor of Richard Levins*, ed. Tamara Awerbuch, Maynard S. Clark, and Peter J. Taylor (Arlington, MA: Pumping Station, 2018), 1–7; Stuart A. Newman, "Marxism and the New Materialism," *Marxism and the Sciences* 1, no. 2 (Summer 2022): 1–12.

88. Mészáros, *Marx's Theory of Alienation*, 162–64.

89. István Mészáros, *Beyond Capital* (New York: Monthly Review Press, 1995), 170–77, 874–77.

90. István Mészáros, *The Necessity of Social Control* (New York: Monthly

Review Press, 2015); John Bellamy Foster, "Mészáros and Chávez: 'The Point from Which to Move the World Today,'" *Monthly Review* 74, no. 2 (June 2022): 26–31.

91. Roy Bhaskar, *Plato Etc.* (London: Verso, 1994), 251, 253.

92. Roy Bhaskar, *Dialectic: The Pulse of Freedom* (London: Verso, 1993), 150–52.

93. Roy Bhaskar, "Critical Realism in Resonance with Nordic Ecophilosophy," in *Ecophilosophy in a World of Crisis*, ed. Roy Bhaskar, Karl Georg Høyer, and Peter Næss (London: Routledge, 2012), 21–22.

94. Roy Bhaskar, *The Order of Natural Necessity*, ed. Gary Hawke (Gary Hawke, 2017), 146.

95. The two works that initiated this analysis were both published in 1999: Paul Burkett, *Marx and Nature* (Chicago: Haymarket, 1999, 2014); John Bellamy Foster, "Marx's Theory of Metabolic Rift," *American Journal of Sociology* 105, no. 2 (September 1999): 366–405.

96. The major contributions of metabolic rift theory are too numerous to enumerate here. A few key works, related especially to the dialectics of nature, include: John Bellamy Foster, *Marx's Ecology* (New York: Monthly Review Press, 2000); John Bellamy Foster, Brett Clark, and Richard York, *The Ecological Rift* (New York: Monthly Review Press, 2010); Ian Angus, *Facing the Anthropocene* (New York: Monthly Review Press, 2016); John Bellamy Foster and Paul Burkett, *Marx and the Earth* (Chicago: Haymarket, 2016); Kohei Saito, *Karl Marx's Ecosocialism* (New York: Monthly Review Press, 2017); Fred Magdoff and Chris Williams, *Creating an Ecological Society* (New York: Monthly Review Press, 2017); Stefano Longo, Rebecca Clausen, and Brett Clark, *The Tragedy of the Commodity: Oceans, Fisheries, and Aquaculture* (New Brunswick, NJ: Rutgers University Press, 2015); Carles Soriano, "Capitalocene, Anthropocene, and Other '-Cenes,'" *Monthly Review* 74, no. 6 (November 2022): 1–29; and Foster and Clark, *The Robbery of Nature*.

1. The Return of the Dialectics of Nature:
The Struggle for Freedom as Necessity

This chapter is the 2020 Deutscher Memorial Lecture, delivered each year by the recipient of the Isaac and Tamara Deutscher Memorial Prize, which was awarded in 2020 to John Bellamy Foster for *The Return of Nature: Socialism and Ecology* (Monthly Review Press, 2020). The lecture was first published in *Historical Materialism* 30, no. 2 (2022): 3–28, and published in revised form in *Monthly Review* (December 2002). It has been further revised for this book.

1. Clive Hamilton and Jacques Grinevald, "Was the Anthropocene Anticipated?," *Anthropocene Review* 2, no. 1 (2015): 59–72.

2.   Karl Marx and Frederick Engels, *Collected Works*, vol. 30 (New York: International Publishers, 1975–2004), 54–66.

3.   Georg Lukács, *History and Class Consciousness*, trans. Rodney Livingstone (London: Merlin, 1971); Roy Bhaskar, *Reclaiming Reality* (London: Routledge, 2011), 131.

4.   Lukács, *History and Class Consciousness*, 24; Martin Jay, *Marxism and Totality* (Berkeley: University of California Press, 1984), 115–18.

5.   Giambattista Vico, *The New Science*, trans. Thomas Goddard Bergin and Max Harold Fisch (Ithaca, NY: Cornell University Press, 1976), 493; John Bellamy Foster, *The Return of Nature* (New York: Monthly Review Press, 2020), 17.

6.   Lukács, *History and Class Consciousness*, 207; Marx and Engels, *Collected Works*, vol. 25, 492.

7.   Georg Lukács, *In Defense of History and Class Consciousness: Tailism and the Dialectic*, trans. Esther Leslie (London: Verso, 2000), 102–7; Georg Lukács, *The Ontology of Social Being*, vol. 3, trans. David Fernbach (London: Merlin, 1980).

8.   Jean-Paul Sartre, *Critique of Dialectical Reason*, vol. 1, trans. Alan Sheridan-Smith (London: Verso, 2004), 32.

9.   Sebastiano Timpanaro, *On Materialism*, trans. Lawrence Garner (London: Verso, 1975); Karl Jacoby, "Western Marxism," in *A Dictionary of Marxist Thought*, ed. Tom Bottomore (Oxford: Blackwell, 1983), 523–26; Lucio Colletti, *Marxism and Hegel*, trans. Lawrence Garner (London: Verso, 1973), 191–92.

10. Max Horkheimer and Theodor Adorno, *The Dialectic of Enlightenment*, trans. John Cumming (New York: Continuum, 1998), 224, 254; Alfred Schmidt, *The Concept of Nature in Marx*, trans. Ben Fowkes (London: New Left, 1971), 156; Max Horkheimer, *The Eclipse of Reason* (New York: Continuum, 2004), 123–27.

11. Herbert Marcuse, *Counter-Revolution and Revolt* (Boston: Beacon, 1972), 59–78.

12. Bhaskar, *Reclaiming Reality*, 132.

13. Foster, *The Return of Nature*, 7.

14. Bhaskar, *Reclaiming Reality*, 115.

15. Lucretius, *On the Nature of the Universe*, ed. Ronald Melville, Don Fowler, and Peta Fowler (Oxford: Oxford University Press, 1999), 93 (III: 869).

16. Anthony Arthur Long, "Evolution vs. Intelligent Design in Classical Antiquity," Townsend Center for the Humanities, 2006, available at https://townsendcenter.berkeley.edu/publications/evolution-vs-intelligent-design-classical-antiquity; Anthony Arthur Long, *From Epicurus to Epictetus* (Oxford: Oxford University Press, 2006); John

Bellamy Foster, Brett Clark, and Richard York, *Critique of Intelligent Design* (New York: Monthly Review Press, 2008), 155–77.

17. John Bellamy Foster, "Marx and the Rift in the Universal Metabolism of Nature," *Monthly Review* 65, no. 7 (December 2013): 1–19.

18. On neo-Kantianism and its consequences for dialectical and materialist philosophy, see Evald Vassilievich Ilyenkov, *Dialectical Logic*, trans. H. Campbell Creighton (Delhi: Aakar Books, 2008), 289–319; Frederick C. Beiser, *After Hegel* (Princeton: Princeton University Press, 2014); Foster, *The Return of Nature*, 264–69. In the words of Lukács, who started out as a neo-Kantian, "according to Kant's theory the world given to us is only appearance, with a transcendental unknowable thing-in-itself behind it." Georg Lukács, *Conversations with Lukács*, ed. Theo Pinkus (London: Merlin, 1974), 76.

19. Marx and Engels, *Collected Works*, vol. 45, 50, 462.

20. Karl Marx, *Letters to Kugelmann* (New York: International Publishers, 1934), 112. Marx was replying to the second edition of Friedrich Albert Lange's *On the Workers' Question* (1870).

21. Karl Marx, *Capital*, vol. 1, trans. Ben Fowkes (London: Penguin, 1976), 102.

22. Bhaskar, *Reclaiming Reality*, 120. Kai Heron, writing from a Lacanian-Hegelian perspective, has recently stated that Marxian ecology based on Marx's theory of metabolic rift is unable "to account for the contingent emergence of ourselves" as "subjects, from nature." This, however, is exactly what the theory of contingent emergence developed in classical historical materialism, which is carried forward by today's dialectical critical realism (including Marxian ecology) is, in the final analysis, all about. To call this "contemplative materialism" thus misses the point: today the issue is the formation of a revolutionary ecological subject, conceived in terms of the "transformative model of social activity," viewed as a contemporary expression of historical materialism. Kai Heron, "Dialectical Materialisms, Metabolic Rifts and the Climate Crisis," *Science and Society* 85, no. 4 (2021): 501–26; Roy Bhaskar, *Dialectic: The Pulse of Freedom* (London: Verso, 1993), 2, 152–73.

23. Georg Lukács, *The Ontology of Social Being*, vol. 2, trans. David Fernbach (London: Merlin, 1978), 6–7, 103. Writing of "the hidden nature speculation in Marx" and Marx's concept of metabolism, Alfred Schmidt observed: "Only in this way"—that is, through the mediation of human activity—"can we speak of a 'dialectic of nature.'" Schmidt's intention was to reduce the notion of the "merely objective dialectic of nature," referred to by Lukács in *History and Class Consciousness*, to the dialectics of nature and society. Alfred Schmidt, *The Concept of Nature in Marx*, trans. Ben Fowkes (London: New Left, 1971), 79.

24. See John Bellamy Foster, "The Dialectics of Nature and Marxist Ecology," in *Dialectics for the New Century*, ed. Bertell Ollman and Tony Smith (Basingstoke: Palgrave Macmillan, 2008), 50–82; Andrew Feenberg, *Lukács, Marx, and the Sources of Critical Theory* (Totowa, NJ: Rowman and Littlefield, 1981); John Bellamy Foster and Paul Burkett, *Marx and the Earth* (Chicago: Haymarket, 2016), 50–66.

25. Marx and Engels, *Collected Works*, vol. 25, 13–14, 503; Lukács, *History and Class Consciousness*, xix.

26. Marx and Engels, *Collected Works*, vol. 25, 461.

27. Marx and Engels, *Collected Works*, vol. 25, 23; Foster, *The Return of Nature*, 254.

28. Leszek Kołakowski, *Main Currents in Marxism*, trans. Paul Stephen Falla (New York: W. W. Norton, 2005), 324–25; Shlomo Avineri, *The Social and Political Thought of Karl Marx* (Cambridge: Cambridge University Press, 1968), 67, 86; Norman Levine, *Dialogue with the Dialectic* (London: George Allen and Unwin, 1984), 10–12.

29. On Hegel's complex, dialectical concept of reflection (and its relation to reflexivity and refraction), see Michael Inwood, *A Hegel Dictionary* (Oxford: Blackwell, 1992), 247–50. For the distinction between the mechanistic and Marxian conceptions of reflection, see Roger Garaudy, *Marxism in the Twentieth Century*, trans. René Hague (New York: Charles Scribner's Sons, 1970) 53–54. Lukács was to relate the origins of dialectical reflection, in the Marxian sense, directly to praxis and production (the metabolism with nature), stating: "The most primitive kind of work, such as quarrying of stones by primeval man, implies a correct reflection of the reality he is concerned with. For no purposive activity can be carried out in the absence of an image, however crude, of the practical reality involved." (Lukács, *History and Class Consciousness*, xxv.) This complex, dialectical view of the concept of "reflection" had roots that went back to Immanuel Kant, who wrote of the "Amphiboly of the Conceptions of Reflection." See Immanuel Kant, *Critique of Pure Reason* (London: J. M. Dent, 1934), 191–208.

30. See Marx and Engels, *Collected Works*, vol. 25, 43, 493–94; G. W. F. Hegel, *The Science of Logic*, trans. A. V. Miller (New York: Humanities, 1969), 399, 405–12, 490–91, 536; Foster, *The Return of Nature*, 244–51; George Lukács, *The Young Hegel*, trans. Rodney Livingstone (Cambridge, MA: MIT Press, 1975), 280; Georg Lukács, *The Ontology of Social Being*, vol. 1, trans. David Fernbach (London: Merlin, 1978), 74–82.

31. Marx and Engels, *Collected Works*, vol. 25, 13, 460. The notion of "biological agency" at all levels of living systems has now attained major significance in the natural sciences. See Sonia E. Sultan, Armin P. Moczek,

and Denis Walsh, "Bridging the Explanatory Gaps: What Can We Learn from a Biological Agency Perspective?" *BioEssays* 44, no. 1 (2021), https://onlinelibrary.wiley.com/doi/epdf/10.1002/bies.202100185.

32. Marx and Engels, *Collected Works*, vol. 25, 110–32, 356–61; Craig Dilworth, "Principles, Laws, Theories, and the Metaphysics of Science," *Synthese* 101, no. 2 (1994): 223–47.

33. Marx and Engels, *Collected Works*, vol. 25, 115–19, 356–61; Hyman Levy, *The Universe of Science* (London: Watts and Co., 1932), 30–32, 117, 227–28.

34. Lukács, *Conversations with Lukács*, 73–75.

35. Bertel Ollman, *The Dance of the Dialectic* (Urbana: University of Illinois Press, 2003), 17; Marx and Engels, *Collected Works*, vol. 25, 120–32; Karl Marx, *Capital*, vol. 1, 929. The notion of the negation of the negation arises out of Hegel's attempts to explain determinate negations that express continuity and change. See G. W. F. Hegel, *The Phenomenology of Spirit*, trans. A. V. Miller (Oxford: Oxford University, 1977), 51.

36. Marx and Engels, *Collected Works*, vol. 25, 313; J. D. Bernal, "Dialectical Materialism," in *Aspects of Dialectical Materialism*, ed. Hyman Levy (London: Watts and Co., 1934), 103–4; Bhaskar, *Dialectic*, 150–52, 377–78; Ernst Bloch, *The Principle of Hope*, vol. 1, trans. Neville Plaice, Stephen Plaice, and Paul Knight (Cambridge, MA: MIT Press, 1986), 9–18, 306–13; Jay, *Marxism and Totality*, 183–86. An account of the dialectic as a spiral form of development was developed by William Morris and E. Belfort Bax, probably in conjunction with Engels, in *The Manifesto of the Socialist League*. See William Morris and E. Belfort Bax, *The Manifesto of the Socialist League* (London: Socialist League Office, 1885), 11. The characterization of the dialectic as a spiral also appears in E. Belfort Bax, *The Religion of Socialism* (Freeport, NY: Books for Libraries, 1972), 2–5.

37. Bloch, *The Principle of Hope*, vol. 1, 71.

38. Kaan Kangal, "Engels's Emergentist Dialectics," *Monthly Review* 72, no. 6 (November 2020): 18–27; John Bellamy Foster, "Engels's Dialectics of Nature in the Anthropocene," *Monthly Review* 72, no. 6 (November 2020): 1–17.

39. Karl Marx, *Early Writings*, trans. Rodney Livingstone and Gregor Benton (London: Penguin, 1974), 260–61.

40. Marx and Engels, *Collected Works*, vol. 25, 459–64; Foster, *The Return of Nature*, 177–215, 273–87.

41. For a standard criticism of Engels in this respect, see Levine, *Dialogue with the Dialectic*, 8–12. For a response, see John L. Stanley, *Mainlining Marx* (New Brunswick, NJ: Transaction, 2002).

42. Benjamin Farrington, *Head and Hand in Ancient Greece* (London: Watts and Co., 1947), 11–15; Aeschylus, *The Oresteia*, trans. George Thomson (New York: Alfred A. Knopf, 2004).

43. Epicurus, *The Epicurus Reader*, trans. Brad Inwood and Lloyd P. Gerson (Indianapolis: Hackett, 1994), 42 (Diog. Laert. 10.34). Epicurus was known for his method of scientific inference as well as his epistemology. A few fragments of his writings have been preserved in the form of letters or collections of maxims. However, all of his three hundred books are lost, except for parts of *On Nature*, which have been recovered from the Herculaneum papyri. Nevertheless, we have a brief summary from Diogenes Laertius of his *Canon*, which was the first distinct epistemological work in the ancient Greek tradition. The most intact Epicurean treatment of the method of scientific inference (retrieved from the Herculaneum papyri) was Philodemus's work on method and signs. See Epicurus, *The Epicurus Reader*, 41–42; Gisela Striker, "Epistemology," in *The Oxford Handbook of Epicurus and Epicureanism*, ed. Philip Mitsis (Oxford: Oxford University Press, 2020), 43–58; Philodemus, *Philodemus: On Methods of Inference*, ed. Philip Howard De Lacey and Estelle Allen De Lacey (Philadelphia: American Philosophical Association, 1941).

44. Foster, *The Return of Nature*, 253.

45. J. D. Bernal, *World Without War* (New York: Prometheus, 1936), 1–2.

46. Marx and Engels, *Collected Works*, vol. 1, 34–107, 403–514. As the Epicurean scholar Cyril Bailey pointed out, Marx was the first figure in modern times to recognize the significance of Epicurus's swerve. Cyril Bailey, "Karl Marx on Greek Atomism," *Classical Quarterly* 22, no. 3–4 (1928): 205–6. Marx drew on a wide body of fragments in writing his dissertation (and his seven *Epicurean Notebooks*) at a time when these had not previously been collected, including one fragment recovered from the charred papyri in the Herculaneum library. Michael Heinrich, *Karl Marx and the Birth of Modern Society* (New York: Monthly Review Press, 2019), 296. On the influence of Epicurus on the British Marxists of the 1930s and '40s, see Foster, *The Return of Nature*, 369–70. Benjamin Farrington, in particular, played a major role in introducing the British Marxian scientists to Epicurus, not only through his own works, but also in facilitating the reading of Marx's doctoral dissertation by thinkers in this tradition. See Lancelot Hogben, *Lancelot Hogben, Scientific Humanist* (London: Merlin, 1998), 105; Benjamin Farrington, *Science and Politics in the Ancient World* (London: George Allen and Unwin, 1939); Benjamin Farrington, *The Faith of Epicurus* (London: Weidenfeld and Nicolson, 1967); George Thomson, *The First Philosophers* (London: Lawrence and Wishart,

1955), 311–14.

47. Joseph Needham, *Time: The Refreshing River* (London: George Allen and Unwin, 1948), 55, 124, 191.

48. See Joseph Fracchia, "Dialectical Itineraries," *History and Theory* 38, no. 2 (1991): 169–97.

49. Ray E. Lankester, *The Kingdom of Man* (New York: Henry Holt, 1911), 159–91; John Bellamy Foster, Brett Clark, and Hannah Holleman, "Capital and the Ecology of Disease," *Monthly Review* 73, no. 2 (June 2021): 1–23.

50. J. B. S. Haldane, *The Science of Life* (London: Pemberton, 1968), 6–11; J. D. Bernal, *The Origin of Life* (New York: World Publishing, 1967), 24–35; Richard Levins and Richard Lewontin, *The Dialectical Biologist* (Cambridge, MA: Harvard University Press, 1985), 277; Vladimir I. Vernadsky, *The Biosphere*, trans. David B. Langmuir (New York: Springer Verlag, 1998).

51. J. B. S. Haldane, *The Marxist Philosophy and the Sciences* (New York: Random House, 1939); Foster, *The Return of Nature*, 383–98.

52. Bernal, "Dialectical Materialism," 103–4; Henri Lefebvre, *Metaphilosophy*, trans. David Fernbach (London: Verso, 2016), 301–2.

53. Needham, *Time*, 233–72.

54. A. G. Tansley, "The Use and Abuse of Vegetational Concepts and Terms," *Ecology* 16, no. 3 (1935): 284–307; Levy, *The Universe of Science*.

55. Foster, *The Return of Nature*, 337–39.

56. J. B. S. Haldane, "Carbon Dioxide Content of Atmospheric Air," *Nature* 137 (1936): 575; Foster, *The Return of Nature*, 397, 612–13.

57. J. D. Bernal, *The Social Function of Science* (New York: Macmillan, 1939).

58. Christopher Caudwell, *Studies and Further Studies in a Dying Culture* (New York: Monthly Review Press, 1971); Foster, *The Return of Nature*, 417–56.

59. Foster, *The Return of Nature*, 489–96; Bernal, *World Without War*; Bernal, *The Origin of Life*, xvi, 176–82.

60. Foster, *The Return of Nature*, 502–26; Foster, Clark, and Holleman, "Capital and the Ecology of Disease"; Helena Sheehan, *Marxism and the Philosophy of Science* (Atlantic Highlands, NJ: Humanities, 1985).

61. Perry Anderson, "Components of the National Culture," *New Left Review* 1, no. 50 (1968): 3–57. Compare Eric Hobsbawm, *Fractured Times* (London: Little, Brown, 2013), 169–83.

62. Perry Anderson, *In the Tracks of Historical Materialism* (London: Verso, 1983), 83.

63. McLellan's *Marxism After Marx* reflected the tendency not only to condemn but also to exclude from the Marxist canon those who were seen as falling outside the narrowly defined Western Marxist tradition.

Thus, of the British Marxists up through the 1930s considered in *The Return of Nature*, including Morris, Hogben, Haldane, Bernal, Levy, Needham, Farrington, Thomson, and Caudwell, only the last is mentioned in the chapter on "British Marxism" in McLellan's work, and this was confined to a mere two sentences. We are told that "Christopher Caudwell was the only really original pre-war British Marxist"—and then only for his treatment of "literature," not his theory of art in general or his analysis of science. See David McLellan, *Marxism After Marx* (Boston: Houghton Mifflin, 1979), 30.

64.   Needham, *Time*, 14–15.

65.   Caudwell, *Studies and Further Studies in a Dying Culture*, 227.

66.   Garaudy, *Marxism in the Twentieth Century*, 61.

67.   Immanuel Kant, *Critique of Judgment*, trans. James Creed Meredith (Oxford: Oxford University Press, 1952), 50–54, 67–74, 77–86.

68.   Systems theory often overlaps with dialectics. See Richard Lewontin and Richard Levins, *Biology Under the Influence* (New York: Monthly Review Press, 2007), 101–24.

69.   Johan Rockstrom et al., "A Safe Operating Space for Humanity," *Nature* 461 (2009): 472–75; Will Steffen et al., "Planetary Boundaries," *Science* 347, no. 6223 (2015): 736–46; Richard E. Leakey and Roger Lewin, *The Sixth Extinction* (New York: Anchor, 1996).

70.   Hamilton and Grinevald, "Was the Anthropocene Anticipated?," 67.

71.   John Bellamy Foster and Brett Clark, "The Capitalinian: The First Geological Age of the Anthropocene," *Monthly Review* 73, no. 4 (September 2021): 1–16; Carles Soriano, "On the Anthropocene Formalization and the Proposal by the Anthropocene Working Group," *Geologica Acta* 18, no. 6 (2020): 1–10.

72.   Marx and Engels, *Collected Works*, vol. 25, 461; Lankester, *The Kingdom of Man*, 159–91.

73.   Marx, *Capital*, vol. 1, 637–38.

74.   Marx, *Capital*, vol. 1, 871; John Bellamy Foster and Brett Clark, *The Robbery of Nature* (New York: Monthly Review Press, 2020), 43–61. Marx strongly preferred the concept of "original expropriation" to "original accumulation," since what was at issue was expropriation, not accumulation. See Karl Marx, *Value, Price, and Profit*, in Karl Marx, *Wage-Labour and Capital/Value, Price and Profit* (New York: International Publishers, 1935), 38.

75.   Foster and Clark, *The Robbery of Nature*, 78–103.

76.   Lewontin and Levins, *Biology Under the Influence*, 103.

77.   Heraclitus, *Fragments*, trans. Brooks Haxton (London: Penguin, 2001), 15.

78.   Bhaskar, *Dialectic*, 115–16; Thomson, *The First Philosophers*, 271–95.

79.   Marx and Engels, *Collected Works*, vol. 5, 141. See also Walter Baier,

Eric Canepa, and Haris Golemis, eds., *Capitalism's Deadly Threat* (London: Merlin, 2021).

80. "The real 'Golden Age' of historical anthropology cannot be conceived of without the just as real 'Golden Age' of a new humanist cosmology." Bloch, *The Principle of Hope*, 138.

2. Marx's Critique of Enlightenment Humanism: A Revolutionary Ecological Perspective

This chapter is based on the closing keynote speech for the international conference on Marx and the Critique of Humanism at the School of Arts and Humanities, University of Lisbon, Portugal, November 8, 2022. It first appeared in the January 2023 issue of *Monthly Review* and has been revised here.

1. Louis Althusser, *For Marx* (New York: Vintage, 1969), 32–39, 221–47. A more compelling and focused interpretation of Marx's "epistemological break" than the one offered by Althusser is provided by Joseph Fracchia in his monumental work *Bodies and Artefacts*. Fracchia sees Marx's emphasis on human corporeal organization in *The German Ideology* as the starting point of his historical materialism. Unlike Althusser's interpretation, Fracchia does not argue that Marx left his humanism behind, but rather he shifted the focus of his materialism to human corporeal existence. See Joseph Fracchia, *Bodies and Artefacts* (Leiden: Brill, 2022), vol. 1, 1–6; vol. 2, 1209–17. This shift, however, was already prefigured in Marx's *Economic and Philosophical Manuscripts*, making it less of a break.

2. Karl Marx, *Dispatches for the* New York Tribune (London: Penguin, 2007), 224.

3. See Ato Sekyi-Otu, *Fanon's Dialectic of Experience* (Cambridge, MA: Harvard University Press, 1996), 16, 20–21, 31, 46, 100, 179, 181, 315; Frantz Fanon, *Black Skin, White Masks* (London: Pluto, 1967), 1; A. James Arnold, Introduction to Aimé Césaire, *The Original 1939 Notebook of a Return to the Native Land,* trans. and ed. A. James Arnold and Clayton Eshleman (Middletown, CT: Wesleyan University Press, 2013), xi–xx; W. E. B. Du Bois, *John Brown* (New York: International Publishers, 2019), 297.

4. Karl Marx and Frederick Engels, *Collected Works*, vol. 30 (New York: International Publishers, 1975–2004), 63.

5. Of course, the irrationalist impulse is not simply to be defined by its opposition to the philosophy of praxis, but rather has a deeper historical significance associated with the imperialist stage of capitalism (and today's late imperialism). Herbert Aptheker wrote that in this context irrationalism can be depicted as a continuum consisting of "the

eclipse of reason, the denial of science, the repudiation of causation. The normal result is cynicism; the abnormal is sadism. The finale is fascism." Herbert Aptheker, "Imperialism and Irrationalism," *Telos* 4 (1969): 168–75.

6. G. W. F. Hegel, *Hegel's Phenomenology of Spirit* (Oxford: Oxford University Press, 1977), 351–53; Marx and Engels, *Collected Works*, vol. 4, 131–32; Marx and Engels, *Collected Works*, vol. 25, 461.

7. Marx and Engels, *Collected Works*, vol. 3, 419; Frederick Engels, *Ludwig Feuerbach and the Outcome of Classical German Philosophy* (New York: International Publishers, 1941), 17, 21.

8. Marx and Engels, *Collected Works*, vol. 3, 162–67; Marx and Engels, *Collected Works*, vol. 5, 410; Karl Marx, *Early Writings* (London: Penguin, 1974), 244; István Mészáros, *Marx's Theory of Alienation* (London: Pluto, 1975), 220–21.

9. Marx and Engels, *Collected Works*, vol. 5, 210. Marx was quoting a real statement by a Yankee slave owner as per the original English.

10. Marx and Engels, *Collected Works*, vol. 19, 209–12.

11. Marx, *Early Writings*, 281, 348, 395; Marx and Engels, *Collected Works*, vol. 4, 7. The term *pseudo-humanism* was used by Jenny Marx in 1846 in a letter to Karl, where she also referred in this connection to "besottedness with progress." This clearly reflected Karl Marx's views as well. Marx and Engels, *Collected Works*, vol. 38, 532.

12. Marx, *Early Writings*, 349–50, 395.

13. Marx, *Early Writings*, 328, 395.

14. Marx and Engels, *Collected Works*, vol. 4, 7, 131. Although Althusser argued that Marx's early humanism was outside of material science and pre-Marxian, he had a harder time dismissing the concept of "real humanism," which Marx used to refer to his transcendence of bourgeois humanism in the form of a materialist analysis focusing on the real, living corporeal human being. See Althusser, *For Marx*, 242–47.

15. Marx and Engels, *Collected Works*, vol. 4, 125, emphases in oriignal.

16. Marx and Engels, *Collected Works*, vol. 4, 126–31.

17. Marx, *Early Writings*, 398–99; Marx and Engels, *Collected Works*, vol. 4, 85, 399–400.

18. Marx and Engels, *Collected Works*, vol. 5, 197.

19. Marx and Engels, *Collected Works*, vol. 5, 7–8.

20. Karl Marx, *The Poverty of Philosophy* (New York: International Publishers 1973), 147.

21. Karl Marx, "The Fetishism of the Commodity and Its Secret," chap. 1 in *Capital*, vol. 1 (London: Penguin, 1976), 172; Engels, *Ludwig Feuerbach*, 41; István Mészáros, *Marx's Theory of Alienation*, 221.

22. Marx, *Early Writings*, 389–90.

23. Marx and Engels, *Collected Works*, vol. 5, 31. The translation here follows Fracchia, *Bodies and Artefacts*, 1–2, emphases in the original.

24. Although *The German Ideology* was co-authored by Marx and Engels, this fundamental discovery was attributed by Engels to Marx. See Frederick Engels, "The Funeral of Karl Marx," in *Karl Marx Remembered*, ed. Philip S. Foner (San Francisco: Synthesis Publications, 1983), 39.

25. Marx, *Early Writings*, 399–400.

26. Marx, *The Poverty of Philosophy*, 98–99, 115, 119–20, 132–44, 155–56, 184; Pierre-Joseph Proudhon, *System of Economical Contradictions* (New York: Arno, 1972), 96–101, 117–18, 126–28, 168, 174–75; John Bellamy Foster, *Marx's Ecology* (New York: Monthly Review Press, 2000), 126–33.

27. In ancient philosophy and during the Enlightenment, Prometheus stood mainly for enlightenment itself, as the bringer of fire to light the darkness. This led to Marx's celebration of Epicurus as the "true radical Enlightener of antiquity," identifying him directly with Prometheus. Later, beginning in the nineteenth century, as represented by Proudhon and Mary Shelley, Prometheanism came to be associated with extreme productivism and extreme industrialism. It was this that Marx took on in his critique of Proudhon. See Marx and Engels, *Collected Works*, vol. 5, 141; John Bellamy Foster, "Marx and the Environment," in *In Defense of History*, ed. Ellen Meiksins Wood and John Bellamy Foster (New York: Monthly Review Press, 1997), 149–62; Walt Sheasby, "Anti-Prometheus, Post-Marx," *Organization and Environment* 12, no. 1 (1999): 5–44.

28. Mészáros, *Marx's Theory of Alienation*, 162–65.

29. John Bellamy Foster, Brett Clark, and Richard York, *Critique of Intelligent Design* (New York: Monthly Review Press, 2008), 86–90; John Bellamy Foster and Brett Clark, *The Robbery of Nature* (New York: Monthly Review Press, 2000), 132–38.

30. Marx, *Capital*, vol. 1, 286.

31. Marx, *Texts on Method*, 190.

32. Marx, *Capital*, vol. 1, 512.

33. Marx, *Early Writings*, 239.

34. Karl Marx, Marx-Engels Archives, International Institute of Social History, Sign. B, 106, 336, quoted in Kohei Saito, "Why Ecosocialism Needs Marx," *Monthly Review* 68, no. 6 (November 2016): 62. Translation altered slightly.

35. See John Bellamy Foster, *Capitalism in the Anthropocene* (New York: Monthly Review Press, 2022), 41–61.

36. Marx, *Capital*, vol. 1, 637.

37. See, for example, John Bellamy Foster, Brett Clark, and Richard York,

The Ecological Rift (New York: Monthly Review Press, 2010); Stefano B. Longo, Rebecca Clausen, and Brett Clark, *The Tragedy of the Commodity* (New Brunswick, NJ: Rutgers University Press, 2015).

38. Kate Soper, "The Humanism in Posthumanism," *Comparative Critical Studies* 9, no. 3 (2012): 368–69.

39. Marx and Engels, *Collected Works*, vol. 5, 8.

40. Marx, *Early Writings*, 398; Kyla Wazana Tompkins, "On the Limits and Promise of New Materialist Philosophy," *Emergent Critical Analytics for Alternative Humanities* 5, no. 1 (2016). Although assemblages have been recognized as crucial to material forms since ancient times, the emphasis on "interlaced assemblages" that deny any hierarchical relations whatsoever in the material world and all forms of emergence or integrative levels, in opposition to material science and dialectics, is peculiar to posthumanism and the new vitalistic materialism.

41. Marx and Engels, *Collected Works*, vol. 1, 413; Foster, *Marx's Ecology*, 52–53. Although Bennett admires Epicurus, she forgets that he was, as Marx explained, the main thinker to insist on the need for "disillusionment" in antiquity, and who is thus at odds with her own criticisms of "demystification" as an approach that leads back to the human. Marx and Engels, *Collected Works*, vol. 5, 141; Foster, *Marx's Ecology*, 2–6, 33–39; Jane Bennett, *Vibrant Matter: A Political Ecology of Things* (Durham, NC: Duke University Press, 2010), 62. Nothing is more absurd than treating Epicurus as a vitalist thinker.

42. Simone Bignall and Rosi Braidotti, "Posthuman Systems," in *Posthuman Ecologies: Complexity and Process After Deleuze*, ed. Braidotti and Bignall (New York: Rowman and Littlefield, 2019), 1; Arie Ben Arie, "The New Materialist Approach to Art and Aesthetics," Well of Faith (blog), July 29, 2021, https://www.well-of-faith.com/post/en/the-%E2%80%9Cnew-materialist%E2%80%9D-approach-to-art-and-aesthetics.

43. Graham Harman, *Bruno Latour: Reassembling the Political* (London: Pluto, 2014), 14.

44. On the critique of vitalism, see John Bellamy Foster, *The Return of Nature* (New York: Monthly Review Press, 2020), 407–9.

45. Timothy Morton, *Humankind: Solidarity with Nonhuman People* (London: Verso, 2019), 155; Timothy Morton, *Ecology without Nature* (Cambridge, MA: Harvard University Press, 2007), 187; Timothy Morton, *Hyperobjects: Philosophy and Ecology After the End of the World* (Minneapolis: University of Minnesota Press, 2013), 1, 41.

46. On the inherent conflict between Marxian ecology and vitalistic new materialism, see SunYoung Ahn, "Magic, Necromancy, and the Posthuman Turn," *Monthly Review* 73, no. 9 (February 2022): 26–37.

47. Timothy Morton, *Dark Ecology: For a Logic of Future Coexistence* (New York: Columbia University Press, 2016), 26, 166; Morton, *Humankind*, 39, 80, 177. On Marx's general ecological perspective, see Foster, *Marx's Ecology*. Bennett also excludes Marx's materialist conception of nature and his ecological materialism, claiming that Marx's materialism was simply a matter of "economic structures and exchanges." Bennett, *Vibrant Matter*, xvi.

48. Morton, *Humankind*, 41-42. See, by way of comparison, Foster and Clark, *The Robbery of Nature*, 130-51.

49. Marx and Engels, *Collected Works*, vol. 3, 419.

50. Morton, *Humankind*, 27-39, 54-56, 70-71, 97-99; Morton, *Dark Ecology*, 24.

51. Morton, *Humankind*, 33, 71, 97. Morton goes so far as to censure Engels for his critique of the occult in *The Dialectics of Nature*, on the grounds that Engels had closed off the paranormal. See Morton, *Humankind*, 166; Marx and Engels, *Collected Works*, vol. 25, 345-55.

52. Marx, *Capital*, vol. 1, 163-77. For the origins and development of Marx's critique of fetishism, see Kaan Kangal, "Young Marx on Fetishism, Sexuality, and Religion," *Monthly Review* 74, no. 5 (October 2022): 46-57; Georg Lukács, *Marxism and Human Liberation* (New York: Dell, 1973), 251.

53. Bruno Latour, "Why Has Critique Run Out of Steam?: Matters of Fact and Matters of Concern," *Critical Inquiry* 30 (2014): 225-48; Bruno Latour, *On the Modern Cult of the Factish Gods* (Durham, NC: Duke University Press, 2010), 9-12; Bruno Latour, *The Politics of Nature* (Cambridge, MA: Harvard University Press, 2004), 20; Harman, *Bruno Latour*, 14, 18, 81, 90, 112-17; Andrew B. Kipnis, "Agency Between Humanism and Posthumanism: Latour and His Opponents," *HAU: Journal of Ethnographic Theory* 5, no. 2 (2015).

54. Bruno Latour, "Love Your Monsters," Breakthrough Institute, February 14, 2012; Bruno Latour, *Down to Earth: Politics in the New Climatic Regime* (Cambridge: Polity, 2018); Bruno Latour, *Facing Gaia* (Cambridge: Polity, 2017).

55. Bennett, *Vibrant Matter*, xiv-xv, 1-4; Marx and Engels, *Collected Works*, vol. 25, 560; Baruch Spinoza, *Ethics* (London: Penguin, 1996), 75, (III, prop. 6); "'From Baruch Spinoza's Letter to G. H. Schuller' (1674)," Explanantia (blog), October 3, 2018, https://explanantia.wordpress.com/2018/10/03/from-baruch-spinozas-letter-to-g-h-schuller-1674/; Richard Manning, "Spinoza's Physical Theory," *Stanford Encyclopedia of Philosophy*, April 24, 2021, https://plato.stanford.edu/entries/spinoza-physics/; Bennett claims not to adhere to strict vitalism. Nevertheless, she relies on the notion of "thing power," Henri Bergson's "critical

vitalism," and metaphysical concepts such as the innate "force" of things (based on a questionable interpretation of Spinoza's concept of *conatus*). See Bennett, *Vibrant Matter*, 63–65.

56. Morton, *Humankind*, 55, 61–63, 166–71.

57. Christopher N. Gamble, Joshua S. Hanan, and Thomas Nail, "What Is New Materialism?" *Angelekai: Journal of the Theoretical Humanities* 24, no. 6 (2019): 119.

58. Morton, *Humankind*, 61, 169–70.

59. Morton, *Humankind*, 56–57. Morton is well aware that from a Marxian dialectical standpoint posthumanism makes no sense: "The logic," of such criticisms, he writes, "goes like this: hypnotized by capitalism, the spiritualist's sin is flat ontology, spirit has become a 'thing among things.'" He is equally aware that the notion that all things have innate agency (and even mind) can be criticized as an attempt to out-reify capitalism itself. Thus, he goes out of his way to insist that "OOO [object-oriented ontology] definitely isn't a manifestation of commodity fetishism." But he bases this on the spurious grounds that commodity fetishism consists not of downplaying the human relations behind objective appearances so much as its opposite: not fully animating the world of things. The argument thus inverts the approach to fetishism introduced by Marx. This follows Latour's notion that the critique of fetishism can be turned any way one wants. See Morton, *Humankind*, 59, 169; Latour, "Why Has Critique Run Out of Steam?".

60. The multiple dangers posed by forms of irrationalist "post" theories, with their flat ontologies, can be seen in their abandonment of revolutionary anticapitalism and anticolonialism. This is powerfully expressed by Oliver W. Baker, in "'Words Are Things': The Settler Colonial Politics of Post-Humanist Materialism in Cormac McCarthy's *Blood Meridian*," *Mediations* 30, no. 1 (2016): 1–24. Similar issues have arisen in relation to Afropessimism, which has been criticized for its flat ontology and regressive erasure of anticolonialism. See Kevin Ochieng Okoth, "The Flatness of Blackness: Afro-Pessimism and the Erasure of Anti-Colonial Thought," *Salvage* 7 (2020); Ato Sekyi-Otu, "Con-Texts of Critique," in *Partisan Universalism: Essays in Honour of Ato Sekyi-Otu* (Quebec: Daraja, 2021), 236–51. Leading posthumanist theorist Rosi Braidotti declares: "What we have learned since 1968 is that capitalism never fails." Given this assumed permanency of capitalism, the message of her new "vital materialism" for feminist, antiracist, and other movements is confined to finding ways to "disassociate and put distance between ourselves" and the "mistaken consumer models," male violence, and white supremacy, which constitute the worst aspects of contemporary capitalism. See Rosi Braidotti, interview by

Iu Andrés, "What Is Necessary Is a Radical Transformation, Following the Bases of Feminism, Anti-Racism, and Anti-Fascism," *Cultural Research and Innovation*, April 2, 2019, https://lab.cccb.org/en/rosi-braidotti-what-is-necessary-is-a-radical-transformation-following-the-bases-of-feminism-anti-racism-and-anti-fascism/; Rosi Braidotti, "A Theoretical Framework for the Critical Posthumanities," *Theory, Culture, and Society* 36, no. 6 (November 2019): 31–61.

61.   Morton, *Humankind*, 6, 30–34, 59, emphasis added; Marx, *Capital*, vol. 1, 311. It is important to recognize that Morton is *not* claiming here that Marx ignored entropy (which could not possibly be claimed on the basis of this quote or in relation to any other statement by Marx or Engels), but rather that Marx simply allowed the coal to go "away" in the sense of ignoring its agency as a "nonhuman person." On Marx, Engels, and thermodynamics, see John Bellamy Foster and Paul Burkett, *Marx and the Earth* (Chicago: Haymarket, 2016), 147–64.

62.   Bennett, in *Vibrant Matter*, goes a step further than Morton and tries animistically to ascribe political agency to all "vibrant matter" (94–109).

63.   Soper, "The Humanism in Posthumanism," 366.

64.   The irrationalism of our time has much in common with the irrationalism of the early twentieth century. It must be combated just as thoroughly. See Georg Lukács, *The Destruction of Reason* (London: Merlin, 1980).

65.   Marx, *Early Writings*, 349; Karl Marx, *Capital*, vol. 3 (London: Penguin, 1981), 949, 959.

3. Engels and the Second Foundation of Marxism

This chapter is based on the Engels Memorial Lecture presented to the Marx Memorial Library in London, England, on November 30, 2022. It was revised in the June 2023 issue of *Monthly Review*, including the addition of a lengthy postscript, from the earlier version first published in the Marx Memorial Library's journal *Theory and Struggle* (May 2023), and has been adapted for this book.

1.   John Bellamy Foster, *The Return of Nature* (New York: Monthly Review Press, 2020), 7, emphasis added. Reference to the "second foundation of Marxist ecological thought" was first introduced twenty years earlier in *Marx's Ecology*. See John Bellamy Foster, *Marx's Ecology* (New York: Monthly Review Press, 2000), 250.

2.   Western Marxism took its point of departure in this respect from the short footnote in Georg Lukács's *History and Class Consciousness*, where he indicated dissatisfaction with Engels's account of the dialectics of nature. Yet, as Lukács indicated on multiple occasions afterward, and as attested by the text of *History and Class Consciousness*, he did

not actually reject the "merely objective dialectics of nature." The distortions of his thought in this respect nonetheless remain dominant. In the translation of his famous *Tailism* manuscript, this went so far as to translate incorrectly what appears as "Dialectics *in* Nature" in the original German in one of the chapter headings as "Dialectics *of* Nature." See Georg Lukács, *History and Class Consciousness* (London: Merlin, 1971), 24, 207; Georg Lukács, *A Defence of History and Class Consciousness: Tailism and the Dialectic* (London: Verso, 2000), 94, 102–7; Kaan Kangal, "Engels' Intentions in *Dialectics of Nature*," *Science and Society* 83, no. 2 (2019): 218; Foster, *The Return of Nature*, 16–21.

3.   Karl Marx and Frederick Engels, *Collected Works*, vol. 25 (New York: International Publishers, 1975–2004), 463–64.

4.   Karl Marx, *Capital*, vol. 1 (London: Penguin, 1976), 279.

5.   Marx and Engels, *Collected Works*, vol. 30, 54–66.

6.   Karl Marx, *Early Writings* (London: Penguin, 1974), 389–90; Marx and Engels, *Collected Works*, vol. 5, 28.

7.   Clive Hamilton and Jacques Grinevald, "Was the Anthropocene Anticipated?," *Anthropocene Review* 2, no. 1 (2015): 67.

8.   Joseph Fracchia, *Bodies and Artefacts*, vol. 1 (Leiden: Brill, 2022), 3.

9.   Marx and Engels, *Collected Works*, vol. 25, 545.

10. Marx, *Early Writings*, 398.

11. Marx and Engels, *Collected Works*, vol. 24, 301; Marx and Engels, *Collected Works*, vol. 25, 633; Marx and Engels, *Collected Works*, vol. 41, 232, 246; Foster, *Marx's Ecology*, 197, 291; Foster, *The Return of Nature*, 251–58.

12. Foster, *Marx's Ecology*, 212–21.

13. Marx and Engels, *Collected Works*, vol. 25, 340; Georg Lukács, *The Destruction of Reason* (London: Merlin, 1980), 403–8.

14. On "dialectical organicism," see Joseph Needham, *Moulds of Understanding* (London: George Allen and Unwin, 1976), 278.

15. Marx and Engels, *Collected Works*, vol. 24, 301; Marx and Engels, *Collected Works*, vol. 25, 23–27, 633; John Bellamy Foster, *The Return of Nature*, 254.

16. Marx and Engels, *Collected Works*, vol. 25, 26–27, 363, 593, 633.

17. On dialectics and integrated levels, see Joseph Needham, *Time: The Refreshing River* (London: George Allan and Unwin, 1943), 233–72; Jean-Pierre Vigier, "Dialectics and Natural Science," in *Existentialism Versus Marxism*, ed. George Novack (New York: Dell, 1966), 243–57.

18. Bertell Ollman, *Dance of the Dialectic* (Urbana: University of Illinois Press, 2003), 11; John Bellamy Foster, *Capitalism in the Anthropocene* (New York: Monthly Review Press, 2022), 304–8; Craig Dilworth, "Principles, Laws, Theories, and the Metaphysics of Science," *Synthese*

101, no. 2 (1994): 223–47; Richard Levins and Richard Lewontin, *The Dialectical Biologist* (Cambridge, MA: Harvard University Press, 1985), 268.

19. A characteristic of much Marxist dialectical thought has been to downplay the negation of negation, or development, evolution, and emergence. This can be seen in Ollman's influential work where "dialectical research" is confined to "four kinds of relations: identity/difference, interpenetration of opposites, quantity/quality, and contradiction." Ollman, *Dance of the Dialectic*, 15. This was even more the case in Soviet Marxism. As Frederick Copleston notes: "In Stalin's time, of course, the law of the negation of the negation was passed over in silence." Frederick C. Copleston, *Philosophy in Russia* (Notre Dame: University of Notre Dame Press, 1986), 327. On Marx and "scientific socialism," see Foster, *The Return of Nature*, 253.

20. Marx and Engels, *Collected Works*, vol. 25, 82, 326.

21. Marx and Engels, *Collected Works*, vol. 25, 126, 324–25.

22. Stephen Jay Gould, *The Structure of Evolutionary Theory* (Cambridge, MA: Harvard University Press, 2002), 479–92; Stephen Jay Gould, *Time's Arrow, Time's Cycle* (Cambridge, MA: Harvard University Press, 1987), 112–15, 133–34; Stephen Jay Gould, *Hen's Teeth and Horse's Toes* (New York: W. W. Norton, 1980), 97–105; Richard York and Brett Clark, *The Science and Humanism of Stephen Jay Gould* (New York: Monthly Review Press, 2011), 21, 28, 40–42.

23. See Helena Sheehan, *Marxism and the Philosophy of Science* (Atlantic Highlands, NJ: Humanities Press, 1985); Foster, *The Return of Nature*, 358–530.

24. V. I. Lenin, *Collected Works*, vol. 14 (Moscow: Progress Publishers, 1977).

25. Sebastiano Timpanaro issued a strong criticism of Western Marxism for abandoning materialism, but since he also rejected the dialectics of nature, his analysis—despite its brilliance—was unable to overcome the limitations he imposed upon it. Sebastiano Timpanaro, *On Materialism* (London: Verso, 1975).

26. The inability of critical theory, due to its shallow materialism and denial of the dialectics of nature, to provide any meaningful ecological analysis is evident in a recent work seeking to promote classical critical theory's contributions to ecology, chiefly that of Theodor Adorno, while at the same time acknowledging that "the classical Frankfurt School critical theorists hardly engaged with natural science" or ecology. Carl Cassegård, *Toward a Critical Theory of Nature* (London: Bloomsbury, 2021), 118.

27. Marx and Engels, *Collected Works*, vol. 25, 460–62. Engels attributed ecological disasters to shortsighted, "unforeseen," "remote natural

consequences," and to the necessary byproducts of a system of production devoted only to immediate gain. In the chapter on "The Revenge of the External" in his *Barbaric Heart*, Curtis White explains that such "unintended consequences" are treated in capitalist economics as externalities, and it is these externalities, *vis-à-vis* natural processes, which are coming back to haunt capitalism. Marx and Engels, *Collected Works*, vol. 25, 461–62; Curtis White, *The Barbaric Heart* (London: Routledge, 2009), 89–107.

28. Marx and Engels, *Collected Works*, vol. 25, 313, emphasis added.

29. Marx and Engels, *Collected Works*, vol. 25, 461.

30. Ray Lankester, *The Kingdom of Man* (New York: Henry Holt and Co., 1911).

31. Lankester's conception of human evolution in its emphasis on the hand was much closer to that of Engels in "The Part Played by Labour in the Transition from Ape to Man" than to either Darwin or Ernst Haeckel. See E. Ray Lankester, *Diversions of a Naturalist* (Freeport, NY: Books for Libraries Press, 1915), 243–44.

32. Lankester, *The Kingdom of Man*, 1–4, 26, 31–33, 184–89.

33. Lankester, *Science from an Easy Chair* (New York: Henry Holt and Co., 1913), 365–79.

34. Carles Soriano, "Anthropocene, Capitalocene, and Other '-Cenes,'" *Monthly Review* 74, no. 6 (November 2022): 1–28.

35. V. I. Vernadsky, *The Biosphere* (New York: Springer-Verlag, 1998); E. V. Shantser, "The Anthropogenic System (Period)," in *The Great Soviet Encyclopedia*, vol. 2 (New York: Macmillan, 1973), 139–44; V. I. Vernadsky, "Some Words About the Noösphere," in *150 Years of Vernadsky*, vol. 2 (Washington, DC: 21st Century Science Associates, 2014), 82. The Anthropogene was initially introduced in the Soviet Union to describe the geological period now known as the Quaternary.

36. Rachel Carson, *Lost Woods* (Boston: Beacon, 1998), 227–45; Barry Commoner, *The Closing Circle* (New York: Bantam, 1971), 60–61, 117, 138–45; Foster, *The Return of Nature*, 502–13; John Bellamy Foster and Brett Clark, "Rachel Carson's Ecological Critique," *Monthly Review* 59, no. 9 (February 2008): 1–17.

37. A. G. Tansley, "The Use and Abuse of Vegetational Concepts and Terms," *Ecology* 18, no. 3 (July 1935): 284–307. In developing the notion of ecosystem, Tansley relied heavily on the systems theory of the Marxist mathematician Hyman Levy. See Hyman Levy, *The Universe of Science* (London: Watts and Co., 1932).

38. Carles Soriano, "On the Anthropocene Formalization and the Report of the Anthropocene Working Group," *Geologica Acta* 18, no. 6 (2020): 1–10.

39. John Bellamy Foster and Brett Clark, "The Capitalinian: The First Geological Age of the Anthropocene," *Monthly Review* 73, no. 4 (September 2021): 1–16.

40. Carles Soriano, "Epistemological Limitations of Earth System Science to Confront the Anthropocene Crisis," *Anthropocene Review* 9, no. 1 (2020): 112, 122; Soriano, "Anthropocene, Capitalocene, and Other '-Cenes,'" 14.

41. Soriano, "Epistemological Limitations of Earth System Science," 121.

42. Kohei Saito, *Marx in the Anthropocene: Towards the Idea of Degrowth Communism* (Cambridge: Cambridge University Press, 2023), 53–55.

43. In Marx's original German as well as in Engels's edition of the third volume of *Capital*, what is presented in the English translation as a single sentence is in fact only a section of a much longer sentence, taking up an entire paragraph. Hence, rather than referring to a "sentence" in the discussion here, the term *passage* is used, particularly as the main issue in dispute concerns only a part of a sentence, even in the English-language edition.

44. Saito, *Marx in the Anthropocene*, 45, 67–68.

45. Karl Marx, *Marx-Engels Gesamtausgabe* (MEGA), II/4.2 (Berlin: Akademie Verlag, 1992), 753; Karl Marx and Friedrich Engels, *Werke*, Band 25 (Berlin: Dietz Verlag, 1964), 822; Saito, *Marx in the Anthropocene*, 53–55, 70; Karl Marx, *Capital*, vol. 3 (London: Penguin, 1981), 949; Karl Marx, *Economic Manuscript of 1864–1865* (Leiden: Brill, 2016), 797–98. Saito also makes the point that Engels's edition of the third volume of *Capital* incorrectly uses the word *life* at the end of the disputed sentence, rather than *soil*. However, both terms essentially convey the same broad meaning in this particular context, while *soil* also appears in the sentence that follows in Engels's edition of the third volume, as well as in Marx's original manuscript. Saito himself said that this discrepancy was probably due to Marx's poor handwriting, in which the words *Boden* and *Leben* look almost identical. Yet, despite acknowledging in his footnote that this could very well have been a result of Marx's poor handwriting, he nonetheless criticizes Engels in his text for substituting the term *life*, claiming that Engels made this change to bring Marx's sentence more in line with Engels's own notion of the "revenge" of nature. Given the penmanship problem and the very problematic nature of Saito's claims about the theoretical significance of the replacement of *soil* by *life*, this whole issue can be set aside in the present discussion. Saito, *Marx in the Anthropocene*, 56, 70.

In correspondence and discussions with me, Joe Fracchia has translated the critical passage in the original German in his *Economic Manuscript of 1864–1865* (as published in MEGA) slightly differently from Saito as: "provoking an irreparable rift in the context of the social and natural metabolism prescribed by the natural laws of the soil." Fracchia's translation is the more literal one mentioned in the text. I owe much of my understanding of these philological problems to Fracchia, who helped me in exploring the differences and nuances in a close comparison of Marx's original German text with his *Economic Manuscript of 1864–1865*, Engels's edited German text of the third volume of *Capital*, and the various English-language translations.

46. Foster, *Capitalism in the Anthropocene*, 41–61; Marx and Engels, *Collected Works*, vol. 30, 54–66.

47. *Marx in the Anthropocene*, 53. On István Mészáros's concept of "second order mediation," see John Bellamy Foster, Foreword to *The Necessity of Social Control*, by István Mészáros (New York: Monthly Review Press, 2015), 16. On Marx's concept of alienated mediation, see Marx, *Early Writings*, 261.

48. Saito, *Marx in the Anthropocene*, 45.

49. Saito, *Marx in the Anthropocene*, 56–57; Marx and Engels, *Collected Works*, vol. 25, 574–76; Justus von Liebig, *Familiar Letters on Chemistry, in Its Relations to Physiology, Dietetics, Agriculture, Commerce, and Political Economy*, 4th ed. (London: Walton and Maberly, 1859), 283–86; John Farley, "The Spontaneous Generation Controversy (1859–1880)," *Journal of the History of Biology* 5, no. 2 (1972): 317; Frederick Engels, *The Housing Question* (Moscow: Progress Publishers, 1979), 92–93.

50. Franklin C. Bing, "The History of the Word 'Metabolism,'" *Journal of the History of Medicine and Allied Sciences* 26, no. 2 (April 1971): 158–80.

51. Marx and Engels, *Collected Works*, vol. 25, 578; J. D. Bernal, *The Freedom of Necessity* (London: Routledge and Kegan Paul, 1949), 363–64; Foster, *The Return of Nature*, 414; Saito, *Marx in the Anthropocene*, 56–57.

52. Julius Robert Mayer, "The Motions of Organisms and Their Relation to Metabolism," in *Julius Robert Mayer: Prophet of Energy*, ed. Robert B. Lindsey (New York: Pergamon, 1973), 75–145; Kenneth Caneva, *Robert Mayer and the Conservation of Energy* (Princeton: Princeton University Press, 1993), 117; Marx and Engels, *Collected Works*, vol. 25, 688.

53. Saito, *Marx in the Anthropocene*, 45, 53.

54. Foster, *The Return of Nature*, 414.

55. Marx, *Capital*, vol. 3, 195, 949, 954.

56. Saito, *Marx in the Anthropocene*, 59, 67.

57. Saito points to Lukács's criticism in *History and Class Conscious* of the validity of scientific experiment as a basis for a dialectical knowledge of the universal metabolism of nature and says that this constitutes the

grounds for Lukács's rejection of Engels's dialectics of nature. Saito fails to note, however, that Lukács later reversed himself on this point in his 1967 preface to his book. Lukács, *History and Class Consciousness*, xix; Saito, *Marx in the Anthropocene*, 85.

58. Marx and Engels, *Marx-Engels Gesamtasugabe* (MEGA) IV/26 (Berlin: Akademie Verlag, 2011), 214–19; Joseph Beete Jukes, *Student's Manual of Geology* (Edinburgh: Adam and Charles Black, 1872), 476–512; Foster, *Capitalism in the Anthropocene*, 51, 270; John Bellamy Foster and Brett Clark, *The Robbery of Nature* (New York: Monthly Review Press, 2020), 143; Saito, *Marx in the Anthropocene*, 65–67.

59. Saito, *Marx in the Anthropocene*, 55, 59.

60. On this, see John Bellamy Foster, Brett Clark, and Richard York, *The Ecological Rift* (New York: Monthly Review Press, 2010).

61. Saito, *Marx in the Anthropocene*, 51; Terrell Carver, *Marx and Engels: The Intellectual Relationship* (Brighton: Wheatsheaf, 1983), 123–25; Foster, *The Return of Nature*, 584. In addition to indicating that he had read the entire manuscript to Marx, Engels said that "it was self-understood between us that this exposition of mine should not be issued without his knowledge." Marx and Engels, *Collected Works*, vol. 25, 9.

62. Oddly, Saito refers elsewhere in his argument to evidence provided by the present author and others pointing to the extent of Marx's involvement in and appreciation of Engels's *Anti-Dühring*. See Saito, *Marx in the Anthropocene*, 48, 241, 253.

63. Saito, *Marx in the Anthropocene*, 67.

64. Frederick Engels, *On Capital* (New York: International Publishers, 1937), 63.

65. Rosa Luxemburg, *Rosa Luxemburg Speaks* (New York: Pathfinder, 1970), 111. An additional factor was that the word *Stoffwechsel* was not originally translated as *metabolism* in the English-language translations of the first and third volumes of *Capital* in 1886 and 1909, but rather as *circulation of matter*.

66. See Foster, *Marx's Ecology*, 21–65; Foster, *The Return of Nature*, 405.

4. Nature as a Mode of Accumulation: Capitalism and the Financialization of the Earth

1. The term *original expropriation* here is used in place of what is often mistakenly referred to as Karl Marx's notion of *primitive accumulation*. This is in line with Marx's own stated preference. See Karl Marx, *Wage Labour and Capital/Value Price and Profit* (New York: International Publishers, 1935), VPP 38-39; Karl Marx, *Capital*, vol. 1 (London: Penguin, 1976), 871; "Ian Angus, *The War Against the Commons* (New

York: Monthly Review Press, 2023), 204–9 Marx carefully distanced himself from this concept of classical-liberal political economy by referring to "so-called primitive accumulation," since, as he insisted, this was not the case of the accumulation of capital, but rather "expropriation" of property. Moreover, *primitive* was a mistranslation of what Marx, following classical political economy, referred to as *original* or *primary*. Capitalism prior to the British Industrial Revolution required such original expropriation to monopolize the means of production, amass start-up capital, and generate a proletarianized labor force. Yet, expropriation of land/nature and thus of the means of production of the workers, as Marx himself indicated, does not stop there, and is continually replicated in the history of capitalism, colonialism, and imperialism, now taking on new dimensions in the twenty-first century. On the expropriation of the English commons, see John Bellamy Foster, Brett Clark, and Hannah Holleman, "Marx and the Commons," *Social Research* 88, no. 1 (2021): 1–30; Ian Angus, *The War Against the Commons.*

2.   Karl Polanyi, *The Great Transformation* (Boston: Beacon, 1944), 178.

3.   William Makepeace Thackeray, *The Newcomes* (London: Penguin, 1996), 488.

4.   Intrinsic Exchange Group, "Solution," https://www.intrinsicexchange.com/en/solution, accessed 10/26/23.

5.   Charles Darwin, *On the Origin of Species* (London: John Murray, 1859), 73. The term *ecosystem services* is usually credited to Paul Ehrlich and Ann Ehrlich, *Extinction: The Causes and Consequences of the Disappearance of Species* (New York: Random House, 1981).

6.   *The State of Natural Capital: Restoring Our Natural Assets* (London: Natural Capital Committee, 2014).

7.   See Erik Gomez-Baggethun, Rudolf de Groot, Pedro L. Lomas, and Carlos Montes, "The History of Ecosystem Services in Economic Theory and Practice: From Early Notions to Markets and Payment Schemes," *Ecological Economics* 69 (2010): 1213. They write mistakenly, in what purports to be a definitive analysis: "Schumacher [in *Small Is Beautiful*] was probably the first author that used the concept of natural capital."

8.   The names here are listed in chronological order in accordance with when they are known to have used the term *natural capital* or the notion of the *earth's capital stock*. A good preliminary treatment of the origins of the term is provided in Antoine Missemer, "Natural Capital as an Economic Concept, History, and Contemporary Issues," *Ecological Economics* 143 (2018): 90–96. However, Missemer misses the roles of Marx, Engels, Waring, Carey, and Liebig in this respect. He also privileges the neoclassical concept of natural capital focusing

on exchange-value, seeing earlier references to natural capital to be of little significance simply because they did not conform to present usage. Thus, despite referring to numerous thinkers who used the term in the nineteenth century, Missemer claims, by sleight of hand, that "the natural capital concept was indeed coined in the 1900s," thereby privileging the neoclassical conception of natural capital as the *only valid concept*. Missemer, "Natural Capital," 93–94. Besides the names mentioned above, a number of other thinkers used the notion of *natural capital* prior to the 1860s. Jean-Baptiste Say's use of the term, where it stood for natural human capital, is highlighted in Pierre-Joseph Proudhon, *What Is Property?* (Cambridge: Cambridge University Press, 1993), 109.

9. The early reference to "natural capital" by Considerant and others was not simply a metaphor, related to commodity capital, but reflected in part the classical recognition that the concept of capital itself had arisen out of a consideration of natural use-values and only took on the primary meaning of capital as accumulated exchange-value with the rise of capitalism. The word *capital* thus arose from *capita*, meaning *heads*, referring to heads of cattle, the entire herd of which was regarded as a *stock*. All of this was in physical or use-value terms. Herman Daly, "The Use and Abuse of the 'Natural Capital' Concept," Center for the Advancement of the Steady State Economy, November 13, 2014.

10. Rondel Van Davidson, "Victor Considerant: Fourierist Legislator, and Humanitarian" (PhD diss., Texas Tech University, December 1970), 68–69; John Cunliffe and Guido Erreygers, "The Enigmatic Legacy of Charles Fourier," *History of Political Economy* 33, no. 3 (2001): 467; Missemer, "Natural Capital," 91–92.

11. Ebenezer Jones, *The Land Monopoly, the Suffering and Demoralization Caused by It, and the Justice and Expediency of Its Abolition* (London: Charles Fox, 1849), 6, 18–21, 27.

12. Jones, *The Land Monopoly*, 10.

13. Jones, *The Land Monopoly*, 19. The complexity of Jones's analysis, which focused on natural capital as the proceeds of nature, defies Missemer's claim that the notion of natural capital was used by Jones simply as a "synonym for land," particularly as land, in classical political economy, was a category that stood for all of nature. Missemer, "Natural Capital," 91.

14. Hal Draper, *The Marx-Engels Chronicle* (New York: Schocken, 1985), 12.

15. Karl Marx and Frederick Engels, *Collected Works*, vol. 5 (New York: International Publishers, 1975), 66–73.

16. On Marx's critique of the naturalization of capital and the treatment of nature divorced from labor as a source of value, see Karl Marx, *Capital*, vol. 3 (London: Penguin, 1981), 953–57. All subsequent references to *Capital*, vol. 3, except as otherwise indicated are to this edition.

17. Karl Marx, "The Value-Form," *Capital & Class* 4 (1978): 134; Karl Marx, "The Commodity," chap. 1 in *Capital*, vol. 1, libcom.org; Karl Marx, *Texts on Method* (Oxford: Blackwell, 1975), 198, 200, 207.

18. John Bellamy Foster, Brett Clark, and Richard York, *The Ecological Rift* (New York: Monthly Review Press, 2010), 53–64; Marx and Engels, *Collected Works*, vol. 37, 732–33.

19. McCulloch, quoted in Marx and Engels, *Collected Works*, vol. 29, 224.

20. Marx, *Capital*, vol. 3, 954.

21. Marx, *Capital*, vol. 3, 756.

22. Karl Marx, *The Poverty of Philosophy* (New York: International Publishers, 1963), 164. *Terre-matière* and *terre-capital* have been inserted here in square brackets to better convey Marx's meaning, as indicated in Marx, *Capital*, vol. 3, 756.

23. Marx, *Capital*, vol. 3, 755–56. In this sentence, Marx uses the term *valorise* to refer to the landlord's realization in exchange-value terms of monopoly rents. It should be noted that the concept of *valorization* (*Verwetung*) is used in two senses in Marx, to refer to: (1) the whole capitalist process of surplus value production, and (2) (more often) the realization of surplus value at the end of the circulation process. Traditionally, *Verwetung* was translated as *realization*, which corresponds to the latter, more limited meaning. However, the 1976 Penguin edition of *Capital* introduced the word *valorization* (which did not at that time exist in the English language) to capture the broader meaning. Here, we are using it in today's more commonplace sense (looser than Marx's second meaning) of conferring value or prices on goods and services. This should not be taken as indicating that land in itself is a *source* of commodity value, which is a product of socially necessary labor and production. Rather, the exchange-value is received by the owner of the land in the form of rent. Valorization is thus used here simply in the sense of conferring titles and exchange-value to land and resources, which generate rents, and are connected to financial markets. Ultimately, this is dependent on the labor and production system. Ernest Mandel, introduction to *Capital*, vol. 1, 36; translator's note in Marx, *Capital*, vol. 1, 252.

24. Karl Marx, *Das Kapital* (Hamburg: Verlag von Otto Meissner, 1894) (*Verwandlung von Surplusprofit in Gundrente*), 158. Translation slightly altered from Marx, *Capital*, vol. 3, 756–57, changing *raw material* to *mere matter*. The correction is in conformity with the 1894 German

edition, which translates *blosser materie* literally as "mere matter" rather than "raw material," and with the French translation, which, in line with the distinction first developed in *The Poverty of Philosophy*, refers to "*la terre-matière une terre-capital.*" Marx, chap. 37 in *Capital*, French translation available at marxists.org. In Marx and Engel's *Collected Works*, vol. 3, 613–14, the entire phrase is unaccountably missing. The Ernest Untermann translation incorporates the phrase but translates the terms as "material land" and "land capital." See Karl Marx, *Capital*, vol. 3 (Chicago: Charles H. Kerr, 1909), 725–26. This then misses the full significance between the earth/land as *mere matter* and the formation of earth-capital. As his references to James Anderson and Henry Carey make clear in the same passage, Marx was concerned here with the ecological issue of the circulation of matter, particularly soil nutrients.

25. Marx, *Capital*, vol. 3, 637–40; Andreas Malm, *Fossil Capital* (London: Verso, 2016), 309–14.

26. Marx, *Capital*, vol. 3, 910–11.

27. Karl Marx, *Capital*, vol. 2 (London: Penguin, 1978), 435; Marx, *Capital*, vol. 3, 213–14; Marx and Engels, *Collected Works*, vol. 30, 63; Paul Burkett, *Marx and Nature* (Chicago: Haymarket, 2014), 141–47; Gómez-Baggethun, Groot, Lomas, and Montes, "The History of Ecosystem Services in Economic Theory and Practice," *Ecological Economics* 69, no. 1 (April 2010): 1211.

28. George E. Waring Jr., "The Agricultural Features of the Census of the United States for 1850," *Bulletin of the American Geological Association* 2 (1857): 189–202.

29. Henry C. Carey, *The Slave Trade, Domestic and Foreign* (Philadelphia: A. Hart, 1853), 199; Karl Marx and Frederick Engels, *Selected Correspondence* (Moscow: Progress Publishers, 1955), 78.

30. Marx, *Capital*, vol. 3, 756–57; John Bellamy Foster, *Marx's Ecology* (New York: Monthly Review Press, 2000), 144–54.

31. Karl Marx, *Critique of the Gotha Programme* (New York: International Publishers, 1938), 1. In *Capital*, Marx wrote of "the dull and tedious dispute over the part played by nature in the formation of exchange-value. Since exchange-value is a definite social manner of expressing the labour bestowed on a thing, it can have no more natural content [separate from labor] than has, for example, the rate of exchange." This did not, however, prevent Marx from constantly insisting that all real *wealth*, as opposed to *value*, stems from nature. Marx, *Capital*, vol. 1, 134, 176.

32. See John Bellamy Foster and Brett Clark, *The Robbery of Nature* (New York: Monthly Review Press, 2020), 12–34.

33. James Maitland, Earl of Lauderdale, *An Inquiry into the Nature and*

Origin of Public Wealth and into the Means and Causes of Its Increase (Edinburgh: Archibald Constable and Co., 1819), 37–59; Foster, Clark, and York, *The Ecological Rift*, 54–58.

34. Irving Fisher, *The Nature of Capital and Income* (New York: Macmillan, 1919), 76; Paul Burkett, *Marxism and Ecological Economics* (Chicago: Haymarket, 2006), 112; Alejandro Nadal, "The Natural Capital Metaphor and Economic Theory," *Real-World Economics Review* 74 (2016): 64–84.

35. Joshua Farley, "Natural Capital," in *Berkshire Encyclopedia of Sustainability*, vol. 5 (Great Barrington, MA: Berkshire, 2012), 264–67; Burkett, *Marxism and Ecological Economics*, 95–101.

36. Robert M. Solow, "The Economics of Resources or the Resources of Economics," *American Economic Review* 64, no. 2 (1974): 146–49.

37. Schumacher, *Small Is Beautiful*, 15–16.

38. Burkett, *Marxism and Ecological Economics*, 95–101, 108–9.

39. Nicholas Georgescu-Roegen, *The Entropy Law and the Economic Process* (Cambridge, MA: Harvard University Press, 1971).

40. Herman Daly, "Toward Some Operational Principles of Sustainable Development," *Ecological Economics* 2 (1990): 1–6.

41. Burkett, *Marxism and Ecological Economics*, 95–101, 108–9.

42. Ibid., 113.

43. The basic elements of Nicholas Georgescu-Roegen's thermodynamic critique of neoclassical economics were accepted from the start by Marxian economists, and viewed as consistent with the classical Marxian tradition, though lacking a social critique. See Paul M. Sweezy, "Ecology and Revolution: A Letter to Nicholas Georgescu-Roegen, July 31, 1974," *Monthly Review* 68, no. 9 (February 2017): 55–57; Elmar Altvater, *The Future of the Market* (London: Verso, 1993); John Bellamy Foster and Paul Burkett, *Marx and the Earth* (Chicago: Haymarket, 2016), 137–64.

44. World Bank, *World Development Report 2003: Sustainable Development in a Dynamic World* (Washington DC/New York: World Bank/Oxford University Press, 2003), 14–15; Burkett, *Marxism and Ecological Economics*, 100.

45. See Burkett, *Marxism and Ecological Economics*, 101–2.

46. See John Bellamy Foster and Hannah Holleman, "The Theory of Unequal Ecological Exchange: A Marx-Odum Dialectic," *Journal of Peasant Studies* 41, no. 1–2 (2014): 223–28.

47. Robert Costanza and Herman E. Daly, "Natural Capital and Sustainable Development," *Conservation Biology* 6, no. 1 (1992): 38.

48. Gómez-Baggethun et al., "The History of Ecosystem Services in Economic Theory and Practice," 1213.

49. Costanza and his coauthors argue: "It is not very meaningful to ask the total value of natural capital to human welfare, nor to ask the value of massive, particular forms of natural capital. It is trivial to ask what is the value of the atmosphere to humankind, or what is the value of rocks and soil infrastructure as support systems. Their value is infinite in total. However, it is meaningful to ask how changes in the quantity and quality of various types of natural capital and ecosystems services may have an impact on human welfare." In practice, then, the analysis is shifted almost entirely to ecosystem services, rather than natural capital. Robert Costanza et al., "The Value of the World's Ecosystem Services and Natural Capital," *Nature* 387 (1997): 255.

50. Robert Costanza et al., "Changes in the Global Value of Ecosystem Services," *Global Environmental Change* 26 (2014): 154; Costanza et al., "The Value of the World's Ecosystem Services and Natural Capital."

51. Herman Daly, "Integrating Ecology and Economics," Center for the Advancement of the Steady State Economy, June 5, 2014.

52. Commodifying ecosystem services—whether in the form of parts of the contemporary economy based directly on the exploitation of natural resources, or through imputing value to ecosystem services—requires "an extreme division (simplification) of nature" antithetical to ecological systems. John Bellamy Foster, *Ecology Against Capitalism* (New York: Monthly Review Press, 2002), 33.

53. Enrique Leff, "Marxism and the Environmental Question," in *The Greening of Marxism*, ed. Ted Benton (New York: Guilford, 1996), 146.

54. Costanza et al., "The Value of the World's Ecosystem Services and Natural Capital."

55. "Methodology: Ecosystem Accounting," UN System of Environmental and Economic Accounting, https://seea.un.org/ecosystem-accounting.

56. Costanza et al., "Changes in the Global Value of Ecosystem Services."

57. Jutta Kill, *The Financialization of Nature* (Amsterdam: Friends of the Earth International, 2015), 3.

58. Gómez-Baggethun et al., "The History of Ecosystem Services," 1214.

59. Nature itself is not strictly a commodity, since it is not produced by human labor. However, it is turned into an economic asset and provides a stream of exchange-value that is valorized or realized by the landlord through rent, constituting one of the forms in which total surplus value is divided. In this way, it becomes part of the general commodity exchange process.

60. "An Inclusive Economy," Intrinsic Exchange Group, https://www.intrinsicexchange.com/en/solution.

61. Sian Sullivan, *Financialisation, Biodiversity Conservation and Equity* (Penang: Third World Network, 2012), 17.

62. Sian Sullivan, "Noting Some Effects of Fabricating 'Nature' as 'Natural Capital,'" *Ecological Citizen* 1, no. 1 (2017): 65–67.

63. Sian Sullivan, "Nature Is Being Renamed 'Natural Capital'—But Is It Really the Planet That Will Profit?" *Conversation*, September 13, 2016; Natural Capital Coalition, *Natural Capital Protocol* (The Hague: Natural Capital Protocol, 2016).

64. "Natural Capital Accounting: In a Nutshell," Economics of Ecosystems and Biodiversity.

65. Sian Sullivan, "Noting Some Effects of Fabricating Nature as 'Natural Capital,' " *The Ecological Citizen* 1 (2017): 65-73.; Tanja Havemann et al., *Levering Ecosystems* (Zürich: Credit Suisse, 2016), 3, 24.

66. Chart on "Creating Natural Asset Companies," in "The Solution," Intrinsic Exchange Group, https://www.intrinsicexchange.com/en/creating-nac-graphic?rq=Creating%20Natural%20Asset%20Companies.

67. "The Solution," Intrinsic Exchange Group.

68. Fabian Huwyler, Jürg Käppeli, and John Tobin, *Conservation Finance: From Niche to Mainstream* (Zürich/New York: Credit Suisse/McKinsey Center for Business and Environment, 2016), 16, 22.

69. Sullivan, "Noting Some Effects of Fabricating 'Nature' as 'Natural Capital,'" 69–70; Havemann et al., *Levering Ecosystems*, 3.

70. Whitney Webb, "New Asset-Class Launch Advances Wall Street's Nature Takeover," *River Cities' Reader*, December 6, 2021.

71. Sullivan, "Nature Is Being Renamed 'Natural Capital.'"

72. On the contradictions of capitalist forestry, see Foster, *Ecology Against Capitalism*, 104–36.

73. "Crazy Carbon Offsets Market Prompts Calls for Regulation," *Bloomberg*, January 6, 2022.

74. Martin Crook, "Conservation as a Genocide: REDD versus Indigenous Rights in Kenya," *Climate and Capitalism*, March 15, 2018.

75. Sullivan, *Financialisation, Biodiversity Conservation and Equity*, 17.

76. Kanyinke Sena, "Recognizing Indigenous Peoples' Land Interests Is Critical for People and Nature," World Wildlife Fund, October 22, 2020.

77. Stefano B. Longo, Rebecca Clausen, and Brett Clark, *The Tragedy of the Commodity: Oceans, Fisheries and Aquaculture* (New Brunswick: Rutgers University Press, 2015).

78. Declan Foraise, "Banks Bankrolling Extinction to Tune of $2.6 Trillion," *Ecosystem Marketplace*, October 29, 2020.

79. "Global Finance Industry Sinks $119bn into Companies Linked to Deforestation," *Financial Times*, October 20, 2021.

80. "Just 100 Companies Responsible for 71% of Global Emissions, Study Says," *Guardian*, July 10, 2017.

81. On the potential negative role of nature derivatives in this respect, see Sullivan, *Financialisation, Biodiversity Conservation and Equity*, 21–23.

82. Schumacher, *Small Is Beautiful*, 46.

83. Marx, *The Poverty of Philosophy*, 164; Marx, *Capital*, vol. 3, 756–57.

84. John Bellamy Foster, *The Ecological Revolution* (New York: Monthly Review Press, 2009), 161–200.

85. Marx, *Capital*, vol. 1, 381.

86. Daly, "The Use and Abuse of the 'Natural Capital' Concept."

87. Foster, Clark, and York, *The Ecological Rift*, 284–87.

88. Sullivan, *Financialisation, Biodiversity Conservation and Equity*, 18.

89. On monopoly-finance capital and current financial crisis tendencies, see John Bellamy Foster, R. Jamil Jonna, and Brett Clark, "The Contagion of Capital," *Monthly Review* 72, no. 8 (January 2021): 1–19.

90. Herman Daly, "Capital, Debt, and Alchemy," Center for the Advancement of the Steady State Economy, April 8, 2012.

91. Robert Solow, "Sustainability: An Economist's Perspective," in *Economics of the Environment*, ed. Robert Dorfman and Nancy S. Dorfman (New York: Norton, 1993), 181.

92. Huwyler, Käppeli, and Tobin, *Conservation Finance*, 17.

93. David Harvey, *Marx, Capital, and the Madness of Economic Reason* (Oxford: Oxford University Press, 2018), 92.

94. John M. Gowdy, "The Social Context of Natural Capital," *International Journal of Social Economics* 21, no. 8 (1994): 43.

95. Marx, *Capital*, vol. 3, 911. On the human economy and nature as complementary, see Herman Daly, "The Return of the Lauderdale Paradox," *Ecological Economics* 25 (1988): 23.

96. Ernst Bloch, *The Principle of Hope*, vol. 2 (Cambridge, MA: MIT Press, 1995), 686, 695.

97. Nicolás Kosoy and Esteve Cobera, "Payments for Ecosystem Services as Commodity Fetishism," *Ecological Economics* 69 (2010): 1228–36.

98. In any rational path of sustainable human development, ecosystems need to be comprehended in their full complexity in terms of natural science, particularly in its more dialectical forms, as in Richard Levins and Richard Lewontin, *The Dialectical Biologist* (Cambridge, MA: Harvard University Press, 1985). Such a rational path requires moving away from the capitalist commodity market and toward social control. For a comprehensive approach based in natural science and political-economic critique, see Fred Magdoff and Chris Williams, *Creating an Ecological Society* (New York: Monthly Review Press, 2017). Material flow analysis and comprehensive energy approaches offer superior alternatives to the natural capital/ecosystem analysis in understanding the changing human relation to nature. Howard

Odum's analysis in particular provides the basis of a deep critique of ecological imperialism. See Friedrich Hinterberg, Fred Luks, and Friedrich Schmidt-Bleek, "Material Flows vs. 'Natural Capital': What Makes an Economy Sustainable?" *Ecological Economics* 23 (1997): 1–14; Howard Odum, *Environment, Power, and Society* (New York: Columbia University Press, 2007), 276–78, 303–5.

99. Marx and Engels, *Collected Works*, vol. 25, 461.
100. Marx, *Capital*, vol. 3, 959.

5. The Defense of Nature: Resisting the Financialization of the Earth

1. John C. Cannon, "Indigenous Leader Sues Over Borneo Natural Capital Deal," *Mongabay*, December 17, 2021; John C. Cannon, "Malaysian Officials Dampen Prospects for Giant Secret Carbon Deal in Sabah," *Mongabay*, February 10, 2022; Chris Lang, "Sabah's Nature Conservation Agreement: A Two Million Hectare Carbon Deal Involving a Fake Director, an Inequitable Agreement, a History of Destructive Logging, Massive Corruption, a Series of Offshore Companies, and a Sprinkling of Neocolonial Racism," REDD-Monitor, December 5, 2021; "'Very Hush-Hush': Borneo's $80bn Carbon Deal Stokes Controversy," *Al Jazeera*, February 2, 2022; Jason Santos, "Not Just US$1000: Hoch Standard Clarifies Involvement with Sabah NCA," *The Vibes*, February 6, 2022.

2. John C. Cannon, "Bornean Community Locked into 2 Million Hectare Deal They Don't Know About," *Mongabay*, November 9, 2021; Lang, "Sabah's Nature Conservation Agreement"; "'Very Hush-Hush.'"

3. Chris Lang, "Whistleblower on Sabah's Nature Conservation Agreement: 'An Obvious Con,'" REDD-Monitor, February 7, 2022; "'Very Hush-Hush.'"

4. Cannon, "Malaysian Officials Dampen Prospects for Giant Secret Carbon Deal in Sabah"; Lang, "Whistleblower on Sabah's Nature Conservation Agreement."

5. As Herman Daly notes, "the term 'natural capital' was introduced in opposition to 'financial capital,' not as an extension of it, or advocacy for 'monetizing nature.'" Herman Daly, "Contribution to GTI Roundtable 'Monetizing Nature,'" Great Transition Initiative, August 2014. For a historically based critique of the concept of *natural capital*, see John Bellamy Foster, "Nature as a Mode of Accumulation," *Monthly Review* 73, no. 10 (March 2022): 1–24.

6. Jill Baker, "Mark Carney's Ambitious $130 Trillion Glasgow Financial Alliance for Net-Zero," *Forbes*, November 8, 2021.

7. Lang, "Whistleblower on Sabah's Nature Conservation Agreement"; Santos, "Not Just US$1000."

8.   "The Growing Case for Conservation Finance," *Environmental Finance*, April 6, 2017, https://www.environmental-finance.com/content/market-insight/the-growing-case-for-conservation-finance.html; World Rainforest Movement, "Growing Speculation: From the Appropriation and Commodification to the Financialization of Nature," *Monthly Bulletin* 181, August 30, 2012; Philip Seufert, Roman Herre, Sofia Monsalva, and Shalmali Guttal, eds., *Rogue Capitalism and the Financialization of Territories and Nature* (Fian International, Transnational Institute, and Focus on the Global South, 2020), https://www.tni.org/en/publication/rogue-capitalism-and-the-financialization-of-territories-and-nature. On the Great Financial Crisis, see John Bellamy Foster and Fred Magdoff, *The Great Financial Crisis* (New York: Monthly Review Press, 2009). On the failure of the 2009 climate negotiations in Copenhagen, see Naomi Klein, *This Changes Everything* (New York: Simon and Schuster, 2014), 8–15.

9.   Corporate Eco Forum and the Nature Conservancy, *The New Business Imperative: Valuing Natural Capital* (Corporate Eco Forum, the Nature Conservancy, 2012), http://www.truevaluemetrics.org/DBpdfs/Metrics/NaturalCapital/The-New-Business-Imperative-Valuing-Natural-Capital-2012.pdf.

10. "Valuing Natural Capital: Accounting for the Benefits that Nature Provides," Conservation International, https://www.conservation.org/projects/valuing-and-accounting-for-natural-capital/.

11. Sian Sullivan, "Nature Is Being Renamed 'Natural Capital'—But Is It Really the Planet That Will Profit?," *Conversation*, September 13, 2016, https://theconversation.com/nature-is-being-renamed-natural-capital-but-is-it-really-the-planet-that-will-profit-65273.

12. Robert Costanza et al., "Changes in the Global Value of Ecosystem Services," *Global Environmental Change* 26 (2014): 152–58; Robert Costanza, Ida Kubiszewski, Natalie Stoekl, and Tom Kompas, "Pluralistic Discounting Recognizing Different Capital Contributions: An Example Estimating the Net Present Value of Global Ecosystem Services," *Ecological Economics* 183 (2021): 1–8.

13. Intrinsic Exchange Group, "Solution," https://www.intrinsicexchange.com/en/solution, accessed 10/26/23. World Rainforest Movement, "Growing Speculation," https://www.wrm.org.uy/bulletin-articles/growing-speculation-from-the-appropriation-and-commodification-to-the-financialization-of-nature/.

14. Barney Dickson et al., *Towards a Global Map of Natural Capital: Key Ecosystem Assets* (Cambridge: UN Environment Programme, 2014), 30–31.

15. Dario Kenner, *Who Should Value Nature? Sustainable Business Initiative—*

Outside Insights (London: Institute of Chartered Accountants in England and Wales, 2014), 9, 12; Mark Bowman, "Land Rights, Not Land Grabs, Can Help Africa Feed Itself," CNN, June 18, 2013.

16. African Forum on Green Economy, "Investing in Natural Capital for a Resilient Africa: 2020 Event Summary," Natural Capital Coalition, June 3, 2020.

17. Karl Marx, *Capital*, vol. 3 (London: Penguin, 1981), 516.

18. Richard Price, *An Appeal to the Public on the Subject of the National Debt* (London: T. Cadell, 1772), 18–19.

19. Karl Marx, *Grundrisse* (London: Penguin, 1973), 375, 842–43; Marx, *Capital*, vol. 3, 516, 519–24; Karl Marx and Frederick Engels, *Collected Works*, vol. 15 (New York: International Publishers, 1975–2004), 512. It was Price's perspective, Marx argued, that led William Pitt the Younger, prime minister of Great Britain from 1804 to 1806, to establish his famous sinking fund to pay off the national debt, in which he "transformed Adam Smith's theory of accumulation into the enrichment of a nation by the accumulation of debts." Marx, *Capital*, vol. 3, 521.

20. Karl Marx, *Capital*, vol. 1 (London: Penguin, 1976), 133.

21. Marx, *Capital*, vol. 1, 126, 131–32; Karl Marx and Frederick Engels, *Selected Correspondence* (Moscow: Progress Publishers, 1975), 180; Elmar Altvater, *The Future of the Market* (London: Verso, 1993), 189–90.

22. Marx and Engels, *Collected Works*, vol. 37, 732–33.

23. On the metabolic rift, see John Bellamy Foster, *Marx's Ecology* (New York: Monthly Review Press, 2000), 141–77; Kohei Saito, *Karl Marx's Ecosocialism* (New York: Monthly Review Press, 2017); John Bellamy Foster and Brett Clark, *The Robbery of Nature* (New York: Monthly Review Press, 2000), 12–34.

24. On Marx and thermodynamics, see John Bellamy Foster and Paul Burkett, *Marx and the Earth* (Chicago: Haymarket, 2016), 137–64.

25. Lucretius, *On the Nature of the Universe* (Oxford: Oxford University Press, 1997), 7–9 (classical citation: I.145–220).

26. *Capital*, vol. 1, 133.

27. John Ryan-Collins, "How Land Disappeared from Economic Theory," *Evonomics*, April 4, 2017; Nicholas Georgescu-Roegen, *The Entropy Law and the Economic Process* (Cambridge, MA: Harvard University Press, 1971), 2–3. Georgescu-Roegen's notion that Marx too fell prey to the same fallacy confused Marx's depiction of the laws of motion of capital in his critique of political economy with his wider *critique*, devoted to the limitations of those laws of motion.

28. Herman Daly, "The Circular Flow of Exchange Value and the Linear Throughput of Matter-Energy," *Review of Social Economy* 43, no. 3 (1985): 282.

29. "Frederick Soddy Facts: The Nobel Prize in Chemistry 1921," Nobel Prize.
     org, https://www.nobelprize.org/prizes/chemistry/1921/soddy/facts/
30. Soddy quoted in Herman Daly, *Beyond Growth* (Boston: Beacon,
     1996), 174.
31. Linda Merricks, *The World Made New: Frederick Soddy, Science, Politics,
     and Environment* (Oxford: Oxford University Press, 1996), 54-55.
32. Merricks, *The World Made New*, 53-54, 78, 86-88; Frederick Soddy,
     *Cartesian Economics: The Bearing of Physical Science Upon State
     Stewardship* (New York: Cosimo Classics, 2012), 43; John W. Evans,
     President, H. Lyster Jameson, Member of the Executive, A. G.
     Church, Secretary, National Union of Scientific Workers, "The British
     Association," *Nature* 106, no. 2664 (November 1920): 373; H. Lyster
     Jameson, *An Outline of Psychology* (N. C. L. C. Publishing Society,
     1938); Stephen A. Craven, "Henry Paul William Lyster Jameson,
     MA, DSc, PhD (1875-1922)—A Polymath: Zoologist, Transvaal
     Educationist, Entrepreneur, Civil Servant and Marxist," *Transactions of
     the Royal Society of South Africa* 67, no. 3 (2012): 127-34.
33. Merricks, *The World Made New*, 115.
34. Soddy quoted in Daly, *Beyond Growth*, 177; John Ruskin, *Unto This
     Last/The Political Economy of Art/Essays on Political Economy* (New
     York: Dutton, 1968), 171, 185, 189-90, 216. For a treatment of Ruskin's
     ecological views, see John Bellamy Foster, *The Return of Nature* (New
     York: Monthly Review Press, 2000), 75-80.
35. Frederick Soddy, *Wealth, Virtual Wealth and Debt* (London: George
     Allen and Unwin, 1926; 1933; repr. Dublin: OMNIA VERITAS, 2021),
     88; John Stuart Mill, *Principles of Political Economy with Some of Their
     Applications to Political Economy* (New York: Longmans, Green, 1920),
     4, 6. On the Lauderdale Paradox, see Foster and Clark, *The Robbery of
     Nature*, 152-72.
36. Money is defined by Keynes as "that by delivery of which debt contracts
     and price contracts are *discharged*, and in the shape of which a store
     of general purchasing power *is held*." It "derives its character from its
     relationship to the money of account" brought "into existence along
     with debts, which are contracts for deferred payment. . . . Money
     proper in the full sense of the term can only exist in relation to a money
     of account." Money and debt are therefore inextricably connected, with
     debt increasingly playing the determining role. John Maynard Keynes,
     A Treatise on Money, in *Collected Writings*, vol. 5 (London: Macmillan,
     1971), 3-4.
37. Soddy, *Wealth, Virtual Wealth and Debt*, 94.
38. Soddy, *Cartesian Economics*, 24, 34; Soddy, *Wealth, Virtual Wealth and
     Debt*, 117.

39. Soddy, *Cartesian Economics*, 12. Soddy's position on imperialism, though seemingly related to a left critique, was left undeveloped. In one sentence in *Cartesian Economics*, he identified finance and usury categorically with "Jews," reflecting a reactionary, anti-Semitic strain in his thought, though this was seldom in evidence. Soddy, *Cartesian Economics*, 31.

40. Soddy, *Wealth, Virtual Wealth and Debt*, 97, 121.

41. Frederick Soddy, *The Role of Money* (London: Routledge, 2003).

42. Soddy, *Wealth, Virtual Wealth, and Debt*, 58.

43. Marx, *Capital*, vol. 1, 255; Karl Marx, *A Contribution to a Critique of Political Economy* (Moscow: Progress Publishers, 1970), 143; Marx and Engels, *Collected Works*, vol. 42, 543.

44. Henry Dunning Macleod, *The Theory and Practice of Banking* (London: Longmans, Green, Reader, and Dyer, 1866), 73, 91.

45. Macleod, *The Theory and Practice of Banking*, 19–20; Henry Dunning Macleod, *The Theory of Credit*, vol. 2, part 1 (London: Longmans, Green, 1894), 594.

46. Soddy, *Wealth, Virtual Wealth and Debt*, 97–100; Ansel Renner, Herman Daly, and Kozo Mayumi, "The Dual Nature of Money: Why Money Systems Matter for an Equitable Bioeconomy," *Environmental Economics and Policy Studies* (2021): 1–12; Kozo Mayumi, *Sustainable Energy and Economics in an Aging Population* (Berlin: Springer Nature, 2020), 118. There are three major theories on the question of whether banks can create money: (1) the credit creation theory of banking, (2) the fractional reserve theory of banking, and (3) the financial intermediation theory of banking. Macleod was the foremost nineteenth-century representative of the credit creation theory of banking by individual banks, or the endogenous theory of money later made famous in Joseph Schumpeter's *The Theory of Economic Development*. The later fractional reserve theory of banking suggested that banks created money not individually through provision of credit but systematically, through the fractional reserve system. The financial intermediation theory of banking, dominant in neoclassical thought but challenged by heterodox thinkers, claims that banks are mere intermediaries like other financial institutions and do not create money, which is exogenous rather than endogenous. Richard A. Werner, "Can Banks Individually Create Money Out of Nothing? The Theories and the Empirical Evidence," *International Review of Financial Analysis* 36 (2014): 1–19; *The Theory of Economic Development* (Oxford: Oxford University Press, 1934), 97. See also Schumpeter, *A History of Economic Analysis* (Oxford: Oxford University Press, 1954), 1115–16. For Soddy, there was no question as to the endogenous creation of money, which

lay behind capitalism's financial excesses. Marx too challenged the orthodox quantity theory of money and suggested that credit money was created endogenously. See Costas Lapavitsas, *Marxist Monetary Theory* (Chicago: Haymarket, 2016). Keynes advanced the credit creation theory of money as early as 1924 in his *Tract on Monetary Reform*. See Werner, "Can Banks Individually Create Money Out of Nothing?." 18.

47. Frederick Soddy, *Money versus Man* (London: Elkin Matthews and Marrot, 1931), 15; Daly, *Beyond Growth*, 187.

48. Frederick Soddy, foreword to *The Frustration of Science* (New York: W. W. Norton, 1935), 7-9.

49. Herman Daly, "Capital, Debt, and Alchemy," Center for the Advancement of the Steady State Economy, April 9, 2012.

50. On the financialization of nature as a response to the Great Financial Crisis and the search for real assets in nature as a basis for the further expansion of the debt economy, see Jenny Simon and Anne Tittor, "The Financialization of Food, Land, and Nature," *Journal für Entwicklungspolitik* 33, no. 2 (2014): 16-45; World Rainforest Movement, "The Growing Case for Conservation Finance."

51. See Foster, "Nature as a Mode of Accumulation," 2-7.

52. The notion that the natural environment, as the commons, in property terms, needs to be internalized within the capitalist economy has its modern expression in Garrett Hardin, "The Tragedy of the Commons," *Science* 162, no. 3859 (1968): 1243-48. For a critique, see Stefano Longo, Rebecca Clausen, and Brett Clark, *The Tragedy of the Commodity: Oceans, Fisheries, and Aquaculture* (New Brunswick, NJ: Rutgers University Press, 2015), 27-38.

53. George Monbiot, "Put a Price on Nature? We Must Stop This Neoliberal Road to Ruin," *Guardian*, July 24, 2014.

54. Robert Costanza et al., "The Value of the World's Ecosystem Services and Natural Capital," *Nature* 387 (1997): 253-60.

55. This paragraph has been adapted from John Bellamy Foster, "The Ecological Tyranny of the Bottom Line," in *Reclaiming the Environmental Debate*, ed. Richard Hofrichter (Cambridge, MA: MIT Press, 2000), 137-39. See also Michael Jacobs, "The Limits to Neoclassicism," in *Social Theory and the Environment*, ed. Michael Redclift and Ted Benton (New York: Routledge, 1994), 69; Barry Commoner, *Making Peace with the Planet* (New York: New Press, 1992), 64-66; Marilyn Waring, *Counting for Nothing* (Toronto: University of Toronto Press, 1999), 216; *National Wildlife* (April–May 1986), 12; Robert Costanza et al., "Twenty Years of Ecosystem Services," *Ecosystem Services* 28 (2017): 9.

56. Costanza et al., "Changes in the Global Value of Ecosystem Services,"
     154.
57. Jonathan Hughes, "The Natural Capital Debt Bubble," Natural Capital
     Forum, June 1, 2015.
58. Guy Standing, *Plunder of the Commons* (London: Penguin, 2019), 121.
59. Monbiot, "Put a Price on Nature?"
60. Foster, *Ecology Against Capitalism*, 33.
61. See Business and Sustainable Development Commission, *Unlocking
     Business Opportunities in Sustainable Land Use with Blended Finance*,
     January 2018; Larry Lohmann, "The Problem Is Not 'Bad Baselines'
     but the Concept of Counterfactual Baselines Itself," REDD-Monitor,
     October 18, 2016, https://redd-monitor.org/2016/10/18/larry-
     lohmann-the-problem-is-not-bad-baselines-but-the-concept-of-
     counterfactual-baselines-itself/.
62. Winston Choi-Schagrin, "Wildfires Are Ravaging Forests Set Aside to
     Soak Up Greenhouse Gases," *New York Times*, August 24, 2021.
63. Guy Duke et al., *Opportunities for UK Business that Value and/or Protect
     Nature's Services*, attachment 1 to final report to the Ecosystem Markets
     Task Force and Valuing Nature Network (London: GHK, 2012), 32.
64. Monbiot, "Put a Price on Nature?"
65. Silvia Favasuli and Vandana Sebastian, "Voluntary Carbon Markets:
     How They Work, How They're Priced and Who's Involved." S&P Global,
     June 10, 2021.
66. Amrei von Hasse and Jan Cassin, *Theory and Practice of "Stacking" and
     "Bundling" Ecosystem Goods and Services*, Business and Biodiversity
     Offsets Programme (Washington, DC: Forest Trends, 2018), 5.
67. Credit Suisse, *Levering Ecosystems* (Zürich: Credit Suisse, 2016).
68. World Rainforest Movement, "The Growing Case for Conservation
     Finance."
69. Peter Burgess, "Building a Platform and Economic Model to
     Actualize the Value of Natural Capital to Incentivize Preservation and
     Conservation over Exploitation of Aboriginal Cultural [*sic*] Their Nature
     Capital Assets and the Nature Capital Assets of the State of Western
     Australia, Northern Territory, and Sabah Malaysia." briefing paper,
     Nature-Capital-Paper 180921, Tierra Australia, https://redd-monitor.
     org/wp-content/uploads/2022/02/Nature-Capital-Paper180921.pdf.
70. Julien Gonzalez-Redin, J. Gareth Polhill, Terence P. Dawson, Rosemary
     Hill, and Iain J. Gordon, "It's Not the 'What,' but the 'How': Exploring
     the Role of Debt in Natural Resource (Un)Sustainability," *Plos One* 13,
     no. 7 (2018): 1–19.
71. John Maynard Keynes, *The General Theory of Money, Interest, and
     Employment* (London: Macmillan, 1973), 159.

72. Herman Daly, "The Return of the Lauderdale Paradox," *Ecological Economics* 25 (1988): 21–23.

73. Karl Marx, *The Poverty of Philosophy* (New York: International Publishers, 1963), 164; Marx, *Capital*, vol. 3, 755–56.

74. Marx, *Capital*, vol. 3, 910–11. On the establishing of titles as the key to "unlocking natural capital assets," see Jake Rostron, "Capitalising on Nature: The Legal Practicalities of Unlocking Natural Capital Assets," Michelmores, November 10, 2018, https://www.michelmores.com/agriculture-insight/legal-practicalities-unlocking-natural-capital-assets/.

75. Ralph Waldo Emerson, *Essays* (London: Arthur L. Humphreys, 1899), 243; Georg Lukács, *The Ontology of Social Being: Marx* (London: Merlin, 1978), 10; Marx and Engels, *Collected Works*, vol. 30, 62–63.

76. Paul Hawken, Amory Lovins, and L. Hunter Lovins, *Natural Capitalism* (Boston: Little, Brown, 1999).

77. Edward B. Barbier, "Natural Capital," in *Nature in the Balance: The Economics of Biodiversity*, ed. Dieter Helm and Cameron Hepburn (Oxford: Oxford University Press, 2014),153–76; Barbier, *Nature and Wealth* (New York: Palgrave Macmillan, 2015), 12, 85–87, 98; Barbier, *Capitalizing on Nature* (Cambridge: Cambridge University Press, 2011).

78. Barbier's term, *ecological capital*, which he then identifies with the world's ecosystems, is related to the view of Costanza and his associates who have defined *natural capital* as meaning ecosystems, insofar as they underlie ecosystem services to the economy. Costanza et al., "Twenty Years of Ecosystem Services," 3. Yet, in referring specifically to *ecological capital*, Barbier goes even further, encompassing under this category every aspect of the earth's ecosystems. Although allowing for the existence of various sub-forms of capital, such as *human capital*, *reproducible capital*, and *natural capital*, his concept of *ecological capital* appears to embrace the Earth System as a whole, such that the planet itself becomes nothing but capital. Barbier, *Nature and Wealth*, 9–13. In Marx's terms, this amounts to the alchemy of the subsumption of "earth matter" by "earth capital." Marx, *Capital*, vol. 3, 756.

79. Marx, *Capital*, vol. 3, 522.

80. The environmental proletariat, in this sense, is not to be identified primarily with the industrial proletariat, but rather with Marx's conception that the proletariat stands for the dispossessed in general, the "focal point of all inhuman conditions" and thus embodies within itself the irrepressible struggles for humanity. See Paul M. Sweezy, *Modern Capitalism and Other Essays* (New York: Monthly Review Press, 1972), 148–49; Karl Marx and Frederick Engels, *The Holy Family* (Moscow: Foreign Languages Publishing House, 1956), 52.

81. Foster and Clark, *The Robbery of Nature*, 46.
82. John Bellamy Foster, *Capitalism in the Anthropocene* (New York: Monthly Review Press, 2022), 483–92.
83. Roman Herre, *Fast Track Agribusiness Expansion in Zambia* (Hands off the Land Alliance, 2013), 7, https://www.tni.org/en/publication/fast-track-agribusiness-expansion-in-zambia; Suefert et al., *Rogue Capitalism*, 25.
84. Bowman, "Land Rights, Not Land Grabs, Can Help Africa Feed Itself," CNN, June 18, 2013.
85. See "Global Campaign for Global Reform," La Via Campesina, viacampesina.org.
86. Brasil de Fato, "Brazil's Natural Resources Are a Target for a Capitalism in Crisis," *Peoples Dispatch*, June 10, 2019, https://peoplesdispatch.org/.
87. Mayank Aggarwal and S. Gopikrishina Warrier, "Environmental Issues in Agriculture a Silent Reason Behind Farmers' Protests," *Mongabay*, December 8, 2020.
88. The Red Nation, *The Red Deal* (Brooklyn: Common Notions, 2021), 141–47; Marx, *Capital*, vol. 3, 949.

6. Ecological Civilization, Ecological Revolution: An Ecological Marxist Perspective

This chapter originated as a lecture delivered to the John Cobb Ecological Academy in Claremont, California, on June 24, 2022, on the topic of ecological civilization and was copublished in 2022 in Chinese in the *Poyang Lake Journal* and in English in *Monthly Review* (October 2022). It has been revised for this book.

1. Karl Marx, *Capital*, vol. 3 (London: Penguin, 1981), 754.
2. Jeremy Lent, "What Does China's 'Ecological Civilization' Mean for Humanity's Future?," Ecowatch, February 9, 2018, ecowatch.com; Xi Jinping, *The Governance of China*, vol. 3 (Beijing: Foreign Languages Press, 2020), 54–56; Xi Jinping, "Full Text of Xi Jinping's Report at the 19th CPC National Congress," *China Daily*, November 4, 2017. An error in Lent's quotes from Xi, where "human and nature" is used instead of "humanity and nature," is corrected here.
3. See Pat Kane, "A New History of Cultural Big Ideas Looks to the East for Solace," *New Scientist*, May 24, 2017; Mark Elvin, *The Retreat of the Elephants: An Environmental History of China* (New Haven: Yale University Press, 2006). Lent's attempt to trace the divergence between humanity and nature, which characterizes the ecological contradiction of the West back to Plato, is not entirely convincing, since Plato himself commented on ecological destruction in his time in his *Critias*, while other ancient thinkers, particularly materialists, such as Epicurus and

his Roman follower Lucretius, evidenced deep ecological values. On Epicurus, see John Bellamy Foster, *Marx's Ecology* (New York: Monthly Review Press, 2000), 2–6, 33–39.

4.   Jeremy Lent, *The Patterning Instinct* (New York: Prometheus, 2017), 264–65. See also Ira E. Kasoff, *The Thought of Chang Tsai 1020–1077* (Cambridge: Cambridge University Press, 1984). Similar views were to be found in Daoism. The ecological character of early Chinese thought was strongly emphasized by the great Marxist scientist, ecological thinker, and leading Sinologist Joseph Needham. See John Bellamy Foster, *The Return of Nature* (New York: Monthly Review Press, 2020), 498–501. The relation of Daoism to ecology is emphasized in P. J. Laska, *The Original Wisdom of Dao De Jing: A New Translation and Commentary* (Green Valley, AZ: ECCS, 2012).

5.   In seeking to demonstrate that Marx was anti-ecological and advanced a Promethean view of the conquest of nature equivalent to that of bourgeois thought, Lent takes Marx's famous statement in *Capital*, volume 3, on the rational regulation of the metabolism between human beings and nature on behalf of the chain of human generations in accord with natural-material conditions and turns it into a flat statement meant to suggest the exact opposite. Thus, using the original English translation, removing the phrase "associated producers" (representing the subject of Marx's statement) and replacing it with "socialism," he writes, "Karl Marx wrote that the goal of socialism was 'rationally regulating [humanity's] material interchange with nature and bringing it under the common control'" — as if this implied a straightforward mastery of nature. In contrast, Marx's statement, quoting from the Penguin translation, reads as follows: "Freedom, in this sphere, can consist only in this, that socialized man, the associated producers, govern the human metabolism with nature in a rational way, bringing it under their collective control, instead of being dominated by it as a blind power; accomplishing it with the least expenditure of energy and in conditions most worthy and appropriate for their human nature." The emphasis here is clearly one of sustainable human development. Lent, *The Patterning Instinct*, 280; Marx, *Capital*, vol. 3 (New York: International Publishers, 1967), 820; Marx, *Capital*, vol. 3 (London: Penguin, 1981), 959.

6.   Lent, "What Does China's 'Ecological Civilization' Mean for Humanity's Future?"; Kane, "A New History of Cultural Big Ideas Looks to the East for Solace."

7.   The analysis in this section draws on John Bellamy Foster, *Capitalism in the Anthropocene* (New York: Monthly Review Press, 2022), 433–56.

8.   John Bellamy Foster, "Late Soviet Ecology and the Planetary Crisis," *Monthly Review* 67, no. 2 (June 2015): 1–20.

9.   A. D. Ursul, ed., *Philosophy and the Ecological Problems of Civilisation* (Moscow: Progress Publishers, 1983).

10. Following the 1983 publication of *Philosophy and the Ecological Problems of Civilisation*, it appears that the vice president of the USSR Academy of Sciences, P. N. Fedoseev (also Fedoseyev), who had written the book's introductory essay on ecology and the problem of civilization, incorporated a treatment of "Ecological Civilization" into the second edition of his *Scientific Communism*. See P. N. Fedoseev (Fedoseyev), *Scientific Communism* (Moscow: Progress Publishers, 1986); Jiahua Pan, *China's Environmental Governing and Ecological Civilization* (Berlin: Springer-Verlag, 2014), 35; Arran Gare, "Barbarity, Civilization and Decadence: Meeting the Challenge of Creating an Ecological Civilization," *Chromatikon* 5 (2015): 167–89; Qingzhi Huan, "Socialist Eco-Civilization and Social-Ecological Transformation," *Capitalism Nature Socialism* 27, no. 2 (2016): 2.

11. N. Fedoseev (Fedoseyev), "The Social Significance of the Ecological Problem," in *Philosophy and the Ecological Problems of Civilisation*, 31; Wang Hui, "Revolutionary Personality and the Philosophy of Victory: Commemorating Lenin's 150th Birthday," Reading the China Dream (blog), (originally published in Chinese April 21, 2020), https://www.readingthechinadream.com/wang-hui-revolutionary-personality.html/.

12. Ivan T. Frolov, "The Marxist-Leninist Conception of the Ecological Problem," in *Philosophy and the Ecological Problems of Civilisation*, 35–42.

13. V. A. Los', "On the Road to an Ecological Culture," in *Philosophy and the Ecological Problems of Civilisation*, 339.

14. Karl Marx and Frederick Engels, *Collected Works*, vol. 46 (New York: International Publishers, 1975–2004), 411, italics in the original; Foster, *Marx's Ecology*, 141–77.

15. Foster, *Capitalism in the Anthropocene*, 73–74.

16. Xi, *The Governance of China*, vol. 3, 6, 20, 25, 417–24.

17. Ibid., 3, 54.

18. Marx and Engels, *Collected Works*, vol. 25, 460–63; see also Cheng Enfu, *China's Economic Dialectic: The Original Aspiration of Reform* (New York: International Publishers, 2019), 150.

19. Xi, *The Governance of China*, vol. 3, 20.

20. Arthur Hanson, *Ecological Civilization in the People's Republic of China: Values, Action, and Future Needs* (Manila: Asian Development Bank, 2019), 3–9. See also John B. Cobb in conversation with André Vltchek, *China and Ecological Civilization* (Jakarta: Badak Merah Semesta, 2019); Paul Burkett, "Marx's Vision of Sustainable Human Development," *Monthly Review* 57, no. 5 (October 2005): 34–62.

21. Chen Xueming, *The Ecological Crisis and the Logic of Capital* (Leiden: Brill, 2017), 573.

22. Hanson, *Ecological Civilization in the People's Republic of China*, 6.

23. See Stephen S. Roach, "China's Growth Sacrifice," Project Syndicate, August 23, 2022.

24. See Joe Scholten, "How China Strengthened Food Security and Fought Poverty with State-Funded Cooperatives," Multipolarista, May 31, 2022.

25. Xiaoying You, "What Does China's Coal Push Mean for Its Climate Goals?," Carbon Brief, March 29, 2022, https://www.carbonbrief.org/analysis-what-does-chinas-coal-push-mean-for-its-climate-goals/.

26. Hanson, *Ecological Civilization in the People's Republic of China*, 6. China has also expanded its coal fired-plants in recent years ostensibly in order to stabilize its power grids, while it expands alternative energy to replace coal. This is all seen as part of a larger strategy of transitioning away from coal. Whether this transition will be realized remains to be seen. See Xiaoying You, "What Does China's Coal Push Mean for Its Climate Goals?," John Bellamy Foster, "Ecological Marxism," *Monthly Review* 75, no. 4 (September 2023): 45.

27. Sit Tsui and Lau Kin Chi, "Surviving Through Community Building in Catastrophic Times," *Monthly Review* 74, no. 3 (July–August 2022): 54–69.

28. Coronavirus Updates by Country, Worldometer, ; Wang, "Revolutionary Personality and the Philosophy of Victory."

29. James Hansen et al., "Young People's Burden: Requirements of Negative CO^2 Emissions," *Earth System Dynamics* 8 (2017): 578; James Hansen, "China and the Barbarians, Part 1," Columbia University, November 24, 2010, https://www.columbia.edu/~jeh1/mailings/2010/20101124_ChinaBarbarians1.pdf; Worldomer CO2 Emissions Per Capita," https://www.worldometers.info/co2-emissions/co2-emissions-per-capita/;"Each Country's Share of CO2 Emissions," Union of Concerned Scientists, July 12, 2023, https://www.ucsusa.org/resources/each-countrys-share-co2-emissions /.

30. Barbara Finamore, *Will China Save the Planet?* (Cambridge: Polity, 2018), 119.

31. Marx, *Capital*, vol. 1 (London: Penguin, 1976), 742; Xi, *The Governance of China*, vol. 3, 55.

32. Ulrich Brand and Markus Wissen, *The Imperial Mode of Living: Everyday Life and the Ecological Crisis of Capitalism* (London: Verso 2021), 5–10.

33. On how the climate legislation in the 2022 Inflation Reduction Act passed by the U.S. Congress, with the backing of the Joe Biden administration, falls short of a Green New Deal, see Jim Walsh and Peter

Hart, "Will the Manchin Climate Bill Reduce Climate Pollution?," Food
and Water Watch, August 10, 2022, https://www.foodandwaterwatch.
org/2022/08/10/will-the-manchin-climate-bill-reduce-climate-
pollution/; Anthony Rogers-Wright, "Why the Inflation Reduction Act
is Less a 'Climate Bill' and More a Poison Pill for Black and Indigenous
Communities and Movements," *Black Agenda Report*, August 24, 2022.

34. John Bellamy Foster, "The Defense of Nature: Resisting the Fin-
     ancialization of the Earth," *Monthly Review* 73, no. 11 (April 2022): 1–22.

35. As a cultural determinist, based on what he calls "cognitive history"
     or the development of worldviews underpinning cultures, Lent seeks
     to weave together what he sees as nonessentialist cultural worldviews
     with postmodernism, and uses this to explain why some cultures are
     more ecologically destructive than others. What this obviates is any
     materialist worldview, leaving these architectonic worldviews hanging
     in the air without foundations. See Jeremy Lent, "Beyond Modernist
     and Postmodernist History," *IAI News*, January 28, 2022.

36. Teodor Shanin, "Marxism and the Vernacular Revolutionary
     Traditions," in *Late Marx and the Russian Road*, ed. Teodor Shanin
     (New York: Monthly Review Press, 1983), 243–79.

37. Marx, *Capital*, vol. 3, 949; John Bellamy Foster, "Marx and the Rift
     in the Universal Metabolism of Nature," *Monthly Review* 65, no. 7
     (December 2013): 1–19.

38. Marx, *Capital*, vol. 3, 911, 949.

39. The above concepts from Marx's ecology and Marxian ecology in
     general are all central to the analysis in John Bellamy Foster, *Capitalism
     in the Anthropocene*.

40. Karl Marx, *Texts on Method* (Oxford: Basil Blackwell, 1975), 195.

7. Marxian Ecology, East and West: Joseph Needham and a Non-Eurocentric View of the Origins of Ecological Civilization

This chapter is based on a talk presented online to the School of Marxism,
Shandong University, Jinan, China, in March 2023. It is revised and
expanded from the original published version in *International Critical
Thought* 13, no. 2 (June 2023): 155–65. It also appears, in a further
revised version, as the "Review of the Month" in the October 2023 issue
of *Monthly Review*, and has been adapted here.

1.   Joseph Needham, *Within the Four Seas: The Dialogue of East and
     West* (Toronto: University of Toronto Press, 1969), 27, 97; Arun Bala,
     "Chinese Organic Naturalism and Modern Science Studies: Rethinking
     Joseph Needham's Legacy," *Culture of Science* 3, no. 1 (2020): 62–63.

2.   Samir Amin, *Eurocentrism*, 2nd ed. (New York: Monthly Review
     Press, 2009), 109. Amin does not specifically mention China in this

context, focusing rather on the Greek tributary mode of production in the pre-Hellenistic age, seen as linked to Egyptian and Phoenician cultures. Amin's argument, though, is complemented by Needham's argument on the simultaneous growth of scientific humanism/ organicist materialism in China, associated with Confucianism and Daoism, which began in the fifth and fourth centuries BCE, thus corresponding in time with the rise of the materialist philosophy of nature in Greece. Needham, *Within the Four Seas*, 97, 212. This thus fits with Amin's general argument on tributary culures, associated with what is often called the axial age.

3.   Needham, *Within the Four Seas*, 66–68.
4.   Ibid., 93.
5.   The foundational role of Epicurean materialism was also a proposition present in most of the other major thinkers of the second foundation of Marxist thought associated with British Red science and cultural materialism, including figures such as Benjamin Farrington, Joseph Needham, J. D. Bernal, J. B. S. Haldane, Lancelot Hogben, Christopher Caudwell, and Jack Lindsay. Other non-Marxian socialists, like Arthur G. Tansley, also drew on Epicurean materialism. John Bellamy Foster, *The Return of Nature* (New York: Monthly Review Press, 2020), 369, 526–30.
6.   On the extraordinary impact of Epicurus and Epicureanism on Marx's thought see John Bellamy Foster, *Marx's Ecology* (New York: Monthly Review Press, 2000), 1–65; Diego Fusaro, *Marx, Epicurus, and the Origins of Historical Materialism* (Oxford: Pertinent Press, 2018).
7.   Jeremy Lent, "What Does China's 'Ecological Civilization' Mean for Humanity's Future," Ecowatch, February 9, 2018, ecowatch.com; John Bellamy Foster, "Ecological Civilization, Ecological Revolution," *Monthly Review* 74, no. 5 (October 2022): 1–11. Lent adopts a culturalist view, which, while seeming to depart from Eurocentrism in his emphasis on the strengths of traditional Chinese philosophy, actually reinforces Eurocentrism by creating what Amin called an "inverted Eurocentrism," which only serves to reinforce Eurocentric views of Europe's own development, while presenting Chinese development as simply an inverse culturalism in relation to Eurocentrism. See Amin, *Eurocentrism*, 2nd ed., 214.
8.   A. D. Ursula, ed., *Philosophy and the Ecological Problems of Civilisation* (Moscow: Progress Publishers, 1983); Foster, "Ecological Civilization, Ecological Revolution," 3–4.
9.   On the problem of imperialism and Marxism in the West see Zhun Xu, "The Ideology of Late Imperialism," *Monthly Review* 72, no. 10 (March 2021): 1–20.

10. Failing to appreciate the Needham thesis that dialectical materialism was an outgrowth of Greek organic materialism for which Chinese organic naturalism had an affinity, mainmstream historians of science have commonly claimed that Needham's thesis "on the relations between Chinese organic materialist science and modern science" were paradoxical, lacking "a coherent philosophical explanation." Bala, "Chinese Organic Naturalism and Modern Science Studies," 73; Wen-yuan Qian, The Great Inertia: Scientific Stagnation in Traditional China (Dover, NH: Croom Helm, 1985), 133.11. This was most clearly articulated in the general introduction to Max Weber's sociology of religion, commonly published as the introduction to Max Weber's The Protestant Ethic and the Spirt of Capitalism (London: Unwin Hyman, 1930), 13–31.

12. Needham, Within the Four Seas, 13.

13. Samir Amin, Eurocentrism, 1st ed. (New York: Monthly Review Press, 1989), vii–xiii.

14. Horace B. Davis, Nationalism and Socialism (New York: Monthly Review Press, 1967), 59–73; Kenzo Mohri, "Marx and 'Underdevelopment,'" Monthly Review 30, no. 11 (April 1979): 32–42, Suniti Kumar Ghosh, "Marx on India," Monthly Review 35, no. 8 (January 1984): 39–53; John Bellamy Foster, "Marx and Internationalism," Monthly Review 52, no. 3 (July–August 2000): 11–22. See also Kevin B. Anderson, Marx on the Margins (Chicago: University of Chicago Press, 2016).

15. Foster, Marx's Ecology, 212–21.

16. John Bellamy Foster, Brett Clark, and Hannah Holleman, "Marx and the Indigenous," Monthly Review 71, no. 9 (February 2020): 1–19; John Newsinger, "The Taiping Peasant Revolt," Monthly Review 52, no. 5 (October 2000): 29–37.

17. Karl Marx, Capital, vol. 1 (London: Penguin, 1976), 90; Kohei Saito, Marx in the Anthropocene (Cambridge: Cambridge University Press, 2022), 184–85; Karl Marx, "The Reply to [Vera] Zasulich" and "A Letter to the Editorial Board of Otechestvennye Zapiski" in Teodor Shanin, Late Marx and the Russian Road (New York: Monthly Review Press, 1983), 124, 134–36.

18. Marx, Capital, vol. 1, 479; Saito, Marx in the Anthropocene, 183–84.

19. Kenneth Pomeranz, The Great Divergence: China, Europe, and the Making of the Modern World Economy (Princeton: Princeton University Press, 2021).

20. David Christian, Maps of Time (Berkeley: University of California Press, 2004), 406–9; Paul Bairoch, "The Main Trends in National Economic Disparities Since the Industrial Revolution," in Disparities in

Economic Development Since the Industrial Revolution, ed. Paul Bairoch and Maurice Lévy-Leboyer (New York: St. Martin's Press, 1981), 7–8.

21. Marx, *Capital*, vol. 1, 916.

22. Needham, *Within the Four Seas*, 27.

23. See Ellen Meiksins Wood, *The Retreat from Class* (London: Verso, 1986); Ellen Meiksins Wood and John Bellamy Foster, eds., *In Defense of History* (New York: Monthly Review Press, 1997).

24. Needham, *Within the Four Seas*, 91; Lucretius, *On the Nature of Things* (New York: E. P. Dutton, 1921), 85–86 (classical citation: II.1090–92). Translation follows Needham's modification of the Leonard text.

25. Joseph Needham, *Science and Civilization in China*, vol. 1 (Cambridge: Cambridge University Press, 1954), 4. On the role of Epicureanism in the development of modern science, see H. Floris Cohen, *How Modern Science Came into the World* (Amsterdam: Amsterdam University Press, 2010), 102–44. Stephen Greenblatt, *The Swerve: How the World Became Modern* (New York: W. W. Norton, 2012).

26. *The Original Wisdom of Dao De Jing: A New Translation and Commentary*, trans. P. J. Laska (Green Valley, Arizona: ECCS Books, 2012), xvii.

27. Karl Marx and Frederick Engels, *Collected Works* (New York: International Publishers, 1975), vol. 1, 413; Foster, *Marx's Ecology*, 52–53; Marx and Engels, *Collected Works*, vol. 5, 141–42.

28. Joseph Needham, "Light from the Orient," *Environment* (New Zealand Environment) 20 (August 1978): 8–11.

29. Joseph Needham, *Science and Civilization in China*, vol. 4, part 1 (Cambridge: Cambridge University Press, 1971), xxvi, 61; Tu Weiming, "The Continuity of Being: Chinese Visions of Nature," in Mary Evelyn Tucker and John Berthrong, eds., *Confucianism and Ecology* (Cambridge, MA: Harvard University Press, 1998), 106; *Dao De Jing*, xi, 80 (verse 63).

30. Joseph Needham, *Time: The Refreshing River* (London: George Allen and Unwin, 1943), 55–56; Epicurus, *The Epicurus Reader* (Indianapolis: Hackett, 1994), 39.

31. Needham, *Time*, 112.

32. Needham, *Within the Four Seas*, 67–68, 94; Joseph Needham, *Science and Civilization in China* (Cambridge: Cambridge University Press, 1956), vol. 2, 55, 484, 567.

33. Needham, *Within the Four Seas*, 63; Bertrand Russell, *The Problem of China* (London: George Allen and Unwin, 1922), 194.

34. Needham, "Light from the Orient," 10–11.

35. Ivan T. Frolov, "The Marxist-Leninist Conception of the Ecological Problem," in Ursul, ed., *Philosophy and the Ecological Problems of Civilisation*, 37–39.

36. V. A. Los', "On the Road to an Ecological Culture," in Ursul, ed., *Philosophy and the Ecological Problems of Civilisation*, 339.

37. Qingzhi Huan, "Socialist Eco-Civilization and Social-Ecological Transformation," *Capitalism Nature Socialism* 27, no. 2 (2016): 51–63; Arran Gare, "Barbarity, Civilization and Decadence: Meeting the Challenge of Creating an Ecological Civilization," *Chromatikon* 5 (2009): 167; Jiahua Pan, *China's Environmental Governing and Ecological Civilization* (New York: Springer, 2016), 35.

38. Wang Wei, "The Marxist Thought on Ecological Civilization," Proceedings of the Second International Conference on Language, Art, and Cultural Exchange, *Advances in Social Science, Education and Humanities Research*, vol. 559 (2021): 617–20; Xiao-pu Wang, Li-min Zhang, Qiu-ying Song, "Marx's Ecological View and Ecological Civilization Construction of China," International Conference on Social Science and Technology Education (Amsterdam: Atlantis Press, 2015): 930–35.

39. Chen Xueming, *The Ecological Crisis and the Logic of Capital* (Leiden/Boston: Brill, 2017), 547–48. The translation has been altered slightly to conform with English usage.

40. Huang Chengliang, "Theoretical Origins of Xi Jinping's Thought in Ecological Civilization," *Chinese Journal of Urban and Environmental Studies* 7, no. 2 (2019): 1–2.

41. Xi Jinping, *The Governance of China*, vol. 3 (Beijing: Foreign Languages Press, 2020), 54–56; Marx and Engels, *Collected Works*, vol. 25, 460–61.

42. Xi Jinping, "Full Text of Xi Jinping's Report at the 19th CPC National Congress," *China Daily*, October 18, 2017; Jeremy Lent, "Can China Really Lead the Way to an 'Ecological Civilization'?," *China Daily*, April 29, 2018; "Xi Jinping Stresses Mobilizing National Resources for Core Technology Breakthroughs in Key Fields," State Council Information Office, People's Republic of China, September 8, 2022.

43. Xi Jinping, "Build an Eco-Civilization for Sustainable Development," in *The Governance of China*, vol. 4 (Beijing: Foreign Languages Press, 2022), 413.

44. Xin Zhou, "Ecological Civilization in China: Challenges and Strategies," *Capitalism Nature Socialism*, 32, no. 3 (2021): 86; *Dao De Jing*, 19 (verse 16), 29 (verse 25).

45. Joseph Needham, *Moulds of Understanding* (London: George Allen and Unwin, 1976), 302–3.

8. Extractivism in the Anthropocene

Originally published in "Bleeding Earth," vol. 25, no. 2 (Autumn 2002) of *Science for the People*. Revised for this chapter.

1. On the Anthropocene, see Jan Zalasiewicz, Colin N. Waters, Mark Williams, and Colin P. Summerhayes, *The Anthropocene as a Geological Time Unit* (Cambridge: Cambridge University Press, 2019); Ian Angus, *Facing the Anthropocene* (New York: Monthly Review Press, 2016).

2. See Zalasiewicz, "Waters, Williams, and Summerhayes," *The Anthropocene as a Geological Time Unit,* 256–57; Angus, *Facing the Anthropocene,* 44–45.

3. Christoph Görg et al., "Scrutinizing the Great Acceleration: The Anthropocene and Its Analytic Challenges for Social-Ecological Transformations," *Anthropocene Review* 7, no. 1 (2020): 42–61.

4. Ulrich Brand and Markus Wissen, *The Imperial Mode of Living* (London: Verso, 2021).

5. Alicia Bárcena Ibarra, in the United Nations Environmental Programme Press Release, "Worldwide Extraction of Materials Triples in Four Decades, Intensifying Climate Change and Air Pollution," July 20, 2016.

6. United Nations Environmental Programme, *Global Material Flows and Resource Productivity* (2016), 5.

7. World Trade Organization, *Trade Profiles 2021.* See also Martin Upchurch, "Is There a New Extractive Capitalism?" *International Socialism* 168 (2020).

8. Eduardo Gudynas, *Extractivisms* (Blackpoint, NS: Fernwood, 2020), 82.

9. Mark Bowman, "Land Rights, Not Land Grabs, Can Help Africa Feed Itself," CNN, June 18, 2013.

10. Guy Standing, "How Private Corporations Stole the Sea from the Commons," *Janata Weekly,* August 7, 2022; Stefano Longo, Rebecca Clausen, and Brett Clark, *The Tragedy of the Commodity* (New Brunswick, NJ: Rutgers University Press, 2015).

11. Vijay Prashad and Taroa Zúñiga Silva, "Chile's Lithium Provides Profit to the Billionaires but Exhausts the Land and the People," *Struggle La Lucha,* July 30, 2022.

12. John Bellamy Foster, "The Defense of Nature: Resisting the Financialization of the Earth," *Monthly Review* 73, no. 11 (April 2022): 1–22.

13. Mohammed Hussein, "Mapping the World's Oil and Gas Pipelines," *Al Jazeera,* December 16, 2021.

14. World Trade Organization, *Trade Profiles 2021,* 22, 70; "USA: World's Largest Producer of Oil and Its Largest Consumer," China Environment Net, July 29, 2022; https://china-environment-news.net/usa-is-worlds-largest-oil-producer/.

15. Donella H. Meadows, Dennis L. Meadows, Jørgen Randers, and William

W. Behrens III, *The Limits to Growth* (Washington, DC: Potomac Associates, 1972); Dennis Meadows, interview by Juan Bordera, "Fifty Years After 'The Limits to Growth,'" MR Online, July 21, 2022.

16. See John-Andrew McNeish and Judith Shapiro, Introduction to *Our Extractive Age: Expressions of Violence and Resistance*, ed. Shapiro and McNeish (London: Routledge, 2021), 3; Christopher W. Chagnon, Sophia E. Hagolani-Albov, and Saana Hokkanen, "Extractivism at Your Fingertips," in *Our Extractive Age*, 176–88; Christopher W. Chagnon et al., "From Extractivism to Global Extractivism: The Evolution of an Organizing Concept," *Journal of Peasant Studies* (May 2022).

17. Alexander Dunlap and Jostein Jakobsen, *The Violent Technologies of Extraction: Political Ecology, Critical Agrarian Studies and the Capitalist Worldeater* (Cham, CH: Palgrave Macmillan, 2020), 34, 100, 120–21.

18. Gudynas, *Extractivisms*, 4, 10.

19. Karl Marx, *Capital*, vol. 1 (London: Penguin, 1976), 287; Marx and Engels, *Collected Works,* vol. 30 (New York: International Publishers, 1975–2004), 145; Marx and Engels, *Collected Works*, vol. 35, 191. Gudyanas attributes the popularization of the term *extractive industry* to international financial institutions such as the World Bank. He rejects the term as connoting that the extractive sector is part of industry and therefore productive. It is important to note that Marx employed the term as part of a sectoral analysis of production as a whole, and thus not separate from production. See Gudynas, *Extractivisms*, 3, 8.

20. Karl Marx, *Capital*, vol 3 (London: Penguin, 1981), 181–82.

21. Marx, *Capital*, vol. 3, 911.

22. Marx, *Capital*, vol. 3, 911, Marx and Engels, *Collected Works*, vol. 30, 62; Marx and Engels, *Collected Works*, vol. 46, 411.

23. Joan Martinez-Alier, "Rafael Correa, Marx and Extractivism," EJOLT, March 18, 2013. See also Eduardo Gudynas, "Would Marx Be an Extractivist?" *Post Development* (Social Ecology of Latin America Center), March 31, 2013.

24. See "Metabolic Rift: A Selected Bibliography," *Monthly Review*, October 16, 2013; Marx, *Capital*, vol. 1, 638.

25. Marx and Engels, *Collected Works*, vol. 20, 129. I am indebted to Ian Angus for drawing my attention to this passage.

26. Marx used the term *expropriation* about thirty times in Part VII of *Capital* on "So-Called Primitive Accumulation," and he used "primitive accumulation"—which he repeatedly prefaced with "so-called" or placed within scare quotes, and used in passages dripping with irony—about ten times. He explicitly indicated in several places that the reality (and historical definition) of "so-called primitive accumulation" was *expropriation*, while the second and third chapters of the sections both

include "expropriation" or "expropriated." See Marx, *Capital*, vol. 1, 871, 873-75, 939-40. For a general discussion of Marx's concepts of appropriation/expropriation, see John Bellamy Foster and Brett Clark, *The Robbery of Nature* (New York: Monthly Review Press, 2020), 35-63.

27. On Polanyi, appropriation, and reciprocity, see Karl Polanyi, *Primitive, Archaic and Modern Economies* (Boston: Beacon, 1968), 88-93, 106-7, 149-56; Foster and Clark, *The Robbery of Nature*, 42-43.

28. Marx, *Capital*, vol. 1, 914-15.

29. Marx and Engels, *Collected Works*, vol. 29, 461.

30. John Bellamy Foster, Brett Clark, and Hannah Holleman, "Marx and the Indigenous," *Monthly Review* 71, no. 9 (February 2020): 1-19.

31. Marx, *Capital*, vol. 1, 638; Marx, *Capital*, vol. 3, 182, 949.

32. Marx and Engels, *Collected Works*, vol. 37, 733, emphasis added.

33. Marx, *Capital*, vol. 1, 638; Marx, *Capital*, vol. 3, 910, 949.

34. Stephen G. Bunker, *Underdeveloping the Amazon: Extraction, Unequal Exchange, and the Failure of the Modern State* (Chicago: University of Chicago Press, 1985), 22.

35. Gudynas, *Extractivisms*, 26-27; Marx and Engels, *Collected Works*, vol. 28, 25; Marx and Engels, *Collected Works*, vol. 29, 461. On current Marxian work on expropriation, see Nancy Fraser, "Behind Marx's Hidden Abode," *Critical Historical Studies* (Spring 2016): 60; Nancy Fraser, "Roepke Lecture in Economic Geography—From Exploitation to Expropriation," *Economic Geography* 94, no. 1; Michael C. Dawson, "Hidden in Plain Sight," 149; Peter Linebaugh, *Stop, Thief!* (Oakland, CA: PM Press, 2014), 73: Foster and Clark, *The Robbery of Nature*.

36. Gudynas, *Extractivisms*, 4-7.

37. Gudynas, "Would Marx Be an Extractivist?"

38. Martin Arboleda, *Planetary Mine: Territories of Extraction Under Late Capitalism* (London: Verso, 2020. *Generalized-monopoly capital* is a term introduced by Samir Amin to designate twenty-first-century world political-economic conditions, in which monopoly capital, with its headquarters for the most part in the imperial triad of the United States/Canada, Western Europe, and Japan, has spread its tentacles across the globe, including the globalization of production under its control. *Late imperialism* is a term directed at how these conditions have promoted new forms of the drain of surplus/value from the periphery to the core of the capitalist system. See Samir Amin, *Modern Imperialism, Monopoly Finance Capital, and Marx's Law of Value* (New York: Monthly Review Press, 2018), 162; John Bellamy Foster, "Late Imperialism," *Monthly Review* 71, no. 3 (July-August 2019): 1-19.

39. Gudynas, *Extractivisms*, 143-44.

40. James Petras and Henry Veltmeyer, *Extractive Imperialism in the Americas* (Leiden: Brill, 2014), 20–48.

41. Paul A. Baran, *The Political Economy of Growth* (New York: Monthly Review Press, 1962), 22–43. In developing his notion of surplus and its relation to the environment, Gudynas declares that Marx's theory of rent is helpful, "but even so the Marxist perspective is limited, particularly because it does not address environmental considerations." His argument here runs into two problems. First, it fails to acknowledge not only the enormous advances in the understanding of Marx's ecological critique in the last several decades, which has generated a vast literature globally. Second, in turning to Baran's analysis of surplus to generate a political-economic and ecological critique of extractivism, Gudynas was drawing his inspiration from one of the leading Marxist economists of the twentieth century.

42. Gudynas, *Extractivisms*, 83. On the relation of Baran's concept of surplus to Marx's concept of surplus value, see John Bellamy Foster, *The Theory of Monopoly Capitalism* (New York: Monthly Review Press, 2014), 24–50.

43. Gudynas, *Extractivisms*, 83–84.

44. Gudynas, *Extractivisms*, 84–85. On how the concept of "natural capital" was converted from a use-value category in classical economics to an exchange-value category in neoclassical economics, see John Bellamy Foster, "Nature as a Mode of Accumulation," *Monthly Review* 73, no. 10 (March 2022): 1–24.

45. Marx and Engels, *Collected Works*, vol. 46, 411; Marx and Engels, *Collected Works*, vol. 30, 62; Marx and Engels, *Collected Works*, vol. 34, 391; Marx, *Capital*, vol. 3, 182, 949. Although Marx and Engels sometimes applied squandering to the destruction of the soil or human bodies, which were also seen as forms of robbery, the destruction of nonrenewable resources was characterized simply as squandering. On the corporeal rift, see Foster and Clark, *The Robbery of Nature*, 23–32.

46. Baran, *The Political Economy of Growth*, 42.

47. Gudynas, *Extractivisms*, 112–13.

48. Clive Hamilton and Jacques Grinevald, "Was the Anthropocene Anticipated?," *Anthropocene Review* 2, no. 1 (2015): 67.

49. The notion of "accumulative society" is taken from Henri Lefebvre, *The Critique of Everyday Life: The One-Volume Edition* (London: Verso, 2014), 622.

50. Carles Soriano, "On the Anthropocene Formalization and the Proposal by the Anthropocene Working Group," *Geologica Acta* 18, no. 6 (2020): 1–10.

51. John Bellamy Foster and Brett Clark, "The Capitalinian: The First Geological Age of the Anthropocene," *Monthly Review* 73, no. 4 (September 2021): 1–16; John Bellamy Foster, "The Great Capitalist

Climacteric," *Monthly Review* 67, no. 6 (November 2015): 1–17.

52.   Marx and Engels, *Collected Works*, vol. 30, 40.

9. Socialism and Ecological Survival

This chapter is a slightly revised version of an article by John Bellamy Foster and Brett Clark with the same title in *Monthly Review* 74, no. 3 (July-August 2002): 1–33.

1.   Epigraph: Barry Commoner, *The Closing Circle: Nature, Man & Technology* (New York: Bantam, 1971), 215.

2.   This is the clear implication of the "Summary for Policymakers" of the IPCC Working Group III report on *Mitigation* in its Sixth Assessment Report, written by scientists and reflecting the scientific consensus. However, the published version of this report, after being redacted by governments—reflecting not the scientific consensus but the governmental consensus—erased all radical social conclusions by the IPCC scientists. On this, see "Notes from the Editors," *Monthly Review* 74, no. 2 (June 2022).

3.   António Guterres, "Mitigation of Climate Change Report 2002: 'Litany of Broken Promises'—UN Chief," United Nations, YouTube video, April 4, 2022, https://www.youtube.com/watch?v=SoJ8S4Khf5E; "Secretary-General Calls Latest IPCC Climate Report 'Code Red for Humanity,' Stressing 'Irrefutable' Evidence of Human Influence," August 9, 2021.

4.   IPCC, *Climate Change 2021: The Physical Science Basis. Working Group I Contribution to the Sixth Assessment Report of the Intergovernmental Panel on Climate Change* (Geneva: IPCC, 2021), 14. The IPCC scenarios are presented in terms of both "best estimates" and "very likely ranges." The "very likely range" for SSP1-1.9 (the most optimistic scenario) for 2021–2040 is 1.2°C–1.7°C. The "best estimate," however, is 1.5°C. The "best estimate" for 2041–2060 is 1.6°C, and for 2081–2100 is 1.4°C.

5.   IPCC, *Climate Change 2022: Mitigation of Climate Change. Working Group III Contribution to the Sixth Assessment Report of the Intergovernmental Panel on Climate Change* (Geneva: IPCC, 2022), 39.

6.   IPCC, *Climate Change 2021: The Physical Science Basis*, 10–12, 15–18, 21–23, 34, 41.

7.   Moira Lavelle, "By 2050, 200 Million Climate Refugees May Have Fled Their Homes. But International Laws Offer Them Little Protection," *Inside Climate News*, November 2, 2021.

8.   Johan Rockström et al., "A Safe Operating Space for Humanity," *Nature* 461, no. 24 (2009): 472–75; Will Steffen et al., "Planetary Boundaries," *Science* 347, no. 6223 (2015): 736–46; John Bellamy Foster, Brett Clark, and Richard York, *The Ecological Rift* (New York: Monthly Review Press, 2010), 13–19.

9. Karl Marx, *Capital*, vol. 3 (London: Penguin Books, 1981), 754.

10. See Stan Cox and Paul Cox, *How the World Breaks: Life in Catastrophe's Path, from the Caribbean to Siberia* (New York: New Press, 2016).

11. Jan Zalasiewicz, Colin N. Waters, Mark Williams, and Colin P. Summerhayes, eds., *The Anthropocene as a Geological Time Unit* (Cambridge: Cambridge University Press, 2019).

12. Ian Angus, *Facing the Anthropocene* (New York: Monthly Review Press, 2016), 42–46; J. R. McNeill and Peter Engelke, *The Great Acceleration* (Cambridge, MA: Harvard University Press, 2014); Clive Hamilton and Jacques Grinevald, "Was the Anthropocene Anticipated," *Anthropocene Review* (2015): 6–7.

13. John Bellamy Foster and Brett Clark, "The Capitalinian," *Monthly Review* 73, no. 4 (September 2021): 12–15.

14. For an interesting reflection on this in the early 1970s, see the chapter on "Nature and Revolution" in Herbert Marcuse, *Counter-Revolution and Revolt* (Boston: Beacon, 1972), 59–78.

15. John Bellamy Foster, *The Return of Nature* (New York: Monthly Review Press, 2020), 502; Sean Walker, "Castle Bravo: Marking the 65th Anniversary of the US Nuclear Disaster," Australian Institute of International Affairs, February 27, 2019.

16. Foster, *The Return of Nature*, 502–3; Richard Hudson and Ben Shahn, *Kuboyama and the Saga of the Lucky Dragon* (New York: Yoseloff, 1965); Ralph E. Lapp, *The Voyage of the Lucky Dragon* (London: Penguin, 1957); Jonathan M. Weisgall, *Operation Crossroads: The Atomic Tests at Bikini Atoll* (Annapolis: Naval Institute Press, 1994).

17. Foster, *The Return of Nature*, 490, 505–8; Commoner, *The Closing Circle*; Virginia Brodine, *Radioactive Contamination* (New York: Harcourt Brace Jovanovich, 1975); Daniel Ford, *The Cult of the Atom: The Secret Papers of the Atomic Energy Commission* (New York: Simon and Schuster, 1982); Lawrence S. Wittner, *Rebels Against War: The American Peace Movement 1933–1983* (Philadelphia: Temple University Press, 1984); Lawrence S. Wittner, *One World or None: A History of the World Nuclear Disarmament Movement Through 1953* (Stanford, CA: Stanford University Press, 1993); Lawrence S. Wittner, *Resisting the Bomb: A History of the World Nuclear Disarmament Movement, 1954–1970* (Stanford, CA: Stanford University Press, 1997); Linus Pauling, *No More War!* (New York: Dodd, Mead, 1983); Linus Pauling, *Linus Pauling in His Own Words: Selected Writings, Speeches, and Interviews* (New York: Touchstone, 1995).

18. Leo Huberman, "Report from Japan," *Monthly Review* 9, no. 2 (June 1957): 49.

19. Foster, *The Return of Nature*, 491–92, 506–7, 509; Barry Commoner,

Science and Survival (New York: Viking, 1966), 119–20; "St. Louis Baby Tooth Survey, 1959–1970," Washington University School of Dental Medicine, http://beckerexhibits.wustl.edu/dental/articles/babytooth. html; Rachel Carson, *Silent Spring* (Boston: Houghton Mifflin, 1994), 277–97; John Bellamy Foster and Brett Clark, "Rachel Carson's Ecological Critique," *Monthly Review* 59, no. 9 (February 2008): 1–17.

20. On socialist thinkers who raised these questions before the publication of *Silent Spring*, see Foster, *The Return of Nature*, 358–416, 457–501.

21. Charles H. Anderson, *The Sociology of Survival: Social Problems of Growth* (Homewood, IL: Dorsey, 1976); Rudolf Bahro, *Socialism and Survival* (London: Heretic Books, 1982). Another important work that addressed the question of survival in the 1970s, though identified with the anarchist tradition, was Murray Bookchin, *Post-Scarcity Anarchism* (New York: Ramparts, 1971).

22. Commoner, *Science and Survival*, 10–11, 125–26; Anderson, *The Sociology of Survival*, 126. Also Bookchin, *Post-Scarcity Anarchism*, 60.

23. Bahro, *Socialism and Survival*, 142, 147–49, 151; E. P. Thompson, *Beyond the Cold War* (New York: Pantheon, 1982), 41–79.

24. See John Bellamy Foster, "'Notes on Exterminism' for the Twenty-First-Century Ecology and Peace Movements," *Monthly Review* 74, no. 1 (May 2022): 1–17.

25. Katherine Richardson, Will Steffen, Wolfgang Lucht, et. al., "Earth Beyond Six of Nine Planetary Boundaries," *Science Advances* 9, no. 37 (September 13, 2023), https://www.science.org/doi/10.1126/sciadv.adh2458.

26. Foster, *The Return of Nature*, 505–6; Michael Egan, *Barry Commoner and the Science of Survival* (Cambridge, MA: MIT Press, 2007), 19–20, 47; Barry Commoner, "What Is Yet to Be Done?" in *Barry Commoner's Contribution to the Environmental Movement*, ed. David Kriebel (Amityville, NY: Baywood, 2002), 75.

27. Commoner, *Science and Survival*, 122.

28. Ibid., 81–83.

29. Ibid., 10–11, 125–26.

30. Commoner, *The Closing Circle*, 215, 230.

31. Barry Commoner, *Making Peace with the Planet* (New York: New Press, 1992), ix.

32. Anderson, *The Sociology of Survival*, 4–5.

33. Anderson, *The Sociology of Survival*, 5, 7, 48–51; Paul A. Baran and Paul M. Sweezy, *Monopoly Capital* (New York: Monthly Review Press, 1966); Herman Daly, "Towards a Steady-State Economy," UK Sustainable Development Commission, April 24, 2008.

34. Richard York, "Asymmetric Effects of Economic Growth and Decline on CO_2 Emissions," *Nature Climate Change* 2 (2012): 762–63.

35. Anderson, *The Sociology of Survival*, 12–13; Barry Weisberg, *Beyond Repair: The Ecology of Capitalism* (Boston: Beacon, 1971), 6.

36. Anderson, *The Sociology of Survival*, 13, 126.

37. Anderson, *The Sociology of Survival*, 153; U.S. Energy Information Administration, "International Energy Outlook, 2021," October 6, 2021.

38. Anderson, *The Sociology of Survival*, 73, 81; Commoner, *Science and Survival*, 127.

39. Anderson, *The Sociology of Survival*, 140. Commoner too drew on this part of Marx's analysis. See Commoner, *The Closing Circle*, 280. For an extended analysis of these issues, see John Bellamy Foster and Brett Clark, *The Robbery of Nature* (New York: Monthly Review Press, 2020).

40. Anderson, *The Sociology of Survival*, 252–69.

41. Ibid., 142.

42. Ibid., 31, 41.

43. Ibid., 235.

44. Ibid., 8, 14, 227, 235, 276–78.

45. Bahro, *Socialism and Survival*, 131.

46. Thompson wrote an introduction to the English edition of Bahro's *Socialism and Survival*, in which he compared the latter favorably to William Morris. Thompson, introduction to *Socialism and Survival*, 8.

47. Rudolf Bahro, *Avoiding Social and Ecological Disaster: The Politics of World Transformation* (Bath: Gateway, 1994), 19.

48. Bahro, *Socialism and Survival*, 28, 157; see also Foster, Clark, and York, *The Ecological Rift*.

49. Bahro, *Avoiding Social and Ecological Disaster*, 29.

50. Bahro, *Socialism and Survival*, 38, 124.

51. Ibid., 16.

52. Bahro, *Avoiding Social and Ecological Disaster*, 5.

53. Bahro, *Socialism and Survival*, 151; Frederick Engels, *The Origin of the Family, Private Property, and the State* (New York: International Publishers, 1975); Foster, *The Return of Nature*, 287–96.

54. Bahro, *Socialism and Survival*, 149.

55. Bahro, *Avoiding Social and Ecological Disaster*, 333.

56. Bahro, *Socialism and Survival*, 149; Bahro, *Avoiding Social and Ecological Disaster*, 323–44.

57. IPCC, *Climate Change 2021: The Physical Science Basis*, 10–12, 15–18, 21–23, 34, 41.

58. IPCC, *Climate Change 2022: Impacts, Adaptation and Vulnerability. Working Group II Contribution to the Sixth Assessment Report of the Intergovernmental Panel on Climate Change* (Geneva: IPCC, 2022), 7–8.

59. Ibid., 8.

60. Ibid., 10–11.

61. Ibid., 11–12.

62. Ibid., 13–15, 18–20, 27.

63. Ibid., 30, 35.

64. Ibid., 29–35. For a discussion of the leaked "Summary for Policymakers" for part 3 on *Mitigation* of the IPCC's Sixth Assessment Report and the relation to the final published report (after being rewritten by governments), see "Notes from the Editors," *Monthly Review* 73, no. 5 (October 2021).

65. Leslie Hook and Chris Campbell, "Climate Graphic of the Week," *Financial Times*, May 13, 2022.

66. For a discussion of the corporeal rift, see Foster and Clark, *The Robbery of Nature*, 23–32.

67. Jeff Masters, "India and Pakistan's Brutal Heat Wave Poised to Resurge," Yale Climate Connections, May 5, 2022.

68. Colin Raymond, Tom Matthews, and Radley M. Horton, "The Emergence of Heat and Humidity Too Severe for Human Tolerance," *Science Advances* 6, no. 19 (2020).

69. Masters, "India and Pakistan's Brutal Heat Wave"; Raymond, Matthews, and Horton, "Emergence of Heat and Humidity"; Tim Andersen, "Wet Bulb Temperature Is the Scariest Part of Climate Change You've Never Heard of," Infinite Universe, July 2, 2021, https://medium.com/the-infinite-universe/wet-bulb-temperature-is-the-scariest-part-of-climate-change-youve-never-heard-of-8d85bef1ca98.

70. Raymond, Matthews, and Horton, "Emergence of Heat and Humidity"; Masters, "India and Pakistan's Brutal Heat Wave."

71. Kim Stanley Robinson, *The Ministry for the Future* (London: Orbit, 2020), 1–12.

72. "Notes from the Editors," *Monthly Review* 73, no. 5.

73. The neoliberal argument is present most prominently in Naomi Klein, *This Changes Everything: Capitalism vs. the Climate* (New York: Simon and Schuster, 2014), 69–73.

74. See Harry Magdoff and Paul M. Sweezy, *Stagnation and the Financial Explosion* (New York: Monthly Review Press, 1987); Joyce Kolko, *Restructuring World Economy* (New York: Pantheon, 1988); John Bellamy Foster and Robert W. McChesney, *The Endless Crisis* (New York: Monthly Review Press, 2012); Grace Blakeley, *Stolen* (London: Repeater, 2019).

75. See Arthur P. J. Mol and Frederick H. Buttel, eds., *The Environmental State Under Pressure* (New York: JAI Publishing, 2002); Magnus Boström and Debra J. Davidson, eds., *Environment and Society* (Cham: Palgrave Macmillan, 2018).

76. For a critique of mainstream ecological modernization theory, including the so-called Environmental Kuznets Curve, see John Bellamy Foster, "The Planetary Rift and the New Human Exemptionalism: A Political-Economic Critique of Ecological Modernization Theory," *Organization and Environment* 25, no. 3 (2012): 211–37.

77. István Mészáros, *The Structural Crisis of Capital* (New York: Monthly Review Press, 2010).

78. John Bellamy Foster, "Absolute Capitalism," *Monthly Review* 71, no. 1 (May 2019): 1–13; Naomi Klein, *The Shock Doctrine* (New York: Picador, 2008).

79. Arthur P. J. Mol, "The Environmental State and Environmental Governance," in *Environment and Society*, ed. Boström and Davidson, 119–42; Mol and Buttel, *The Environmental State Under Pressure*.

80. John Bellamy Foster, "Nature as a Mode of Accumulation," *Monthly Review* 73, 10 (March 2022): 1–24; John Bellamy Foster, "The Defense of Nature," *Monthly Review* 73, no. 11 (April 2022): 1–22.

81. Mol, "The Environmental State and Environmental Governance," 119–42.

82. Marx and Engels, *Collected Works*, vol. 25 (New York: International Publishers, 1975–2004), 145–46, 153, 270.

83. UN Environmental Programme, *Emissions Gap Report 2021: The Heat Is On: A World of Climate Promises Not Yet Delivered* (Nairobi: UNEP, 2021); Gaurav Ganti, Carl-Friedrich Schleussner, Claire Fyson, and Bill Hare, "New Pathways to 1.5°C," *Climate Analytics*, May 25, 2022.

84. Kjell Kühne, Nils Bartsch, Ryan Driskell Tate, Julia Higson, and André Habet, "'Carbon Bombs'—Mapping Key Fossil Fuel Projects," *Energy Policy* 166 (2022): 1.

85. Markus Wacket and Kate Abnett, "G7 Pledges to Phase Out Coal But No Date," *Canberra Times*, May 27, 2022, https://www.canberratimes.com.au/story/7757002/g7-pledges-to-phase-out-coal-but-no-date/.

86. "Notes from the Editors," *Monthly Review* 73, no. 5; "Notes from the Editors," *Monthly Review* 74, no. 2. Oxfam and the Stockholm Environment Institute have shown that the wealthiest 10 percent of the world's population are responsible for 52 percent of carbon emissions, while the poorest 50 percent account for only 7 percent. "Confronting Carbon Inequality," Oxfam Media Briefing, September 21, 2020. See also Jess Colarossi, "The World's Richest People Emit the Most Carbon," *Our World*, December 5, 2015; John Bellamy Foster, "On Fire This Time," *Monthly Review* 71, no. 6 (November 2019).

87. United Nations, "Secretary-General Warns of Climate Emergency, Calling Intergovernmental Panel's Report 'a File of Shame,' While Saying Leaders 'Are Lying,' Fueling Flames," Meetings Coverage and Press Release, April 4, 2022.

88. Marx and Engels, *Collected Works*, vol. 4, 394; Foster, *The Return of Nature*, 184, 196.

89. Julia Rock, "Biden Is Preparing to Crush a Historic Climate Change Lawsuit," *Lever News*, May 26, 2022; Mary Christina Wood, "On the Eve of Destruction: Courts Confronting the Climate Emergency," *Indiana Law Journal* 97, no. 1 (2022): 277.

90. Rock, "Biden Is Preparing to Crush a Historic Climate Change Lawsuit."

91. Christopher Flavelle, "FEMA's Director Wants Capitalism to Protect Us from Climate Change," *Bloomberg*, December 12, 2016.

92. Editors, "Fix Disaster Response Now," *Scientific American*, September 1, 2021; Junia Howell and James R. Elliott, "Damages Done: The Longitudinal Impacts of Natural Hazards on Wealth Inequality in the United States," *Social Problems* 66, no. 3 (2018): 448–67.

93. Thomas Joseph Dunning, *Trades' Unions and Strikes* (London: London Consolidated Society of Bookbinders, 1873), 42. Quoted slightly differently in Karl Marx, *Capital*, vol. 1 (New York: International Publishers, 1976), 926.

94. Foster, "The Defense of Nature."

95. The dominant ideology in the capitalist world has manufactured an unbalanced and one-sided environmental history of state socialism as one of continual ecological destruction pervading all of society, said to exceed the ecological destruction wrought by the West. However, a new book by Salvatore Engel-Di Mauro has shown that the record of environmental policy in Soviet-type societies was much more mixed. Actually existing "socialist states scored . . . many environmental successes," comparing favorably in many respects, especially when historically contextualized and placed in a global context, with the nations of the capitalist core. This is even more the case in the twenty-first century when examining the results of socialist-directed ecological planning in states such as China and Cuba. See Salvatore Engel Di-Mauro, *Socialist States and the Environment* (London: Pluto, 2021), 198. The great advantage of socialist-type societies is that it is possible to engage in social and ecological planning that puts people and the long-term environmental good ahead of profits. See Paul M. Sweezy and Harry Magdoff, "Socialism and Ecology," *Monthly Review* 41, no. 4 (September 1989): 1–8.

96. Don Fitz, "Cuba Prepares for Disaster," *Resilience*, March 24, 2022.

97. Fitz, "Cuba Prepares for Disaster"; see also Richard Stone, "Cuba Embarks on a 100-Year Plan to Protect Itself from Climate Change," *Science*, January 10, 2018.

98. Rey Santos quoted in Fitz, "Cuba Prepares for Disaster."

99. Fitz, "Cuba Prepares for Disaster." On the Jevons Paradox, see Foster, Clark, and York, *The Ecological Rift*, 169–82.

100. "As World Burns, Cuba No. 1 for Sustainable Development: WWF,"
     teleSUR, October 27, 2016; Matt Trinder, "Cuba Found to Be the
     Most Sustainable Country in the World," *Green Left*, January 10, 2020;
     Mauricio Betancourt, "The Effect of Cuban Agroecology in Mitigating
     the Metabolic Rift," *Global Environmental Change* 63 (2020): 1–10;
     Rebecca Clausen, Brett Clark, and Stefano B. Longo, "Metabolic
     Rifts and Restoration: Agricultural Crises and the Potential of Cuba's
     Organic, Socialist Approach to Food Production," *World Review of
     Political Economy* 6, no. 1 (2015): 4–32; Christina Ergas, *Surviving
     Collapse* (New York: Oxford University Press, 2021); Sinan Koont,
     "The Urban Agriculture of Havana," *Monthly Review* 60, no. 8 (January
     2009): 44–63.
101. "Notes from the Editors," *Monthly Review* 73, no. 4 (September 2021);
     "Soaring International Prices Aggravate Cuba's Food Crisis," Reuters,
     May 20, 2021.
102. Manolo de los Santos and Vijay Prashad, eds., "Cuba's Revolution
     Today: Experiments in the Grip of Challenges," *Monthly Review* 73, no.
     8 (January 2022).
103. Ricardo Vaz, "Maduro Orders Asset Transfers as Grassroots Groups
     Aim to Boost Production," *Venezuela Analysis*, May 28, 2022;
     Pablo Gimenez, "Venezuela: Fighting the Economic War 'People to
     People,'" interview by Fight Racism! Fight Imperialism!, *Venezuela
     Analysis*, October 29, 2019; Frederick B. Mills and William Camacaro,
     "Venezuela and the Battle Against Transgenic Seeds," *Venezuela
     Analysis*, December 11, 2013; John Bellamy Foster, "Chávez and the
     Communal State," *Monthly Review* 66, no. 11 (April 2015): 1–17; Chris
     Gilbert, "Red Current, Pink Tide," *Monthly Review* 73, no. 7 (December
     2021): 29–38; Chris Gilbert, "A Commune Called *Che*," *Monthly Review*
     73, no. 10 (March 2022): 28–38.
104. On ecological civilization, see Arthur Hanson, *Ecological Civilization in
     the People's Republic of China: Values, Action and Future Needs* (Manila:
     Asian Development Bank, 2019); John Bellamy Foster, *Capitalism in
     the Anthropocene* (New York: Monthly Review Press, 2022), 433–56.
105. Joe Scholten, "How China Strengthened Food Security and Fought
     Poverty with State-Funded Cooperatives," *Multipolarista*, May
     31, 2022; "Reported Cases and Deaths Per Country: COVID-19,"
     Worldometer, https://www.worldometers.info/coronavirus/; Liu Min
     and Hu Angang, "How China Became the World's Largest Contributor
     to Forest Growth," *China Environment News*, March 22, 2022; Barbara
     Finamore, *Will China Save the Planet?* (Cambridge: Polity, 2018); John
     Bellamy Foster, "The Earth-System Crisis and Ecological Civilization,"
     International Critical Thought 7, no. 4 (2017): 449–53; Lau Kin Chi,

Sit Tsui, and Yan Xiaohui, "Tracing a Trajectory of Hope in Rural Communities in China," *Monthly Review* 72, no. 5 (October 2020): 32–34; Lau Kin Chi, "Revisiting Collectivism and Rural Governance in China," *Monthly Review* 72, no. 5 (October 2020): 35–49.

106. Kali Akuno, "Build and Fight: The Program and Strategy of Cooperation Jackson," in *Jackson Rising: The Struggle for Economic Democracy and Black Self-Determination in Jackson, Mississippi*, ed. Kali Akuno and Ajamu Nangwaya (Wakefield, Québec: Daraja, 2017), 3, 11.

107. "About Us," Cooperation Jackson, https://cooperationjackson.org/; Akuno, "Build and Fight," 3.

108. Akuno, "Build and Fight," 4.

109. Akuno, "Build and Fight," 5–7. On human needs versus capital needs, see Michael A. Lebowitz, *Build It Now* (New York: Monthly Review Press, 2006).

110. Akuno, "Build and Fight," 15–29.

111. Red Nation, *The Red Deal: Indigenous Action to Save Our Earth* (Brooklyn, NY: Common Notions, 2021), 21–22.

112. James Connolly, *Songs of Freedom: The James Connolly Songbook* (Oakland, CA: PM, 2013), 59.

10. Planned Degrowth: Ecosocialism and Sustainable Human Development—An Introduction

1. Epigraph: Herman E. Daly, *Beyond Growth* (Boston: Beacon, 1996), 2.

2. In Marxist terms, *degrowth* stands for a shift from expanded reproduction in terms of material throughput to simple reproduction. See Paul M. Sweezy, *The Theory of Capitalist Development* (New York: Monthly Review Press, 1970), 75–95. The preeminent theorist of a steady-state economy (aimed at simple reproduction in the context of a full-world economy) is the late Herman E. Daly in works such as *Beyond Growth* and *Steady-State Economics*. Daly was a sharp critic of the existing capitalist economy and frequently made use of Marx in his analysis. However, his approach to steady-state economics was originally inspired by John Stuart Mill's conception of the "stationary state" and like Mill sought, in Marx's words, to "reconcile the irreconcilables" of capital and labor, seeing a no-growth economy as compatible with capitalism or at least a market system, and implemented by government policy, licensing, and caps. The irrealism of this was partly recognized by Daly, who dealt with the implementation of a no-growth economy as a matter of faith, ending his great work *Beyond Growth* with God and a "Creation-centered economy." Nevertheless, his analysis was at its core deeply critical and even radical. See Daly, *Beyond Growth*, 216–24; Herman E. Daly, *Steady-State Economics* (Washington, DC: Island Press,

1991); Herman E. Daly and John B. Cobb Jr., *For the Common Good* (Boston: Beacon, 1989). For a criticism of attempts to reconcile a no-growth economy with capitalism, see John Bellamy Foster, *Capitalism in the Anthropocene* (New York: Monthly Review Press, 2022), 363–72.

3. Herman E. Daly, "Economics in a Full World," *Scientific American* (September 2005): 100–7.

4. Howard T. Odum and Elisabeth C. Odum, *A Prosperous Way Down* (Boulder: University Press of Colorado, 2001).

5. Jason Hickel, *Less Is More* (London: Windmill, 2020), 30.

6. For ecological critiques of national income accounting, see Daly and Cobb, *For the Common Good*, 64–84, 401–55; John Bellamy Foster and Brett Clark, *The Robbery of Nature* (New York: Monthly Review Press, 2020), 260–61; Marilyn Waring, *Counting for Nothing* (Toronto: University of Toronto Press, 1999).

7. For a discussion of waste in capitalism, see Victor Wallis, *Red-Green Revolution: The Politics and Technology of Ecosocialism* (Toronto: Political Animal Press, 2022), 24–30.

8. Karl Marx and Frederick Engels, *Collected Works*, vol. 37 (New York: International Publishers, 1975–2004), 732–33.

9. Waring, *Counting for Nothing*, 153–81.

10. Johan Rockström et al., "A Safe Operating Space for Humanity," *Nature* 461, no. 24 (2009): 472–75; Will Steffen et al., "Planetary Boundaries," *Science* 347, no. 6223 (2015): 736–46; Sadrine Dixson-Declève et al., *Earth for All* (Gabriella, BC: New Society Publishers, 2022): 13–19.

11. Carles Soriano, "Anthropocene, Capitalocene, and Other '-Cenes,'" *Monthly Review* 74, no. 6 (November 2022): 1.

12. United Nations Intergovernmental Panel on Climate Change, *Sixth Assessment Report, Working Group I: The Physical Science Basis* (2021), 14; Andrea Januta, "Explainer: The U.N. Climate Report's Five Futures Decoded," Reuters, August 9, 2021; International Energy Agency, "Net Zero by 2050 Scenario (MZE)," Global Energy and Climate Model, October 2022, www.iea.org.

13. Kevin Anderson, "IPCC's Conservative Nature Masks True Scale of Action Needed to Avert Catastrophic Climate Change," The Conversation, March 24, 2023; see also David Spratt, "Faster, Higher, Hotter: What We Learned About the Climate System in 2022," part 1, Resilience.org, February 20, 2023.

14. "Global Temperatures Set to Reach New Records in Next Five Years," World Meteorological Organization, May 17, 2023.

15. Leaked Scientist Consensus Report on Mitigation, AR6, part 3, section B4.3, https://mronline.org/wp-content/uploads/2021/08/summary_ draft1.pdf; "Notes from the Editors," *Monthly Review* 74, no. 2 (June

2022). On low-energy solutions to climate change, see Joel Milward Hopkins, Julia K. Steinberger, Narasimha D. Rao, and Yannick Oswald, "Providing Decent Living with Minimum Energy: A Global Scenario," *Global Environmental Change* 65 (November 2020); Jason Hickel et al., "Urgent Need for Post-Growth Climate Mitigation Scenarios," *Nature Energy* 6 (2021): 766–68.

16. Anderson, "IPCC's Conservative Nature"; Hickel, *Less Is More*, 126–64.

17. John Kenneth Galbraith, *Economics and the Public Purpose* (New York: New American Library, 1973), 77–204; Paul M. Sweezy, "Utopian Reformism," *Monthly Review* 25, no. 6 (November 1973): 1–11.

18. Jacques Sapir, "Is Economic Planning Our Future?" *Studies on Russian Economic Development* 33, no. 6 (2022): 583–97.

19. Karl Marx, *Capital*, vol. 1 (London: Penguin, 1976), 99; Frederick Engels, *The Housing Question* (Moscow: Progress Publishers, 1975), 97.

20. Karl Marx and Frederick Engels, *The Communist Manifesto* (New York: Monthly Review Press, 1964), 40, 74.

21. Karl Marx, *Grundrisse* (London: Penguin, 1973), 173; Michael A. Lebowitz, *The Socialist Imperative* (New York: Monthly Review Press, 2015), 70–71.

22. Karl Marx and Frederick Engels, *Selected Correspondence* (Moscow: Progress Publishers, 1975), 186–87; Marx and Engels, *Collected Works*, vol. 3, 375–76, 418–43.

23. Marx, *Capital*, vol. 1, 172–73.

24. Karl Marx and Frederick Engels, *Writings on the Paris Commune*, ed. Hal Draper (New York: Monthly Review Press, 1971), 77.

25. Karl Marx, *Capital*, vol. 3, 959. Most current ecosocialist approaches to degrowth rely heavily on Marx's notions of social metabolism and metabolic rift. See Mattias Schmelzer, Andrea Vetter, and Aaron Vansintjan, *The Future Is Degrowth* (London: Verso, 2022), 84–86, 122–23, 237–44; Kohei Saito, *Marx in the Anthropocene* (Cambridge: Cambridge University Press, 2022).

26. Marx to Engels, March 25, 1868, Marx and Engels, *Selected Correspondence*, 190; John Bellamy Foster, "Capitalism and the Accumulation of Catastrophe," *Monthly Review* 63, no. 7 (December 2011): 3–5.

27. Marx and Engels, *Collected Works*, vol. 25, 281–82; Engels, *The Housing Question*, 92.

28. Marx and Engels, *Collected Works*, vol. 25, 279, 282–83.

29. Ibid., 219, 282.

30. Ibid., 294–95.

31. Ibid., 277–82; Jasper Bernes, "The Belly of the Revolution," in *Materialism and the Critique of Energy*, ed. Brent Ryan Bellamy and Jeff Diamanti (Chicago: MCM´ Publishing, 2018), 340–42.

32. Marx and Engels, *Collected Works*, vol. 25, 463–64.

33. Ibid., 460–63.

34. Walt Rostow, *The World Economy* (Austin: University of Texas Press, 1978), 47–48, 659–62; William R. Catton, *Overshoot* (Urbana: University of Illinois Press, 1982).

35. Marx and Engels, *Collected Works*, vol. 25, 269–70.

36. Michał Kalecki argued for "a synthesis of central planning and workers' control." Michał Kalecki, *Selected Essays on Economic Planning* (Cambridge: Cambridge University Press, 1986), 31. Marta Harnecker stressed the participatory planning system developed in Kerala State in India as a viable model. Marta Harnecker, *A World to Build* (New York: Monthly Review Press, 2015), 153–57. She also provided a guide for the implementation of participatory planning: Marta Harnecker and José Bartolemé, *Planning from Below: A Decentralized Participatory Planning Proposal* (New York: Monthly Review Press, 2019). For a critical Marxist work on the role of the direct producers in "real socialism," see Michael A. Lebowitz, *The Contradictions of "Real Socialism"* (New York: Monthly Review Press, 2012).

37. Marx and Engels, *Collected Works*, vol. 24, 519; Karl Marx, *On the First International* (New York: McGraw Hill, 1973), 11; Marx, *Grundrisse*, 159, 171–72; Paul Burkett, "Marx's Vision of Sustainable Human Development," *Monthly Review* 57, no. 5 (October 2005): 43; Ernest Mandel, "In Defense of Socialist Planning," *New Left Review* 159 (September–October 1986): 7.

38. Marx, *Capital*, vol. 1, 448–49; Lebowitz, *Contradictions of "Real Socialism,"* 21. The concept of *social metabolic reproduction* was developed by István Mészáros based on Marx's use of the concept of social metabolism in the *Grundrisse*. See István Mészáros, *Beyond Capital* (New York: Monthly Review Press, 1995), 39–71.

39. Karl Marx, *Theories of Surplus Value*, vol. 3 (Moscow: Progress Publishers, 1971), 309–10, emphasis in original; John Bellamy Foster and Paul Burkett, *Marx and the Earth* (Chicago: Haymarket, 2016), 149. The Greek word δίναμις, as used by Aristotle, refers to *power* as a source of change, thus a causal power. William Charlton, "Aristotelian Powers," *Phronesis* 32, no. 3 (1987): 277–89.

40. Marx and Engels, *Communist Manifesto*, 2.

41. Marx and Engels, *Collected Works*, vol. 25, 460–61; Jean-Paul Sartre, *Critique of Dialectical Reason*, vol. 1 (London: Verso, 2004), 164. Marx and Engels utilized the notion of *extermination* in the nineteenth-century sense of both death and removal in the context of the ecological ruination of Ireland under British colonialism. See Foster and Clark, *The Robbery of Nature*, 64–77. On the dialectic of

exploitation, expropriation, and exhaustion in Marx and Sartre, see Alberto Toscano, "Antiphysics/Antipraxis: Universal Exhaustion and the Tragedy of Materiality," in *Materialism and the Critique of Energy*, ed. Bellamy and Diamanti, 480–92; Michael A. Lebowitz, *Between Capitalism and Community* (New York: Monthly Review Press, 2020), 176–77.

42. Marx and Engels, *Collected Works*, vol. 25, 460–61; Sartre, *Critique of Dialectical Reason*, vol. 1, 164–66. Engels vividly described how deforestation in Russia "destroyed the stocks of subsoil water," so that "the rain and snow water flowed quickly along the streams and rivers without being absorbed, producing serious floods," while "in summer the rivers became shallow and the ground dried out. In many of the most fertile areas of Russia the level of subsoil water is said to have dropped a full metre, so that the roots of the corn crops can no longer reach it and wither away. So that not only are the human beings ruined, but in many areas so is the land itself for at least a generation." Marx and Engels, *Collected Works*, vol. 27, 387. Such ecological observations were to impact later socialist thinkers. V. I. Lenin specifically noted these passages in Engels on deforestation and impoverishment of the soil in Russia. V. I. Lenin, *Collected Works*, vol. 39 (Moscow: Progress Publishers, 1974), 501.

43. John Bellamy Foster, *The Return of Nature* (New York: Monthly Review Press, 2020), 137–38.

44. Burkett, "Marx's Vision of Sustainable Human Development," 34–62; Saito, *Marxism in the Anthropocene*, 232–42.

45. Paul A. Baran, *The Longer View* (New York: Monthly Review Press, 1969), 151.

46. Andrew Zimbalist and Howard J. Sherman, *Comparing Economic Systems* (Orlando: Academic Press Inc., 1984), 130.

47. Alec Nove, *An Economic History of the U.S.S.R.* (London: Penguin, 1969), 101.

48. Ibid., 74, 80; Zimbalist and Sherman, *Comparing Economic Systems*, 132.

49. Zimbalist and Sherman, *Comparing Economic Systems*, 130.

50. Tadeusz Kowalik, "Central Planning," in *Problems of the Planned Economy*, ed. John Eatwell, Murray Milgate, and Peter Newman (London: Macmillan, 1990), 43.

51. Tamás Krausz, *Reconstructing Lenin* (New York: Monthly Review Press, 2015), 335–38; Moshe Lewin, *Lenin's Last Struggle* (London: Pluto, 1975), 26–28, 115–16; Nove, *An Economic History of the U.S.S.R.*, 52, 58; Alfred Rosmer, *Moscow Under Lenin* (New York: Monthly Review Press, 1972), 131–33.

52. Nove, *An Economic History of the U.S.S.R.*, 100–101, 134; Fyodor I.

Kushirsky, *Soviet Economic Planning, 1965–1980* (Boulder, CO: Westview, 1982), 6–8; Zimbalist and Sherman, *Comparing Economic Systems*, 147.

53. Nove, *An Economic History of the U.S.S.R.*, 120; V. I. Lenin, *Collected Works*, vol. 32 (Moscow: Progress Publishers, 1973), 429–30.

54. Nikolai Bukharin, *The Politics and Economics of the Transition Period* (London: Routledge, 1979), 108–13; E. A. Preobrazhensky, *The Crisis of Soviet Industrialization* (White Plains, NY: M. E. Sharpe, 1979), 63; Harry Magdoff and Paul M. Sweezy, "Perestroika and the Future of Socialism—Part Two," *Monthly Review* 41, no. 11 (April 1990): 2; Nicholas Spulber, *Soviet Strategy for Economic Growth* (Bloomington: Indiana University Press, 1964), 102–3.

55. Nove, *An Economic History of the U.S.S.R.*, 124–28, 132, 147; Spulber, *Soviet Strategy for Economic Growth*, 66–68, 72.

56. Nove, *An Economic History of the U.S.S.R.*, 137; Harry Braverman, *Labor and Monopoly Capital* (New York: Monthly Review Press, 1998), 8–12; Gregory Grossman, "Command Economy," in *Problems of the Planned Economy*, 58–62.

57. Moshe Lewin, *Russia/USSR/Russia* (New York: New Press, 1995), 95–114. See also Alec Nove, *The Economics of Feasible Socialism* (London: George Allen and Unwin, 1983), 79–81; Michael Ellman, "Socialist Planning," in *Problems of the Planned Economy*, 14.

58. Lewin, *Russia/USSR/Russia*, 112, 95–108; Magdoff and Sweezy, "Perestroika and the Future of Socialism—Part Two," 2; Spulber, *Soviet Strategy for Economic Growth*, 126.

59. Lewin, *Russia/USSR/Russia*, 108–9.

60. Ernest Mandel, *Marxist Economic Theory*, vol. 2 (New York: Monthly Review Press, 1968), 557–59.

61. Lewin, *Russia/USSR/Russia*, 114. For a listing of the main structural characteristics of the Soviet planned economy, see Paul Cockshott, *How the World Works* (New York: Monthly Review Press, 2019), 209–10.

62. Joseph Stalin, quoted in Baran, *The Longer View*, 179.

63. "Invasion of the Soviet Union, June 1941," *Holocaust Encyclopedia* (Washington, DC: United States Holocaust Memorial Museum).

64. David Kotz, "The Direction of Soviet Economic Reform," *Monthly Review* 44, no. 4 (September 1992): 15.

65. Lewin, *Russia/USSR/Russia*, ix, 142; Moshe Lewin, "Society and the Stalinist State in the Period of the Five-Year Plans," *Social History* 1, no. 2 (May 1976): 172–73; Paul M. Sweezy, *Post-Revolutionary Society* (New York: Monthly Review Press, 1980), 144–45; Harry Magdoff and Fred Magdoff, "Approaching Socialism," *Monthly Review* 57, no. 3 (July–August 2005): 40–41.

66. Elena Veduta, "Some Lessons on Planning from the World's First Socialist

Economy," *Monthly Review* 74, no. 5 (October 2022): 23–36; Lebowitz, *Contradictions of "Real Socialism,"* 115–20. The notion promoted by the "Austrian School" of economics, including figures such as Ludwig von Mises, Friedrich Hayek, and Lionel Robbins, that central planning was impossible because it would require simultaneously solving millions of equations, was wrong from the start, as adequately demonstrated by Oskar Lange. Today the bulk of goods are not produced on the basis of market signals but are the product of internal, intrafirm corporate planning. Nevertheless, computerization of the inputs and outputs in the planning system would have greatly aided overall efficiencies. Oskar Lange and Fred M. Taylor, *On the Economic Theory of Socialism* (New York: McGraw-Hill, 1938), 57–98; Ernest Mandel, "In Defense of Socialist Planning," *New Left Review* 1/159 (September–October 1986), 11; P. Cockshott, A. Cottrell, and J. Dapprich, *Economic Planning in an Age of Climate Crisis* (London: Cockshott, Cottree, and Dapprich, 2022).

67. Magdoff and Sweezy, "Perestroika and the Future of Socialism—Part Two," 6; Magdoff and Magdoff, "Approaching Socialism," 44.
68. Sweezy, *Post-Revolutionary Society*, 140–41.
69. Helen Yaffe, *Che Guevara: The Economics of Revolution* (New York: Palgrave Macmillan, 2009), 38–39; Michael Löwy, *The Marxism of Che Guevara* (New York: Rowman and Littlefield, 1973), 7–51, 440–41. On Soviet enterprises, see Spulber, *Soviet Strategy for Economic Growth*, 119–29; Magdoff and Magdoff, "Approaching Socialism," 44; Galbraith, *Economics and the Public Purpose*, 108–17.
70. Zimbalist and Sherman, *Comparing Economic Systems*, 24–25.
71. Magdoff and Sweezy, "Perestroika and the Future of Socialism—Part Two," 3–7; János Kornai, *The Socialist System* (Princeton: Princeton University Press, 1992).
72. For a comparison of U.S. and Soviet growth rates, see David M. Kotz with Fred Weir, *Russia's Path from Gorbachev to Putin* (London: Routledge, 2007), 35–36.
73. Stephen F. Cohen, *Soviet Fates and Lost Alternatives* (New York: Columbia University Press, 2011), 136–40; Stanislav Menshikov, "Russian Capitalism Today," *Monthly Review* 51, no. 3 (July–August 1999): 81–99; Kotz, *Russia's Path from Gorbachev to Putin*, 105–25; Gordon M. Hahn, *Russia's Revolution from Above, 1985–2000* (New Brunswick, NJ: Transaction Publishers, 2002).
74. Magdoff and Magdoff, "Approaching Socialism," 49.
75. On China's land reform, see William Hinton, *Through a Glass Darkly* (New York: Monthly Review Press, 2006), 37–84.
76. Fred Magdoff, preface to *The Unknown Cultural Revolution: Life and*

Change in a Chinese Village, by Dongping Han (New York: Monthly Review Press, 2008), x.

77.   Rostow, *World Economy*, 522, 536.

78.   Chris Bramall, *In Praise of Maoist Economic Planning: Living Standards and Economic Development in Sichuan Since 1931* (Oxford: Oxford University Press, 1993), 335–36.

79.   Samir Amin, "China 2013," *Monthly Review* 64, no. 10 (March 2013): 20.

80.   Yi Wen, "The Making of an Economic Superpower: Unlocking China's Secret of Rapid Industrialization," working paper, Federal Reserve Board of St. Louis, Economic Research, August 2015, 2; John Ross, *China's Great Road* (Glasgow: Praxis, 2021), 13, 178.

81.   Lowell Dittmer, "Transformation of the Chinese Political Economy in the New Era," in *China's Political Economy in the Xi Jinping Epoch*, ed. Lowell Dittmer (Singapore: World Scientific Publishing, 2021), 8; Gang Chen, "Consolidating Leninist Control of State-Owned Enterprises: China's State Capitalism 2.0," in *China's Political Economy in the Xi Jinping Epoch*, 44.

82.   Chen, "Consolidating Leninist Control of State-Owned Enterprises," 59.

83.   Chen, "Consolidating Leninist Control of State-Owned Enterprises," 45–49, 59; Tian Hongzhi and Li Hui, "How Does the Five-Year Plan Promote China's Economic Development?," *Hradec Economic Days* (2021), diglib.uhk.cz.

84.   Cheng Enfu, *China's Economic Dialectic* (New York: International Publishers, 2021), 48–49, 66–67, 143, 295–310.

85.   Wen, "The Making of an Economic Superpower," 9.

86.   China's seeming ability to avoid major business-cycle swings does not mean that the society is free from crises in a larger transformational sense. See Wen Tiejun, *Ten Crises: The Political Economy of China's Development (1949–2020)* (New York: Palgrave Macmillan, 2021); John Ross, "Why China's Socialist Economy Is More Efficient than Capitalism," MR Online, June 6, 2023.

87.   "Wealth and Inequality in the U.S. and China," University of Southern California US-China Institute, November 19, 2020, china.usc.edu; Cheng Enfu, *China's Economic Dialectic*, 287–93; Marc Blecker, "The Political Economy of Working Class Re-formation," in *China's Political Economy in the Xi Jinping Epoch*, 87–105; John Bellamy Foster and Robert W. McChesney, *The Endless Crisis* (New York: Monthly Review Press, 2012), 155–83.

88.   Magdoff and Sweezy, "Perestroika and the Future of Socialism—Part Two," 1; Mandel, "In Defense of Socialist Planning," 9.

89.   See Martin Hart-Landsberg, "Planning an Ecologically Sustainable

and Democratic Economy: Challenges and Tasks," *Monthly Review* 75, no. 3 (July–August 2023): 114-25. On British wartime planning, see Cockshott, Cottrell, and Dapprich, *Economic Planning in an Age of Climate Crisis*, 63-75.

90.   "Rosie the Riveter: More than a Poster Girl," U.S. Army Ordnance Corps, goordnance.army.mil; "Rosie the Riveter," History.com, March 27, 2023.

91.   Magdoff and Magdoff, "Approaching Socialism," 53-54.

92.   Kalecki, *Selected Essays on Economic Planning*, 27.

93.   Fred Magdoff and Chris Williams, *Creating an Ecological Society* (New York: Monthly Review Press, 2017), 290.

94.   Marx, *Capital*, vol. 1, 742.

95.   Magdoff and Magdoff, "Approaching Socialism," 54-55.

96.   Lange, "On the Economic Theory of Socialism," 72-73. The term *whole-process people's democracy* is intrinsic to contemporary Chinese conceptions of how democracy might be made more meaningful. Despite limitations on how this has been applied in China itself, the concept is critically important in the development of socialist democracy. Xi Jinping, *The Governance of China*, vol. 4 (Beijing: Foreign Languages Press, 2022), 299-301.

97.   Mandel, "In Defense of Socialist Planning," 6-8, 13-17, 22, 25; Karl Marx, *Texts on Method* (Oxford: Basil Blackwell, 1975), 195; Gregory Grossman, "Material Balances," in *Problems of the Planned Economy*, 178.

98.   A key work in the ideological attack on the Soviet environmental record was Murray Feshbach and Arthur Friendly Jr., *Ecocide in the USSR* (New York: Basic Books, 1992). The technique utilized was to play up Soviet ecological destruction, while ignoring the fact that many of the same ecocidal conditions existed and often on a larger scale in per capita terms and global impact in the West.

99.   Salvatore Engel-Di Mauro, *Socialist States and the Environment* (London: Pluto, 2021), 115; Foster, *Capitalism in the Anthropocene*, 328.

100. Foster, *Capitalism in the Anthropocene*, 316-37.

101. Engel-Di Mauro, *Socialist States and the Environment*, 120-24, 139.

102. John Bellamy Foster, "Ecological Civilization, Ecological Revolution," *Monthly Review* 74, no. 5 (October 2022): 1-11.

103. P. G. Oldak, "Balanced Natural Resource Utilization and Economic Growth," *Problems in Economics* 28, no. 3 (1985): 3; P. G. Oldak, "The Environment and Social Production," in *Society and the Environment: A Soviet View* (Moscow: Progress Publishers, 1977), 56-68; P. G. Oldak and D. R. Darbanov, "A Bioeconomic Program," *Soviet Studies in Philosophy* 13, no. 2-3 (1974): 68-73.

104. Engel Di-Mauro, *Socialist States and the Environment*, 129-31, 141-42.

105. Paul M. Sweezy, "Socialism and Ecology," *Monthly Review* 41, no. 4 (September 1989): 1–8.

106. Engel Di-Mauro, *Socialist States and the Environment*, 170–94; "As World Burns, Cuba Number 1 for Sustainable Development: WWF," *teleSUR*, October 27, 2016; Matt Trinder, "Cuba Found to Be the Most Sustainable Country in the World," *Green Left*, January 10, 2020; Mauricio Betancourt, "The Effect of Cuban Agroecology in Mitigating the Metabolic Rift: A Quantitative Approach to Latin American Food Production," *Global Environmental Change* 63 (2020): 1–10; Rebecca Clausen, Brett Clark, and Stefano B. Longo, "Metabolic Rifts and Restoration: Agricultural Crises and the Potential of Cuba's Organic, Socialist Approach to Food Production," *World Review of Political Economy* 6, no. 1 (2015): 4–32.

107. "Comparing the United States and China by Economy," Statistics Times.com, May 15, 2021.

108. Foster, "Ecological Civilization, Ecological Revolution"; Barbara Finamore, *Will China Save the Planet?* (Cambridge: Polity, 2018), 156–58.

109. Ana Felicien, Christina M. Schiavoni, and Liccia Romero, "The Politics of Food in Venezuela," *Monthly Review* 70, no. 2 (June 2018): 1–19; Owen Schalk, "Venezuela's Seed Law Should Be a Global Model," *Canadian Dimension*, January 16, 2023. On Venezuela and degrowth, see Chris Gilbert, "'Where Danger Lies . . .': The Communal Alternative in Venezuela," *Monthly Review* 75, no. 3 (July–August 2023). See also John Bellamy Foster, "Chávez and the Communal State," *Monthly Review* 66, no. 11 (April 2015): 1–17.

110. One of Lenin's last articles was "Better Fewer, but Better." Baran later wrote an essay titled "Better Smaller but Better." Both had to do with strategic political retreats. But they also both reflected a way of thinking that recognized that qualitative changes are often more important than quantitative changes in achieving meaningful progress. See V. I. Lenin, "Better Fewer, but Better," in Lewin, *Lenin's Last Struggle*, 156–76; Baran, *The Longer View*, 203–9.

111. Odum and Odum, *A Prosperous Way Down*, 139.

112. Erald Kolasi, "The Ecological State," *Monthly Review* 72, no. 9 (February 2021): 23–36; Tom Athanasiou and Paul Baer, *Dead Heat: Global Justice and Global Warming* (New York: Seven Stories, 2002).

113. The leaked original scientists' consensus report on mitigation, prior to its being censored by governments before publication, indicated that the scale-up of carbon capture and sequestration, bioenergy with carbon capture and sequestration, and nuclear technologies were all impractical and unable to play anything but a minor role in mitigating climate

change. See Leaked Scientist Consensus Report on Mitigation, AR6, part 3, B4.3, https://f3b9m7v4.rocketcdn.me/wp-content/uploads/2021/08/summary_draft1.pdf. See also Mathilde Fajardy, Alexandre Köberle, Niall MacDowell, and Andrea Fantuzzi, "BECCS Deployment: A Reality Check," Briefing Paper no. 28, Grantham Institute, Imperial College London, January 19, 2019; Julian Allwood, "Technology Will Not Solve the Problem of Climate Change," *Financial Times*, November 16, 2021.

114. On the ecological and economic waste of monopoly capital, see Foster, *Capitalism in the Anthropocene*, 373–89.

115. On the environmental proletariat, see Foster, *Capitalism in the Anthropocene*, 483–92.

116. Harry Magdoff, "A Note on Market Socialism," *Monthly Review* 47, no. 1 (May 1995): 12–18.

117. Anthony Giddens, *The Politics of Climate Change* (Cambridge: Polity, 2011), 95; Andreas Malm, *Fossil Capital* (London: Verso, 2016), 382; On the various ways of combining plan and market, see Alec Nove, "Planned Economy," in *Problems of the Planned Economy*, 195–97.

118. Fred Magdoff and John Foster, "Grand Theft Capital," *Monthly Review* 75, no. 1 (May 2023): 19–20; Carter C. Price and Kathryn A. Edwards, "Trends in Income from 1975 to 2018," RAND Corporation Working Paper WR-A156-1, Santa Monica, 2020, 12 (fig. 2), 40; "U.S. Housing Market Has Doubled in Value Since the Great Recession, Gaining $6.9 Trillion in 2021," Cision PR Newswire, January 27, 2002.

119. On the calculation of economic surplus, see Michael Dawson and John Bellamy Foster, "The Tendency of the Surplus to Rise, 1963–1988," in *The Economic Surplus in Advanced Economies* (Brookfield, VT: Edward Elgar, 1992): 42–70.

120. William Morris, *Signs of Change* (London: Longmans, Green, 1896), 141–73; Foster, *The Return of Nature*, 103–5.

121. Schmelzer, Vetter, and Vansintjan, *The Future Is Degrowth*, 240.

122. Kurt Vonnegut Jr., *Player Piano* (New York: Dell, 1974). The plot of Vonnegut's dystopian novel focuses on the negative effects of a totally automated society where human labor is no longer necessary.

123. Leo Huberman and Paul M. Sweezy, "The Triple Revolution," *Monthly Review* 16, no. 7 (November 1964): 422; Robert W. McChesney and John Nichols, *People Get Ready* (New York: Nation Books, 2016), 80–81; Giorgos Kallis, "The Degrowth Alternative," Great Transition Initiative, February 2015, https://greattransition.org/publication/the-degrowth-alternative.

124. See the critique offered in Foster and Clark, *The Robbery of Nature*, 269–87.

125. See, for example, Noam Chomsky and Robert Pollin, *Climate Crisis and the Global Green New Deal* (London: Verso, 2020). Pollin, whose

views are somewhat distinguished from Chomsky's in this respect, is a strong opponent of degrowth alternatives, insisting that absolute decoupling on the scale required can be achieved at minimal cost without economic growth contracting through an "industrial policy" framework with green taxes, state financing, and market incentives.

126. Max Ajl, *A People's Green New Deal* (London: Pluto, 2021).
127. William J. Baumol and William G. Bowen, *Performing Arts: An Economic Dilemma* (Cambridge, MA: MIT Press, 1968).
128. Varun Ganapathi, "Understanding Baumol's Cost Disease and Its Impacts on Healthcare," *Forbes*, April 8, 2022; Aaron Benanav, *Automation and the Future of Work* (London: Verso, 2020), 57–60.
129. Magdoff and Williams, *Creating an Ecological Society*, 251–57; Herman Daly, Postscript to *Economics, Ecology, Ethics: Essays Toward a Steady State Economy*, ed. Herman Daly (San Francisco: W. H. Freeman, 1980), 366.
130. Paul A. Baran, *The Political Economy of Growth* (New York: Monthly Review Press, 1957), 42.
131. Noam Chomsky, *The End of Organized Humanity*, Climate Damage, YouTube video, 19:24 min., April 12, 2023.
132. Marx and Engels, *Collected Works*, vol. 25, 145–46, 153, 270; Marx and Engels, *The Communist Manifesto*, 2; Karl Marx and Frederick Engels, *Ireland and the Irish Question* (Moscow: Progress Publishers, 1971), 142. See also Walter Benjamin, *Selected Writings*, vol. 4 (Cambridge, MA: Harvard University Press, 2003), 402; Michael Löwy, *Fire Alarm* (London: Verso, 2016), 66–67; John Bellamy Foster, "Engels's Dialectics of Nature in the Anthropocene," *Monthly Review* 72, no. 6 (November 2020): 1–3.
133. Diogenes of Oeananda, *The Fragments*, edited by C. W. Chilton (Oxford: Oxford University Press, 1971), 11. Translation follows Benjamin Farrington, *The Faith of Epicurus* (London: Weidenfeld and Nicolson, 1967), 134.

Index

Saito, Kohei, 96–102, 175–76
Santos, Orlando Rey, 232
Sapir, Jacques, 245
Sartre, Jean-Paul, 34, 44, 251
Schmalhausen, I. I., 30–32
Schmidt, Alfred, 23–24
Schmidt, Oscar, 86
Schopenhauer, Arthur, 86
Schumacher, E. F., 106, 114, 127
science, 85; alienation of, 211–12; Red
    science, 89
scientific humanism, 182
Second Congress on International
    History of Science (1931), 25–27
second foundation of Marxism, 38-40,
    45, 82-103, 178
Sena, Kanyinke, 125
Shirokov, Mikhail, 28–29
Shute, Nevil, 205
Skvortsov-Stepanov, Ivan Ivanovich,
    19, 20
slavery, 63, 67-68, 104, 177, 191, 230,
    238
Smith, Adam, 139
social Darwinism, 86
socialism: ecosocialism, 35–41, 168–
    70, 207–16; Marx and Engels on
    planning in, 249–50; as pathway to
    ecological survival, 202; planning
    essential for, 263–64; second foun-
    dation of, 82–90; socialist states and
    environment, 264–67
socialist ecology, 43
social labor, 269
social metabolism 12, 24, 27, 32-33,
    39-40, 44, 49, 54, 66, 73-76, 81,
    89-90, 94, 98, 128, 131, 139, 157,
    163, 169, 192, 198, 212, 231-33, 244,
    250, : Bhaskar on, 40; Engels on, 54,
    96–103; Lukács on, 33; Marx on, 47,
    49–50, 73; Mészáros on, 39.
social metabolic reproduction, 39,
    231, 250
social reproduction, 155, 193, 213

society, as emergent form of nature, 8
Soddy, Frederick, 138, 141–45
Solow, Robert, 114, 129–30
Soper, Kate, 81
Soriano, Carles, 62, 94–96, 197
Soviet Union: central planning in,
    252–58; ecological civilization
    in, 168, 181; ecology in, 161–62;
    environmental policies of, 264–66;
    Marxism in, 16–22, 83; suppression
    of materialist dialectics and science
    in, 29–32
Spencer, Herbert, 56
Spinoza, Baruch, 17, 79
Stalin, Joseph, 21; central plan-
    ning under, 256; "Dialectical and
    Historical Materialism," by, 29; on
    dialectics of nature, 24–25; in Great
    Debate, 255; Western Marxism's
    opposition to, 22
Standing, Guy, 148
standing natural capital, 124
state capitalism, 253
state socialism, 260
Stedile, João Pedro, 156
substantive equality, 9, 15, 64, 94, 157,
    240, 263
Sukachev, Vladimir Nikolayevich, 30, 32
surplus, 194–96
sustainability: Solow on, 130; weak-
    and strong-, 116–17
sustainable human development, 5, 9,
    16, 41, 64, 157, 161-62, 164, 169-70,
    182-83, 231-34, 237-40, 251, 267;
    Burkett on, 251-52
Sweezy, Paul M., 266, 270
System of Environmental-Economic
    Accounting, 120

Tans, Peter, 220
Tansley, Arthur George, 27, 57, 94
temperature: United Nations report
    on, 225–26, 241–42; wet-bulb and
    dry-bulb, 221–22